Linear Induction Accelerators for High-Power Microwave Devices

Linear Induction Accelerators for High-Power Microwave Devices

Igor Vintizenko

CISP

CRC Press
Taylor & Francis Group
Boca Raton London New York

CRC Press is an imprint of the
Taylor & Francis Group, an **informa** business

CRC Press
Taylor & Francis Group
6000 Broken Sound Parkway NW, Suite 300
Boca Raton, FL 33487-2742

First issued in paperback 2020

ISBN 13: 978-0-367-57138-2 (pbk)
ISBN 13: 978-1-138-59527-9 (pbk)

Visit the Taylor & Francis Web site at
http://www.taylorandfrancis.com

and the CRC Press Web site at
http://www.crcpress.com

Contents

Introduction

The monograph presents the analysis of the published works and the results of the studies conducted at the Tomsk Polytechnic University (TPU) during the period since the beginning of 80's of last century till present time. The studies are united by the theme of the application of linear induction accelerators (LIAs) for the generation and amplifying of microwave radiation by relativistic devices.

The possibility of designing the relativistic microwave devices with enhanced power arose at the end of 60's because of the emergence of high-current nanosecond electron accelerators which form electron beams with the energy of 0.5–2 MeV and a current of 1–100 kA. Electron beams with the indicated characteristics are used in different areas of science and technology: for heating of plasma to the thermonuclear temperature, developing new methods of collective acceleration of ions, the study of phase transformations in solid bodies and the properties of materials, and also for the generation and amplification of high-power electromagnetic radiation. The formation of relativistic electron beams is accomplished by this accelerator with the discharge of the forming line at a high-voltage diode. The charging of the forming lines originates from the pulse voltage generators, which have significant overall sizes, low efficiency and a single operating mode. Further, the problem of the electric strength of elements of generators with respect to the total accelerating voltage leads to the complication of installation and increase in their weight and dimension characteristics by the growth in the acceleration energy.

In the monograph LIAs utilized as power supples of the devices of relativistic high-frequency electronics are described. This branch of electronics is dedicated to the use of high-current relativistic electron beams for the amplification, generation and conversion of electromagnetic oscillation. Relativistic effects appear, as a rule, in velocities of the electrons, commensurate with the speed of light, i.e. with sufficiently high values of accelerating voltage.

As show by the analysis of physical processes in the relativistic devices with the vacuum electrodynamic structures, an increase in the efficiency in such devices and the possibility of the formation of powerful flows of coherent radiation are closely related with the formation of the electron beams which possess small angular and energy spreads. In this case, for relativistic electronics with the largest possible energy of acceleration with comparatively small overall sizes and energy reserve of accelerator are promising. Special prospects in this region are related to the use of devices whose output voltage acts only along the accelerating circuit, and the electric field is excited between two sections of the internal surface of the closed conducting cavity – the linear induction accelerators. The monograph describes the design and the output parameters of LIA developed specially or used with specific modifications of relativistic devices.

In 1978 Yu.P. Bakhrushin and A.I. Anatskii [12] published their monograph which is still the only book dedicated to LIA. Problems related with thr magnetic reversal of ferromagnetic induction systems, forming of the pulse of accelerating voltage, the spatial distribution of the magnetic field, the shaping and structure of the electron beam are studied in detail and designs of the LIA developed at that period are analyzed. In the past 40 years the technology of LIA has changed, for example "non-iron" LIAs and LIAs on the magnetic elements are now available. The element base of accelerators has changed substantially. From simple spark gas dischargers multichannel dischargers with the forced division of the current between the channels and magnetic commutators in the form of saturated chokes have appeared. A significant number of LIAs have been developed on the basis of a new element base. And finally, LIAs were adopted actively for relativistic devices. All the above-mentioned facts are reflected in the monograph based on the experience acquired at the TPU in the design, development, simulation and the application of LIAs for different studies, including the field of relativistic high-frequency electronics. So, some of the first in the world theoretically and experimentally investigated generators like relativistic magnetrons and klystrons, reflex triodes, antenna amplifier are developed by the colleagues of TPU. LIAs were adapted as the power supply source, on the basis of original layout diagram and element base were prepared.

The methods of resolving various physical and engineering problems which appear during calculation, construction, simulation and operation of the linear induction accelerators, intended for the power supply of relativistic devices are also analyzed in the monograph. It should be noted that the pulse-periodic regime of operation of relativistic devices is most promising for the practical application provided that the compact radiation complexes are created. Microwave sources with the high average power are used in the following directions: microwave sources for linear resonance electron accelerators with the high rate of acceleration; in radars including non-linear; the electromagnetic compatibility studies of radio-electronic equipment; for sterilization. For solving above mentioned problems it is necessary to develop power sources with the parameters corresponding to the requirements of relativistic devices and also other systems, which allow prolonged operation of devices with a high pulse repetition rate.

It should be noted that when using LIAs with various output characteristics of power, duration and pulse repetition frequency different factors that restrict the output parameters of relativistic generators appear. In this connection there is a need for development and research of accelerators which are most suitable for some of microwave devices. On the whole, the LIAs require more detailed studies and specific generalization. This concerns mainly the study of physical processes in the elements of the accelerator, the calculation of output characteristics, the development of the original design of LIA (including the power sources) and their experimental testing.

The first chapter of the monograph is a review. Its aim is to describe the operating principles of LIA, their basic elements and their advantages in comparison with the high-current electron accelerators (section 1.1). In this chapter two modifications of LIA – with the ferromagnetic induction systems (section 1.2) and the so-called 'non-iron' LIA (section 1.3) are examined. The LIA induction system usually contains cores from a ferromagnetic material, which enable to considerably reduce the current in the primary winding of inductors in comparison with the 'non-iron' version of accelerators are used. This, in turn, enables to increase the efficiency of accelerator to higher values.

However, the 'non-iron' linear induction accelerators make it possible to achieve energies above 10 MeV with retention of the high currents characteristic of the nanosecond electron accelerators. Both modifications techniques of the LIA are fairly complicated, which required resolving a number of electrophysical problems, in particular: 1) the development of the multiunit accelerating system, which uses hundreds of the dischargers switched on low-impedance forming lines ($\rho \approx 1$ Ohm) per a predetermined program with sub-nano-accuracy per second; 2) the development of methods of forming and transporting high-current electron beams (with the current of tens of kilo-amperes) to significant distances (of the order of 10 m). This chapter describes the basic elements of the LIA: pulse voltage generators, induction systems, forming lines commutators, electron guns, magnetic systems and other elements.

The second chapter describes the LIAs developed by specialists at the TPU that made a significant contribution to the accelerator technology. The accelerators have an original layout diagram and element base: low-impedance strip forming lines; multichannel spark dischargers with forced division of current between the channels; non-linear saturated core devices. The accelerators are used for the formation of electron beams, as a supply source of relativistic generators (undulators, klystrons, magnetrons, reflex triodes), pumping of gas lasers and so forth.

The distinctive special feature of the LIA is the use of magnetic elements which makes it possible to ensure the high frequency of repetition and also high reproducibility of the amplitude and time characteristics of pulses. In these accelerators the magnetic commutators are used for the commutation of the forming lines and they are capable to commutate current with unlimited resources into hundreds of kilo-amperes with a frequency of kilohertz. For the realization of this switch with the minimum overall size and the corresponding minimum inductance the forming lines must be charged within hundreds of nanoseconds by magnetic pulse generators (MPG). The LIAs on magnetic elements developed in the recent past are the injector modules, intended for forming the electron beams of a comparatively low energy (300–500 keV with the current of 3–6 kA). They are used for the supplying power to relativistic magnetron and reflex triodes.

The procedures of technical/engineering calculations and the results of computer simulation of the LIA on the basis of equivalent circuits are described in this chapter. The effect of overlapping of the phases of the charge and discharge of compression stages, the unbalance of the capacitance of the capacitor of the last stage of compression of MPG and the forming lines that enables to increase the efficiency of the accelerator with simultaneous reduction in the mass of the ferromagnetic material of the cores are also outlined. This material can be useful in the design of MPG for different applications. Furthermore, the power supply diagrams of the LIA are given and the principle of their operation is described. One of the diagrams, used for the power supply of the LIA located on the mobile platform, enables to form current in the magnetic system, ensures charging of the forming lines, shaping of demagnetization field and synchronous start of the channels of the multichannel discharger. The power supply diagram with power recuperation and stabilization of the level of the charge of primary storage that ensures the pulse repetition rate of the LIA to 320 Hz is represented. Important characteristic like the maximum pulse repetition rate of the LIA on the magnetic elements related to the magnetic reversal of ferromagnetic cores of the saturated core devices of MPG, the pulse transformer and damping of inter-pulse fluctuations are examined.

Processes in the 'power supply–relativistic device' system with a strong feedback are investigated by computer simulation. The simulation of such a complex system enables to study in more detail the physical processes, which take place in different elements of system and to analyze the effects of the mutual coupling of the load and power source. The load (relativistic magnetron) is represented in the form of non-linear parametric resistance, which at each step of the integration of differential equations is defined as the result of a solution in the model of the relativistic magnetron. This method makes it possible to operationally conduct tuning of the device for extreme power output, electron and total efficiency.

Third chapter presents information about the operating principle and design of relativistic devices with the rectilinear electron beams formed by the LIA. The design of devices like analogous and the classical sources of coherent emission (a klystron in section 3.2 and a Cherenkov generator in section 3.5) and those based on new principles (undulator, laser on free electrons in section 3.1, and also the orotron and the gyrotron in section 3.3) are described.

One of the drawbacks in the instruments based on the principles of classical microwave electronics is the fact that the operating current cannot exceed the limiting current of transport in the drift tube. The generators with the virtual cathode are promising for using in contemporary LIA with all possibilities. Their distinctive special feature is the possibility to generate radiation only by the current that exceeds the maximum vacuum current which is a condition for the formation of virtual cathode. The purpose of the work, carried out at TPU, was to create a compact microwave radiating complex (see section 3.4). The LIA was selected as the power source for triodes due to the possibility of operation in the periodic regime and also due to high efficiency and compact size. The reflex triode differs from other devices by the simplicity of design, the absence of the magnetic system which substantially increases energy input, weight and overall size of the entire installation. For use as the power source for the reflex triode low-resistance load changes were introduced in the design of LIA which made it possible to match the internal resistance of the accelerator with the impedance of the triode and to realize the periodic regime of pulses for the first time. Some deficiencies in the reflex triode were removed in the original axially symmetrical construction. It was possible to reduce the inductance of current-leads to the grid of the triode, to decrease the duration of the front of the voltage pulse, and to also ensure the symmetry of the grid voltage supply.

One of the interesting ideas in the development of microwave generators with the virtual cathode is the idea of the hybrid generators, in which the transient current of electrons is modulated with the help of a virtual cathode. The virtual cathode enters the circuit which is tuned to the regime of the travelling-wave tube. The wave which is excited in the tube returns to the virtual cathode. Thus the feedback is achieved. The output of microwave radiation is produced near the virtual cathode. Thus, this hybrid generator, which was called 'virtod' is the 'vircator + travelling-wave tube' system. The results of experiments with a similar instrument are described in section 3.4.3.

The potential applications of the sources of powerful microwave pulses include information- telecommunication systems, such as the radar systems of survey, probing of the atmosphere and the earth's surface, systems of distant space communication which are still held up because of the insufficient progress in resolving the problem of controllability. In these cases it is necessary to provide the possibility of the variation of the largest possible number of characteristics of output emission – power, frequency and phase for the monochromatic signals, the spectrum for the multi-frequency signals and the high frequency pulses and also guiding the microwave ray output. Section 3.6 describes the results of studies of antenna-amplifiers on a compact module of the LIA conducted at the TPU. The final goal of the studies is the experimental demonstration of amplifying the control capabilities of the output parameters of the microwave pulse in a wide range of the power level in the order of tens of megawatts and the duration in the order of tens of nanoseconds in the three-centimeter wavelength range. The concept of the antenna-amplifier includes the idea of developing a compact controlled source of powerful radiation by the association of the electron accelerator and the electrodynamic system of interaction and the radiating antenna.

The super-reltron is described in section 3.7; its operating principle is similar to the principle of the klystron. Initial studies of this device were carried out with the power source of pulse voltage generators with all its inherent deficiencies. Improved design of the super-reltron based on the LIA is proposed in this section. In the super-reltron the design of elements of the power source and the device increases the pulse repetition rate, total efficiency rises and the weight and dimension characteristics are reduced considerably. This is achieved by using the LIA original layout.

The fourth chapter outlines the development, calculation, design and testing of the separate units of the relativistic magnetron (RM), intended for operating with the high pulse repetition rate. Section 4.1 presents the results of the studies of the RM in the pulse-periodic regime using LIAs with multichannel dischargers.

The experimental data obtained in the analysis of the first tests of the RMs and LIAs with multichannel dischargers are used to explain the need for design differences in the elements of installationd based on the LIAs on magnetic elements which allow operation of the RM with a high pulse repetition rate during prolonged time intervals. First of all they are connected with the use of DC magnetic systems, powerful systems of vacuum evacuation and water-cooled anode blocks (see sections 4.2–4.7). Further, the results of experimental studies of the relativistic magnetron generators with the power supply from the LIA on magnetic elements carried out at TPU (section 4.7) and the Physics International Company, USA (section 4.8) are represented in this chapter.

The fifth chapter (sections 5.1–5.4) examines the operating principles, electrical circuits and design of generators of microsecond pulses (GMP) which use the technology of LIA on magnetic elements. The output parameters of these generators are: voltage 450–1000 kV; current 1–2 kA; the duration of the flat part of the pulse 1 µs; the pulse repetition rate up to 1 kHz. The basis of GMP design is the original idea to realize the sequential discharge of several magnetic pulse generators of the primary winding of high-voltage pulse transformer synchronized in a specific manner. By selecting the elements it is possible to shape voltage pulses as linearly increasing

or linearly decreasing. The generators of microsecond pulses allow
an operational change in the polarity of the output pulses with
the retention of the amplitude and time characteristics. Similar
parameters of power sources are unique. They cannot be obtained
in the traditional pulse-generating circuits. The latter cannot ensure
the rectangular form of the microsecond voltage pulse duration and
the high repetition frequency.

The design-engineering problems related with the development of
GMPs are investigated in this chapter. Formulas for the selection of
the cross section of the windings of pulse transformers and chokes
(section 5.5) are given and the calculation of the thermal condition
of the elements of a generator (section 5.6) and computer simulation
on the basis of equivalent circuits (section 5.7) are carried out. One
of the section is dedicated to the estimation of inductance leakage,
magnetization inductance, dynamic capacity, ohmic equivalents of
losses. Such material can be useful for designing of the high-voltage
devices which use magnetic elements, including the linear induction
accelerators. The detailed description of the original scheme of the
power supply of the generator which allows operation with a pulse
repetition rate up to 1 kHz in the regime of their long packets (to
5000) is also presented.

The generators of microsecond pulses with similar unique output
characteristics can find application in the formation of electron
beams of relativistic devices of both O- and M-types. In this case
with relatively low levels of the output power it is possible to obtain
longer high-energy microwave pulses. As a result the average power
of the radiation of the installation increased and the prospects for
its practical use emerge. In this case the limitations related with the
heating of collectors of the relativistic devices of O-type, the anode
blocks of relativistic magnetrons and their possible destruction are
partially removed.

On the basis of the results presented in the book it can be
concluded that the linear induction accelerators can be successfully
adapted as sources of power for numerous relativistic microwave
devices.

Principle of operation, construction and parameters of linear induction accelerators

1.1. Introduction

The development of high-voltage pulse accelerating technology started in the 60s of the last century and by now has achieved significant success. Intensive use is made of high-current electronic accelerators (HCEA), which allow producing high-energy particles at a sufficiently high efficiency of energy transfer from the power source to the beam. Beams of relativistic electrons are used in developing new methods of acceleration, searching for ways to implement controlled thermonuclear fusion, and also for generating bremsstrahlung and microwave radiation. At the same time, the maximum electron energy in HCEA is limited by the electrical strength of the system with respect to the total accelerating voltage.

To obtain powerful beams of charged particles, linear induction accelerators (LIA) are also actively used. For the first time the idea of LIA in the form of a system of successively installed impulse transformers was proposed in the early 1930s by A. Bouversey [1]. However, this idea was realized only after forty years, when the level of the development of physics and technology allowed us to achieve the required magnitude of the accelerating field. The first accelerators were created in the early 1960s in the USA under the guidance of N. Christofilos [2] in connection with the work on controlled thermonuclear fusion. One of the first large installations was an 'Astron' injector for 3.7 MeV energy and a beam current of

350 A in a 300 ns pulse [3]. Later, the accelerator was reconstructed, the energy was increased to 6 MeV, and the current to 500 A in a 300 ns pulse. The technique of LIA was especially successful in the 1970s and early 1980s. At this time, a number of large accelerators, such as ERA, ETA, FXR and ATA, for energy from 4 to 50 MeV and a pulsed beam current from 1 to 10 kA were built. In the USSR, in 1967, under the leadership of V. I. Veksler, the LIA-3000 for 3 MeV and a beam current of 200 A in a pulse was constructed to implement the collective acceleration method [4, 5] at The Joint Institute of Nuclear Physics (JINP) (Dubna) was built in cooperation with the NIIEFA (Leningrad) for the duration of 500 ns. Later on, a fundamentally new SILUND induction accelerator (a high-current induction linear accelerator of the nanosecond range) was created at Dubna, and in 1981 the SILUND-20 accelerator with a beam current of 1 kA, an electron energy of 2 MeV and a pulse repetition frequency of 50 Hz was put into operation. The pulse duration was approximately 20 ns, the emittance of the electron beam was about 3π cm · mrad for 60% of the current. In the 1970s, LIA-5000 (5 MeV, 2 kA, 50 ns) and the first section of the LIA 30/250 (3 MeV, 250 A, 500 ns) were constructed. The accumulated experience made it possible to proceed to the construction of technological installations. The first of them, LIA 1,25-200 and LIA 1-5, retained the main features of previous accelerators. They were made on the basis of small induction modules, powered from generators of high-voltage pulses, built on hydrogen thyratrons without the use of industrial transformers. They differed from each other in the type of material of the inductor cores (permalloy or ferrite) and the duration of the accelerating voltage pulse. Such accelerators began to be developed for research on the creation and retention of hot plasma, for experiments on the investigation of new methods for accelerating particles. In the 1980s, compact LIAs based on generators with magnetic compression were developed, which made it possible to increase the pulse repetition frequency up to several kHz [6]. It is hoped that, along with the increasing application of LIA in scientific research, they will be used in industry. For example, the LIA can be used as a basis for constructing mobile, cheap and easy-to-use installations for X-ray analysis, gamma-ray logging, flaw detectors, capable of operating in field or factory conditions can be created. Since the end of the 1980s linear induction accelerators have been applied as power sources for relativistic high-frequency devices [7, 8].

A few words should be said about the collective methods of acceleration, since in the USSR they served as the basis for the development of the technique of LIAs. The collective methods of acceleration occupy a leading place among the new effective methods of accelerating the contaminated particles. Many scientific centres of the world are carrying out experimental studies on various modifications of the collective method, facilities are being created, new concepts of acceleration are emerging, and accelerators of the future are being designed for higher energies [5]. The incentives for the development of collective methods were the proposals of Soviet physicists G.I. Budker, V.I. Veksler, and Ya.B. Fainberg, presented at the 1956 conference in Geneva. Each of these proposals served as a basis for determining the main directions for the development of collective methods.

G.I. Budker suggested using the intrinsic electromagnetic field of a self-focusing electron–ion ring for sustaining the ions in cyclic orbits and focusing them during acceleration to high energies (accelerator with collective focusing). V.I. Veksler proposed to apply for the acceleration of ions the fields arising from the interaction of an ion bunch with an electron beam or a flux of electromagnetic radiation, or the intrinsic fields of a two-component bunch. The proposal made by Ya.B. Fainberg on the use of space-charge waves for the acceleration of ions in the plasma was subsequently intensively developed. It underlies controlled schemes of collective acceleration in high-current relativistic electron beams.

In comparison with the traditional systems, the collective accelerators have a higher rate of ion acceleration, and therefore, greater economy and smaller dimensions. These advantages are achieved as a result of using the intrinsic electromagnetic field for accelerating the electron bunch. The electric field strength associated with the space charge inside the bunch reaches values of (10^5–10^6 V/cm) at a comparatively low electron charge density. High intrinsic electric fields lead to strong repulsion of electrons. To reduce this effect, it is suggested to form a bunch in the form of a ring of relativistic electrons. In this case, the repulsive forces acting on the electrons are weakened γ^2 times (where γ is the relativistic factor) and they can be compensated for by comparatively small external fields or fields of the ions themselves. When an electron ring accelerates in an external field in a direction perpendicular to the plane of the ring, the non-rotating ions are entrained by the strong intrinsic field of the electrons. The simultaneous motion of electrons and ions leads

to a significant gain in the rate of acceleration in comparison with
the direct acceleration of ions in an external field. At the same time,
having identical velocities, ions acquire significantly more energy
than electrons ($AM/(m\gamma)$ times, where m and M are the masses of the
electron and ion, respectively, and A is the mass number of the ion).

To obtain electron rings with the required parameters (electron
energy 2 MeV, electron density $\sim 10^{13}$ cm^{-3}, the average radius of
the ring is 3.5 cm; ring thickness 0.2 cm), their compression in a
magnetic field growing in time is proposed in an installation called
an adgezator (adiabatic generator of charged toroids). A rectilinear
relativistic electron beam is introduced into the adgezator chamber,
which is folded into a ring in a weakly focusing magnetic field. The
magnetic field grows in time so that the ring contracts radially. In
this case, the dimensions of the cross-section of the ring decrease
(approximately as the average radius), and the electron energy
increases inversely proportional to the radius of the ring.

Analysis of the beam parameters shows that the most suitable
injector for a collective accelerator is a linear induction accelerator
giving a high current of electrons with a short pulse duration and
a small energy spread. Therefore, LIAs were used earlier and still
used as relativistic electron injectors in many experiments on the
acceleration of ions by electron rings in the former USSR/Russia
(Dubna JINR, Moscow ITEP, Tomsk Polytechnic University), USA
(Maryland), Germany (Karlsruhe, Garshing), Japan and Italy.

1.2. Principle of operation and parameters of linear induction accelerators

The principle of the action of LIA is based on the use for the
acceleration of charged particles of a vortex electric field excited
in a system consisting of several ring-shaped transformers (Fig.
1.1). With simultaneous feeding to all magnetizing coils (turns or
windings) a voltage pulse from the pulse generator, an alternating
magnetic flux dB/dt is excited in the inductor cores, and an axial
electric field E is generated on the axis of the system, the value of
which is given by the expression

$$\mathrm{rot}\, E = -\frac{dB}{dt}. \tag{1.1}$$

With the help of the body of the LIA and diaphragms, the vortex
field is evenly distributed along the accelerating tube. The role of the

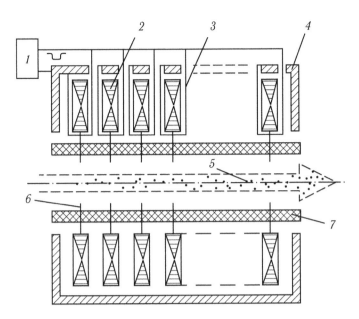

Fig. 1.1. LIA scheme: *1* – pulse generator; *2* – ferromagnetic core; *3* – magnetizing coil; *4* – accelerator body; *5* – the electron beam; *6* – diaphragms; *7* – accelerator tube body

secondary winding for all inductors is performed by an accelerated beam. Thus, the accelerating field in the LIA is distributed along the entire accelerator, the gap to which the total voltage corresponding to the total energy of the beam is applied, is absent. This greatly simplifies the element base of the accelerators, increases the service life and reduces the weight and dimensions.

To obtain a beam of charged particles with a uniform charge density and the same increase in energy along the length of the accelerator, it is necessary to ensure a linear change of the induction in time. Then the expression (1.1) can be rewritten in the form

$$E = -\frac{n}{l} S \frac{\Delta B}{\tau}, \qquad (1.2)$$

here *n* is the number of inductors; *l* is the length of the inductor system; *S* is the cross section of the steel of one inductor; ΔB is the induction increment in the core; τ is the pulse duration. The condition of constant rate of change in the induction of the magnetic field can be satisfied by feeding the primary winding (magnetizing coil) of the inductor with a rectangular voltage of duration τ.

It is seen from expression (1.2) that the energy transferred to the charged particle beam during the pulse time is proportional to the change in the magnetic flux in the inductor core and does not depend on the pulse duration. To fully utilize the core material, before applying a voltage pulse to the magnetizing winding the core is transferred to the region of negative saturation (it is demagnetized), which is almost always done.

This method allows one to create an accelerator with an electron beam current of hundreds and thousands of amperes, which at the same time has the advantages of waveguide accelerators, in particular, the ease of input and output of a beam. The particle energy is increased by growth of the number of consecutively connected inductors. In a linear induction accelerator, beam focusing is substantially simplified and capture is ensured in the regime of acceleration of practically all particles, irrespective of the injection energy. The maximum value of the beam current is limited only by the difficulty of the transportation of a high-current electron beam. The high quality of the accelerated beam, which is caused by the video-pulse mode of operation, also belongs to the main advantages of LIA. The energy spread may be less than 1%, emittance not more than 0.02π cm ·rad. This property makes the linear induction accelerator attractive for use as an injector of many research installations.

Table 1.1 shows the parameters of some linear induction accelerators, created in different years and having various applications. As can be seen from the table, the pulse duration of the accelerating voltage is in the range of 20–500 ns. For a longer duration, the induction system of the accelerator becomes too cumbersome, and at a lower duration the efficiency factor decreases due to the increase in energy losses for magnetization reversal of the core. For the specified pulse duration range one can be use 'non-iron' LIAs. Recently, it has become possible to advance accelerators in the microsecond range of duration of the voltage pulse durations. Chapter 5 of this monograph will be devoted to the description of the generator of microsecond pulses, made using the LIA technology.

All the installations listed in Table 1.1 are characterized by a pulsed beam current that is 2–3 orders of magnitude greater than the current in linear waveguide accelerators. At the same time, the energy of the accelerated electrons is limited to 30 MeV (LIA 30/250) and 47.5 MeV (ATA). It is also realistic to create accelerators with much higher energies. Recently, the project of LIA for accelerating to an

Table 1.1. Parameters of LIA produced by scientific organizations (electron energy E, beam current I, pulse duration τ, pulse repetition rate F), application areas, characteristics and design features of accelerators (FEL – free-electron lasers)

LIA Organization	E, MeV	I, A	τ, ns	F, Hz	Design features
Astron Lawrence LivermoreNational Laboratory, USA	4.2	800	300	60	The energy spread is less than 2%. Emittance 25 cm mrad.
ETA-2 Lawrence LivermoreNational Laboratory, USA	6-7	2000	50	5000	For FEL 140 GHz for heating the plasma in a tokamak. It consists of an injector of 1.5 MeV and 60 accelerating inductors.
ATA Lawrence LivermoreNational Laboratory, USA	50	10000	70	1000.	Batch mode of operation (10 pulses) For FEL 140 GHz for heating the plasma in a tokamak.
ERA Berkeley, United States	4.25	500	45	1	The energy spread is less than 0.5%. Emittance 70 cm mrad.
LIA 3000 JINR Dubna, NIIEFA Leningrad, USSR	3	200	350	25	
LIA 30/250 JINR Dubna, NIIEFA Leningrad, USSR	30	250	500	50	
LIA-30 VNIIEF, Arzamas-16, the USSR	40	100,000	20		The "non-iron" LIA

Table 1.1. (continued)

Name / Location					Notes
RHEPP-1 RHEPP-2 Sandia National Laboratories, Albuquerque, USA	1 2.5	25000 25000	60 60	120 120	Average power of electron beam is 100 kW for RHEPP-1. Average power of electron beam is 300 kW for RHEPP-2
LAX-1. National Laboratory of KEK, Japan	2	2000	120		For FEL 10 GHz. Strip forming lines with adjustable capacity. The change in capacitance is effected by the motion of dielectric plates between the electrodes of the lines.
AIRIX France	20 (project)	3500	60		For radiographic measurements
I-3000 VNIIEF, Russia	3	10 000	16		The 'non-iron' LIA.
LIA for industrial applications China	5 (project)	200	1000	100	Mobile LIA for radiation technological applications. Artificial DFL.
National Bureau of Standards USA	750	750	2000		Sectioned inductors
SNOMAD-1 SNOMAD-4 Science Research Laboratory, USA	0.6 1.5	600 600	60 50	5000 5000	For a copper vapour laser. Thermocathode. The acceleration rate is 3 MeV / m. Injector 0.5 MeV and accelerating module 1 MeV. The average power of the electron beam is 500 kW.
LIA with multichannel arresters Tomsk Polytechnic Institute	300 300	3 3.3	80 70	160 20	For relativistic magnetrons Packet mode of operation (3 pulses). Mobile version.

Table 1.1. (continued)

LELIA Center d	Etudes Scientifiques et Techniques d'Aquitaine Comissariat a l'Energie Atomique, France	3 (project) 2.1	3000 2500	50	1000	For FEL. Forming lines are coaxial cables of 100 Ohm. Dispenser osmium cathode
Compact LIA Physics International Company, Olin Corporation Aerospace Division, San Leandro, USA	0.75	10000	60	200	For relativistic magnetrons and relativistic klystrons
LIA 4/2 Tomsk Polytechnic University	2.4	1000	110	3300	Packet mode operation (5 pulses)
LIA "Corvette" VNIIEF, Russia	1.5	35 000	40		The 'non-iron' LIA for the vircator.
3.4 MeV LIA China Academy of Engineering Physics, Chengdu, China .	3.4	2000	90		For FEL
LIA 4/6 Tomsk Polytechnic University	400	3.6	170	200	For relativistic magnetrons

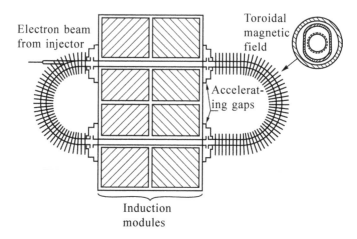

Fig. 1.2. Scheme of the Racetrack induction accelerator.

energy of 1.6 GeV of the electron ring was considered. The average intensity of the accelerating field should be 5 MeV/m, which required an accelerator length of 320 m [9]. In the paper [10], the design of the Racetrack induction accelerator is proposed and the principle of its operation is described (Fig. 1.2). It is a cyclic accelerator capable of functioning in the kiloampere current range. A high-current electron beam circulates through linear induction modules with a pulse duration of the electric field ~2 µs. In the modules the electrons accelerate. The increase in energy is

$$V = \frac{V_m T_m}{\tau_{rev}}, \tag{1.3}$$

where V_m – increase in energy per revolution; T_m – pulse duration of the module; τ_{rev} – time of beam revolution. Each straight section has a length L with a total length of the cores L_c with a radius R_c. Each waveguide is a circle of radius R. Assuming that the electrons move at a velocity close to the speed of light c, and the magnitude of the induction amplitude in any section of the core is the same, the expression for the energy gain can be written as

$$V_\tau = V_m T_m = 2\Delta B R_c L_c. \tag{1.4}$$

In this case, the expression for the travel time τ_{rev} takes the form

$$\tau_{rev} = \frac{(2L + 2\pi R)}{c} = \frac{2\pi R_a (R/R_0)^2}{c((R/R_0) - 1)},$$

$$(1.5)$$

where

$$R_0 = \frac{V}{\Delta Bc} \frac{LR}{L_c R_c}.$$

This expression is interesting in that it gives the minimum travel time τ_{min}, which determines the size of the system. The minimum τ_{rev} is reached at $R = 2R_0$. The volume of the cores is equal to

$$V_c = 2\pi R_c^2 L_c = 16\pi^2 \left(\frac{V}{\Delta Bc}\right)^3 \left(\frac{RL}{R_c L_c}\right)^2. \qquad (1.6)$$

An important element of the system is the electron beam transport system. It is possible to design a system of a magnetic field that will ensure the weakening of the forces affecting propagation of the beam in the entire energy range. A similar configuration is provided by a quadrupole field, which, like in a stellarator, is untwisted into a spiral with $\Omega \sim 2$. The simplest configuration of the coil for the formation of such a field is shown in Fig. 1.2. All particles with an energy close to the maximum appear in this field, and particles with higher energy have such an orbit that they do not enter the aperture of the accelerator. The stellarator winding in conjunction with the induction module forms an accelerator with constant fields. Since focusing and confining fields may not depend on time, an arbitrary pulse shape of the induction module is permissible. To accelerate particles to energies exceeding the maximum values at which the particles are held by a constant field, a time-varying vertical magnetic field can be applied. This field can be chosen so that the Larmor radius of rotation of the electrons is equal to the radius of the transport path. Thus, for a magnetic field of a stellarator with an induction of 1 T at a waveguide length of 1 m, an electron beam with energies up to 10 MeV can be obtained at 40 electron revolutions. With the same stellarator field together with an external vertical field with an induction of 0.24 T, an electron beam with an energy of up to 100 MeV is possible. Synchronizing a vertical magnetic field with the energy of electrons, i.e., making it variable in time, it is theoretically possible to create a Racetrack induction accelerator for the energy up to 1 GeV.

The main problem of Racetrack is injection into the toroidal accelerator. Closed magnetic field lines serving to hold the transverse dimensions of the electron beam complicate the design of the injector. An important advantage of the stellarator magnetic field is the admissibility of injecting a high electron energy spread. In the case of Racetrack, a high energy gain and a short acceleration time reduce synchrotron radiation losses. Thus, the described design of the installation, based on the use of the LIA technique, makes it possible to advance into the region of high electron energies.

The maximum pulse repetition rate in the developed LIA is still small. It is determined by the quality of the beam transportation along the accelerator and the characteristics of the switching elements in the pulsed power system. Almost all LIAs presented in Table 1.1 are devices of the stationary type, that is, they are rather cumbersome. The weight of the accelerators from the calculation of an energy gain of 1 MeV, reduced to a pulse duration of 50 ns and a beam current of 1 kA, remains in the range 800–1200 kg at a repetition rate of 50 Hz. In general, the accelerators presented in the table require very qualified service, fine tuning, strict preventive maintenance. Only the recently constructed linear induction accelerators on magnetic elements favourably differ from the others in the listed indicators. This class of accelerators will be discussed in detail in Chapter 2.

1.3. Linear induction accelerators with ferromagnetic cores

In general, the LIA consists of a set of modules: injectors, forming charged particle beams, and accelerating, in which the particles are accelerated. The injector modules contain high-voltage electrodes located along their axis on which the vortex emf applied to the cathode (anode) of the electron beam forming system is summed. Inside the accelerating modules there are tracts for transporting particle beams with magnetic systems to hold the transverse dimensions of the beam. A block diagram of the LIA module is shown in Fig. 1.3. It includes a charging pulse generator (CPG) for charging the forming lines (FL), and also a switch (S) connecting the FL to the induction system (IS). The forming line defines the amplitude and duration of the output pulses of the installation. The induction system raises the voltage from low (LV) to high (HV) values on the cathode holder of the injector module or forms a vortex

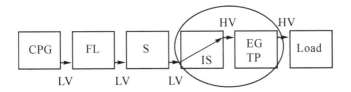

Fig. 1.3. Block diagram of a linear induction accelerator.

electric field on the axis of the accelerating module. Note, that the induction system is structurally integrated with the electron gun (EG) in the injector modules or with the transport path (TP) of the beam in the LIA acceleration modules, while the CPG, FL and S form a pulse generator (*1* in Fig. 1.1).

Triggering switch S (a gas spark gap or a magnetic commutator in the form of a saturation choke with a core of a ferromagnetic material) connects the FL through the induction system and the electron gun to a load (Load), for example, to the electron beam formation system of a relativistic microwave generator. The electron gun consists of a high-voltage insulator designed to separate the oil-filled volume of the accelerator and the vacuum volume of the vacuum tract and a system of electrodes for forming an electron beam. All these elements are a common high-current circuit and determine the shape and magnitude of the accelerating voltage pulse.

1.3.1. Generators of charge pulses based on capacitive energy storage devices

The limiting value of the electric field in modern capacitors reaches scales of $\sim 10^6$ V/cm, which determines the density of energy stored in them $\sim 10^7$ J/m³. Based on the successive addition of voltage on the pulse capacitors, it is possible to build a pulse voltage generator (PVG). One of the possible options for implementing such an addition was proposed in 1925 by Arkadiev and Marks.

The Arkadiev–Marks generator consists of N identical cells, each of which contains a series-connected capacitor and a spark gap (Fig. 1.4). Initially, all the capacitors are charged in parallel through the resistance R from an external high voltage source (typical charging voltage is 50–100 kV). Then, with the help of spark gaps, all the cells are switched in series, which makes it possible to obtain on the output of the generator a voltage equal to the charging voltage multiplied by the number of cells. Advantages of such a scheme are the simplicity of the design and the fact that only a few first (three

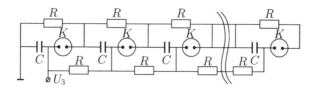

Fig. 1.4. Scheme of the Arkadiev–Marks generator.

to four) spark gaps should be manageable, while the rest are triggered by overvoltage on them. Its drawbacks include the non-rectangular shape of the output voltage pulse, a large number of dischargers (one for each cell), the criticality of the amplitude of the voltage pulse with respect to the time variation in the operation of the spark gaps (this can be partially compensated by the setting of the peaker), and also the flow through the spark gaps of the same operating current as in the load.

To double the voltage on the capacitors of the PVG cells and, respectively, to double the output voltage, a two-way charge circuit can be used. In this case, both half-periods of the rectified voltage are used to charge the capacitors.

Analysis of the operation of the PVG shows that the amplitude of the output pulse, as well as its duration and the duration of the front, are determined not only by the inductance (L_k), capacitance (C_k) and resistivity (R_k) of the discharge circuit, but also by the spark gap time constant (Θ). The smaller Θ, the greater the amplitude, the shorter the duration of the output pulse and its front. At a constant voltage, $U_0 = pd_{el}$ (where p is the gas pressure in the spark gap [atm], d_{el} – interelectrode gap [cm]), the quantity $\Theta \sim p^{-1}$. Consequently, the higher the gas pressure in the spark gap, the lower the value Θ. Thus, at high pressure the parameters of the pulse will be determined only by the parasitic parameters of the discharge circuit – the inductance of the load. If the influence of C_k can be neglected, the length of the front between the levels 0.1 and 0.9 will be $t_f = 2.2L_k$. However, if $R_k \gg 2\sqrt{L_k / C_k}$, then the pulse duration is $\tau = 0.7R_kC_k$. Typical parameters of PVGs used to feed LIA forming lines are: energy reserve ~0.1–1 MJ; the output voltage 0.5–1 MV; the pulse duration is 1–100 μs; internal wave resistance from one to tens of ohms.

Another method of voltage multiplication is known. This is the Fitch scheme proposed in 1964 (Fig. 1.5). Each cell of PVG, built in this circuit, consists of a pair of series-connected capacitors the polarities of which are opposite in sign during the charging process. Then, with the help of an additional circuit from the switch and

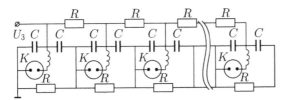

Fig. 1.5. Fitch scheme.

the inductance, the polarity of one of the capacitors in the cascade changes to the opposite one. Thus, the voltage at each stage increases in a time equal to half the oscillation period in the *LC* circuit, from zero to almost twice the charging voltage, so that the total voltage on the PVG is close to the charge voltage multiplied by the number of capacitors. The main advantages of this scheme include: half the number of dischargers; the form of the output voltage is less critical to the spread of the operation of the spark gaps; the current of the spark gaps can be substantially lower than the operating voltage; it is possible to correct the shape of the voltage pulse on the load when selecting the response time of the output spark gap.

The Fitch scheme also has the following drawbacks: all dischargers must be manageable; to obtain a steep voltage front, it is necessary to use an output discharger (sharpener); a rapid inversion of the voltage at the capacitors significantly reduces their service life; many additional elements are used, including control circuits.

In addition to the above schemes, magnetic pulse generators are used to generate high voltage pulses to charge the forming lines. The principle of operation of such generators will be discussed in detail in Chapter 2 when describing the LIA, developed by the Tomsk Polytechnic Institute.

Generator of charge pulses based on inductive energy storage devices. The advantage of inductive energy storage devices is a high specific energy content (up to 10^9 J/m^3), significantly exceeding the analogous index of capacitive storage devices. In addition, the use of generators on such energy storage devices allows increasing the output power and eliminating the influence of the prepulse on the load operation. They were considered as power sources for "non-iron" LIAs, so it makes sense to briefly introduce the principle of their operation.

A typical scheme of a charging pulse generator based on an inductive storage device is shown in Fig. 1.6. In the inductance of storage, for example, by using a capacitance discharge, the current

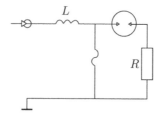

Fig. 1.6. The CPG based on inductive storage and current breaker

rises and then flows through the breaker I. The voltage on the load does not exceed the voltage of the power source. When the current reaches the level at which the breaker operates, the current in the inductance circuit drops sharply. As a result, after a breakdown of the switch K, a high-voltage pulse is formed on the load and the capacitor is discharged to the load. The voltage on the load,increases jumpwise. At the same time, the power developed on the load exceeds the power of the power source. With an active constant load, the current in it falls exponentially with a time constant determined by the values of R and L. The creation of an inductance, a power source and a switch does not cause technical difficulties. However, in order to transfer energy into the load in a short time, a fast current breaker is necessary. The problem of creating a multiple and reliable breaker is still not solved. The existing powerful circuit interrupters are built either on the basis of explosion of the conductors in heating with current or the detonation of explosives, or on the basis of a plasma current breaker (PCB). The former work only once, the latter are very unreliable. The typical efficiency of energy transfer in such schemes is about 25% for the inductive load and not more than 50% for the active resistance.

Current breakers based on electrical explosion of conductors. The qualitative description of the processes occurring during interruption of current in this type of breaker is based on introducing the characteristic opening time τ_s and the exponential current drop in time. Equations for the electrical circuit after the opening of the breaker have the form

$$\left[\begin{array}{l} I_s + I_R = I_L; \\ -L\dot{I}_L = I_R R, \end{array} \right. \tag{1.7}$$

where I_R is the current through the load R; I_L is the current in the inductive storage. Here it is assumed that during the opening time τ_s, which is much less than the characteristic time of the rise of current in the inductance, the voltage drop at the capacitance is insignificant. Solving the system of differential equations, one can obtain

$$I_R = I_0 \frac{\tau_0}{\tau_s - \tau_0} \left(e^{-t/\tau_s} - e^{-t/\tau_0} \right), \tag{1.8}$$

where $\tau_0 = L/R$ is the characteristic time of the circuit. At the time

$$t_0 = \frac{\tau_0 \tau_s}{\tau_0 - \tau_s} \ln\left(\frac{\tau_0}{\tau_s} \right) \tag{1.9}$$

current reaches the maximum value

$$I_{max} = I_0 \left(\frac{\tau_s}{\tau_0} \right)^{-\tau_s/(\tau_s - \tau_0)}. \tag{1.10}$$

The maximum of the current in a load close in magnitude to I_0 is possible only if the condition $\tau_s/\tau_0 \ll 1$ is fulfilled. Having integrated the power in time, the energy released on the load is calculated as follows:

$$Q = \frac{L I_0^2}{2} \frac{\tau_0}{\tau_s + \tau_0}. \tag{1.11}$$

It is seen that with decreasing τ_s/τ_0 it also tends to the energy stored in the inductance.

Plasma current breakers (PCB). The principle of operation of a plasma current breaker is as follows. Near the load of the pulse generator, a plasma channel is created between the grounded and high-voltage electrodes. The generator current initially flows through this channel. In this case, a partial or complete transfer of energy from the capacitive storage to the inductive one takes place. When the conductivity of the plasma bridge decreases, a vortex emf is generated and the energy flow accumulated in the inductance is switched to the load. In experiments with inductive energy storage with PCB, the possibility of generating pulses of duration ≤ 10 ns with a voltage amplitude of several MV at a current of 1 MA was demonstrated [11].

To calculate the output characteristics of the generators, information is needed on the parameters of the current breaker, which have been only partially obtained so far. However, some general principles of constructing generators can be formulated from the analysis of the circuit shown in Fig. 1.6. In this circuit, the current breaker is modelled by the active resistance, which at the initial instant of time sharply increases. The load appears to be an ohmic constant resistance. This representation of the PCB is based on the fact that the resistance of the PCB increases to its maximum value during the time in which the current in the inductive storage does not have time to decrease noticeably. The maximum power mode is realized when the resistance of the current breaker and load equals. In this case, the power in the load reaches $P = 0.25\ (I_L)^2 R$. If the current I_L of the inductance L is obtained by discharging the capacitance C charged to the voltage U_0, then $I_L = U_0/\rho_0$, wherein $\rho_0 = (L/C)^{1/2}$. Then

$$P = \frac{U_0^2}{4\rho_0} \frac{I_L R}{U_0}. \tag{1.12}$$

It follows from (1.12) that in order to obtain high power values in the load, it is necessary to have a primary storage device with a high output voltage and a low wave resistance. To obtain the same maximum power increase factor equal to $I_L R/U_0$, the voltage of the primary energy storage should be reduced.

To fill the interelectrode gap of the PCB with plasma, coaxial plasma guns are used with electrical breakdown over the surface of the dielectric washer. The surface sources of multicomponent plasmas (H^+, C^+, C^{++}), which are a chain of discharge gaps and allow the generation of a large number of plasma fluxes for uniform filling of the interelectrode space of the PCB, are widely used. To obtain single-component plasma fluxes, it is possible to use systems with a gas inlet (H_2, D_2, He, Ar, Kr). Usually several plasma sources are located evenly along the azimuth on the surface of the outer electrode in which there are longitudinal slits. The distance to the central high-voltage electrode is chosen to be sufficient for the spatiotemporal separation of the plasma bunch to the neutral component produced by the operation of the plasma source. The power sources of the plasma guns are low-inductance capacitor banks with an energy capacity from one to tens of kJ, depending on the number of guns and the required amplitude of the current through the PCB.

1.3.2. Induction systems

The induction system of LIAs usually contains cores made of a ferromagnetic material, which makes it possible to significantly reduce the current in the primary winding of inductors in comparison with the so-called 'non-iron' version of the accelerator. For example, for cores made of 50 NP alloy strips (permalloy) with a thickness of 20 and 10 μm, the average value $\mu = \Delta B/(\mu_0 H)$ for a pulse duration $\tau = 0.5$ μs is 2000 and 4000, respectively [12]. This makes it possible to improve the efficiency of the accelerator.

The induction system is assembled from sections whose dimensions are determined by the requirements for the focusing system, the vacuum system, and purely constructive considerations. Focusing lenses are installed in the gaps between the sections, as well as nozzles for connecting vacuum pumps and placing diagnostic equipment. The sections, in turn, consist of a number of identical elements (inductors). When choosing the type and thickness of the insulating material installed between the inductors, as well as the inductor steel and the magnetizing coil, account should be taken of the corresponding electric field strength distributions shown in Fig. 1.7.

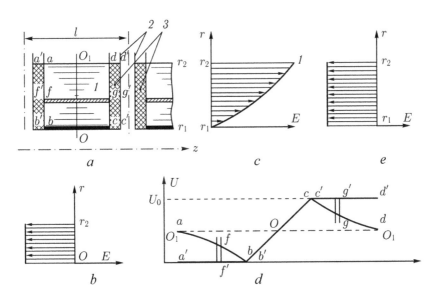

Fig. 1.7. Diagrams of the electric fields of the inductor (*1* – core, *2* – magnetizing coil, *3* – insulation): *a*) sketch of part of the section; *b*) the diagrams of the field inside the section; *c*) the field diagrams inside the inductor; *d*) the field between the inductors; *e*) potential difference.

One of the most important moments in the design of LIA is the correct choice of the basic geometric dimensions of inductors, mainly the ferromagnetic core, depending on the necessary rate of acceleration and the magnitude of the accelerated charge. Proceeding from the condition that the supply voltage of the magnetizing coil U of the accelerating vortex emf acting along the axis of the accelerator is equal to the acceleration rate, the quantity $T = eU/l$ (where l is the axial length of the inductor and e is the electron charge) is called the acceleration rate. To excite a vortex emf with a voltage U for a time τ, it is necessary to have a ferromagnetic material cross section of one inductor (in the case of an inductor consisting of several ferromagnetic cores, Fig. 1.8), equal to

$$S = \frac{1}{\Delta B}\int_0^\tau U\,dt = (r_2 - r_1)l\frac{k}{1+(l-a_f n_1)/l} \approx \frac{U\tau}{\Delta B}, \tag{1.13}$$

where k is the coefficient of steel filling of the core volume in the radial direction; $a_f n_1$ is the total width of the ferromagnetic toroids of the inductor; n_1 is the number of ferromagnetic cores in the inductor; r_1 and r_2 are the inner and outer radius of the ferromagnetic core.

The volume of the ferromagnetic material of one inductor is

$$V = 0.5(r_2^2 - r_1^2)l\,k_2 = \frac{U\tau(r_2 + r_1)}{2\Delta B}, \tag{1.14}$$

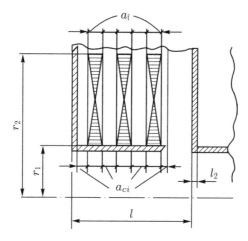

Fig. 1.8. Schematic representation of the inductor.

where k_2 is the total filling factor ($k_2 = kl/(a_f n_1)$). If we proceed from minimizing the ratio of the total volume of the ferromagnetic material to the maximum possible rate of acceleration, then the optimum occurs at $r_2/r_1 = 2.2$. The LIA operation experience shows that it is at a ratio $r_2/r_1 = 2.2$–2.5 that the minimum value of the inductor impedance components is observed. The last condition, as well as the equations (1.13) and (1.14), are the starting points for calculating electromagnetic processes in the accelerating system, the choice of its geometry, etc. Although the LIA system is actually a set of single-turn pulse transformers, the calculation method used in the design of pulse transformers operating in the microsecond range [13, 14] proves to be unacceptable here, especially when using a ferromagnet with a rectangular hysteresis loop (RHL). The fact is that the pulse transformer theory focuses on processes in the cores, made of electrical steel which is characterized by a nearly linear dependence $\Delta B(H)$. Comparative experiments have shown that a higher efficiency in LIA can be obtained by using precision soft magnetic alloys with a narrow hysteresis close to a rectangular loop. In particular, many iron–nickel 50NP alloys (15, 16] are used in many accelerators, including those in the Tomsk Polytechnic Institute. The magnetization reversal processes for such alloys have their own peculiarities, considered below. The magnetization reversal of the 50NP alloy with a 500 ns pulse is shown in Fig. 1.9 *b*. The magnetizing hysteresis loops of cores from other ferromagnetic alloys are similar in form to those discussed here.

In addition, with pulse lengths typical for LIA of tens and hundreds of nanoseconds, the thickness of the tape from which the cores are made is selected in the range of 10–20 μm. With this thickness, the action of eddy currents is commensurate with the effect of magnetic viscosity, which is not taken into account in the theory of pulse transformers. The magnetization reversal of a core from a ferromagnet with RHL taking into account eddy currents and magnetic viscosity is considered in a number of articles (see, for example, [12]). With the accuracy sufficient for practice it described by the equation

$$H(t) = H_0 + g(B)\frac{dB}{dt}. \tag{1.15}$$

Here H_0 is the start field, the value of which depends on the material grade and is 1.5–2.5 times higher than the coercive force H_c; $g(B)$ is given by

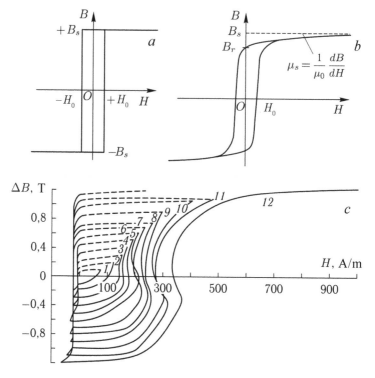

Fig. 1.9. The rectangular (*a*) and idealized (*b*) hysteresis loops, as well as the reversal of the magnetization of the 50 NP alloy by a pulse with the 500 ns duration 500 (*c*) [12].

$$g(B) = \left(\frac{\sigma\delta^2}{8B_s}\right)(B+B_r) + \left(R_m\left(1-B^2B_s^{-2}\right)\right)^{-1}, \qquad (1.16)$$

where B_r and B_s are the residual induction and induction of saturation; δ is the thickness of the tape from which the core is curved; σ is the electric conductivity of a ferromagnetic; R_m is a constant depending on the material type [Ohm/m]. The first member of the right-hand side of (1.16) characterizes the action of eddy currents (for ferrites it is zero). Its second term is due to magnetic viscosity and has a place both for ferrites, and for metallic ferromagnetics.

Calculations and experiments show that for metallic cores the first term in (1.16) can be neglected with a tape thickness <5 μm, and the second with a tape thickness >30–40 μm. The LIA uses tapes with a thickness of 10–20 μm, since the use of a tape thinner than 5 μm is associated with technological difficulties, besides, it reduces the filling factor of the core volume with steel, and the cost of such a tape is relatively high. With a tape thickness of more than 20 μm, the efficiency falls. Thus, in the calculations it is necessary to take into account both terms in the expression (1.16).

Integrating (1.15) with respect to time, we obtain

$$\int_0^\tau [H(t) - H_0]dt = \int_{B_{ini}}^{B_s} g(B)dB. \tag{1.17}$$

The integral on the left-hand side of (1.17) is called the field impulse. With full reversal of magnetization, its value does not depend on the duration and shape of the magnetization reversal voltage and current, since

$$\int_0^\tau [H(t) - H_0]dt = \int_{-B_s}^{B_s} g(B)dB = g_1(+B_s) - g_1(-B_s) = S_\omega, \tag{1.18}$$

where τ is the pulse duration with full reversal of magnetization. The field pulse at full reversal of magnetization is called the switching coefficient. It represents the amount of electricity necessary for a complete magnetization reversal of the core, per unit of its length. After integrating (1.17) with allowance for (1.18), we find the expression for the field momentum:

$$\int_{-B_s}^{B_s} g(B)dB = \frac{\sigma\delta^2}{16B_r}\left\{(B(t) + B_r)^2 + \frac{B_s}{R_m}\left[\operatorname{arcth}\frac{B_r}{B_s} + \operatorname{arcth}\frac{B(t)}{B_s}\right]\right\}. \tag{1.19}$$

As can be seen from Fig. 1.9 at $B > 0.75\ B_s$, the slope of the magnetization reversal curves changes abruptly. Therefore, it is usually not advisable to remagnetize the core to $B > B_r$. Recalling that in the interval $-B_r \leq B \leq B_r$ the function arcth can be approximated by a linear function; from (1.19) we obtain the field impulse:

$$Q(\lambda) = \frac{B_r\sigma\delta^2}{4}\lambda^2 + \frac{2B_s}{R_m}\operatorname{arcth}\frac{B_r}{B_s}\lambda, \tag{1.20}$$

where $\lambda = \Delta B(t)/2B_r$. For $\lambda = 1$, from (1.20) it follows that $S_\omega = S_{\omega e} + S_{\omega 0}$, where $S_{\omega e}$ is the switching factor due to the action of eddy currents, and $S_{\omega 0}$ is its component due to the action of viscosity. Further, we have

$$Q(\lambda) = S_{\omega e}\lambda^2 + S_{\omega 0}\lambda. \tag{1.21}$$

Differentiating (1.20) with respect to time, taking into account that for a rectangular voltage impulse $\dfrac{d\lambda}{dt} = \dfrac{1}{\tau}$, we obtain

$$H(t) = H_0 + \frac{S_{\omega 0}}{\tau} + \frac{S_{\omega e}}{B_r}\frac{\Delta B(t)}{\tau}. \tag{1.22}$$

It can be seen that, up to the accepted assumptions, the dynamic hysteresis loop expands due to the start field and magnetic viscosity and has an inclination due to eddy currents.

Table 1.2 shows the calculated values of S_ω for the most common materials [12], as well as some of their characteristics.

The real magnetization reversal curve (see Fig. 1.9 *b*) differs somewhat from that calculated from (1.22). Therefore, the exact values of the magnetization reversal field must be in accordance with the experimental characteristics. The non-linearity of the

Table 1.2. Characteristics of the most common materials

Alloy	Thickness, mm	Magnetic permeability mH / m	Switching factor μK / m	Coercive force A / m	Saturation induction T	Coefficient of square-ness
50 NP	0.01	25	110	32		0.83
	0.02	50	160	20	1.50	0.85
	0.05					
	0.10	50		18		0.85
34 NKMP	0.01	44	145	24		0.92
	0.02	50		16	1.50	0.90
	0.05	75		12		0.87
	0.10	125		8		0.85
35NKKhSP	0.01	38		24		0.85
	0.02	50		16	1.30	0.85
	0.05	75		12		0.85
	0.10	125		8		0.80
68NMP	0.02	125		8.0		0.90
	0.05	250		5.6	1.15	0.90
	0.10	280		4.0		0.90

magnetization reversal characteristic complicates the analytical consideration of processes in the induction system.

Typically, induction systems of the LIA contain a demagnetization current generator:

$$i_m(t) = \pi D_{av} H(t), \tag{1.23}$$

where D_{av} is the average diameter of the ferromagnetic core. As already mentioned above, with the aid of a demagnetization system, the cores are transferred to a state of negative saturation. For materials with RHL, after the core is transferred to the saturation state, the demagnetizing field can be reduced to zero. In this case, pulses of a current of opposite polarity (in particular, a half-wave of a sinusoid) are used for demagnetization. When using materials with a small rectangularity, for example ferrites of the nickel–zinc group, the core must be placed under the action of a demagnetizing field before the operating pulse is applied in order to have the greatest increment in induction.

The energy loss in the core, corresponding to the area bounded by the magnetic hysteresis loop on a particular cycle, is determined as follows:

$$W_m = \int_0^t u i_m dt. \tag{1.24}$$

From (1.24), taking into account (1.21) and (1.22), we obtain

$$W_m = VH_0\Delta B + \left(\frac{2VB_s}{\tau}\right)\left(S_{\omega 0}\lambda^2 + S_{\omega e}\lambda^3\right), \tag{1.25}$$

where V is the volume of the core. (It is assumed that the material has RHL and $B_r \approx B_s$.) The first term on the right-hand side of (1.25) is proportional to the energy accumulated in the core, which, generally speaking, can be used again, but this is not done in the LIA. This term is always proportional to the volume of the core and inversely proportional to the time of its magnetization reversal. The second term represents dynamic losses in eddy currents and viscosity. The dependence on induction differential ΔB is more complicated and is determined by the relation between the $S_{\omega 0}$ and $S_{\omega e}$. The quantity $S_{\omega 0}$ is the characteristic of the material itself and does not depend on the thickness of the rolled product. The value of $S_{\omega e}$ is proportional to the square of the thickness of the tape. With a small tape thickness (<5 μm), $S_{ue} \ll S_{\omega 0}$. If $S_{\omega 0} \ll S_{\omega e}$, then the eddy current energy losses predominate, and the energy loss is proportional to ΔB^3.

In view of the fact that the energy losses in the core are proportional to its volume, and the acceleration applied to the particles is proportional to the area of its cross section, in order to obtain a greater efficiency of the accelerator at its low weight, it is necessary to minimize the diameter of the cores and increase their axial size.

The value $W_n/(W_n + W_m)$, which is the ratio of the energy transferred to the beam (W_n), to the total energy spent on demagnetizing the cores, is the main component of the overall efficiency of the accelerator. The experimental data obtained in [12] allow us to estimate this value. When using a strip of 50NP alloy with a thickness of 10 microns, it has values of 0.12, 0.6, and 0.95 for a pulse duration of 500 ns and beam currents of 100, 1000, and 10 000 A, respectively. With a pulse duration of 50 ns and the same values of the beam current, this value is 0.02, 0.15, and 0.6. Thus, the efficiency of the LIA grows with the increase in the pulse duration and the rise in the current of the accelerated beam. The latter circumstance makes it possible to maintain a sufficiently high efficiency even at small pulse durations.

The paper [17] describes a LIA with a microsecond duration of the output current pulse. The program for creating such an accelerator was started in the 1970s. By that time, LIAs were operated with a current of several hundred amperes with a pulse duration of 40 and 300 ns and an electron energy of up to 4–5 MeV. Then the National Bureau of Standards of the USA made a decision to produce LIA with an output pulse of microsecond duration for some technological applications of electron beams. The efforts of developers were aimed at using LIA technology in applying cheaper ferromagnetic materials. The accelerators operating at that time used ferrite in the inductor cores (for example, LIA ERA), which does not provide the required pulse duration. Therefore, the properties of three magnetic materials (permalloy, silicon steel (Si–Fe) and low-carbon steel containing 3.5% Si) were analyzed. Of all the materials listed, permalloy has the smallest coercive force, and hence the lowest losses due to the magnetizing current. This material can be rolled into a thin tape to reduce the eddy current losses. However, the developers found its use extremely expensive, especially when taking into account the required significant amount of steel. Therefore, tape cores made of an alloy of silicon and low-carbon steels were tested. The latter is produced in the United States in large quantities for industrial transformers. The thickness of its tape can reach a value of 5 μm, sufficient to ensure small losses in eddy currents. The following considerations also served in favour of the choice of low-carbon steel. First, the saturation induction of low-carbon steel is only 10% lower than that of the silicon steel. Secondly, the use of a low-carbon steel tape is twice as thin as the silicon steel makes it possible to practically equalize the losses in the cores during magnetization. Also the low-carbon steel is much cheaper.

In order to achieve a high acceleration rate in the LIA, it is necessary to apply high voltage to the magnetization coils of the inductors. Preliminary experiments showed that in order to prevent electrical breakdowns, it is sufficient to have a 0.63 μm interlayer insulation with mylar between the turns of the steel strip of the core. Core winding is carried out using liquid polyester resin, which subsequently hardens, giving the core sufficient mechanical strength. Note that the use of mylar and polyester as an interlayer insulation does not allow for annealing of the core, as is usually done in the case of ferromagnetic cores of LIA. Therefore, in the material of the core there remain the stresses and distortions of the structure formed during the manufacture of the magnetic tape and winding. As a result, it is possible to increase the energy losses during magnetization reversal of the core and to reduce the achievable induction range.

To form electron beams of the microsecond duration with a high rate of acceleration, cores with a large steel cross section are required. In this case, to maintain a reasonable length of the accelerating path, cores with a large ratio of the outer radius to the inner radius are needed. The maximum use of magnetic material takes place when the steel of the entire core is saturated simultaneously. At the same time, it is known that the material on the inner radius is saturated earlier than the peripheral part. Thus, most of the core is magnetized only partially, which reduces the induced voltage.

This problem was resolved by dividing the core into segments (Fig. 1.10). With the parallel connection of the magnetizing turns of the segments it was possible to increase the utilization rate of steel from 50 to 90%. Such an improvement has led to another positive effect associated with an increase in the rate of acceleration of the LIA. Since the secondary turn of the induction system covers all segments of the core, the accelerating voltage is the sum of the induced voltages from each segment. Increasing the voltage reduces

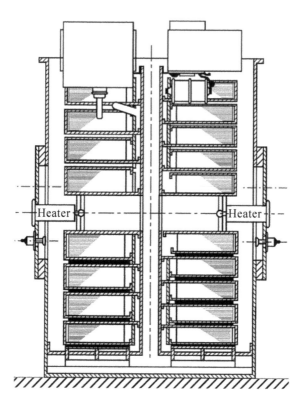

Fig. 1.10. Inductors from four and five segments of LIA constructed by the National Bureau of Standards USA.

the output voltage of the pulse generator, which in turn reduces its cost and improves reliability.

Figure 1.10 shows the cross sections of the experimental cores consisting of four and five radial segments. The turns of magnetization of each segment of the core are made of a copper sheet 0.51 mm thick and connected in parallel by means of two wide copper strips. Together with the grounded planes, they form a compact low-inductance conductive system with a high admissible current. The weight of the cores is 400 kg. When placed in an oil-filled tank, they are able to withstand a voltage pulse of 200 kV with a duration of 2 μs. The current flowing through the magnetizing turns of the cores is created when the forming line with lumped parameters is discharged when the spark gap is switched on. The output impedance of the system is 1.29 Ohm; output voltage 100 kV. The accelerator of the National Bureau of Standards USA used 4 cores. Created on the basis of the described technology, the LIA generates current pulses with an amplitude of 750 A and a duration of 2 μs at an electron energy of 750 keV.

Induction LIA systems of the Tomsk Polytechnic University. There are several variants of induction systems. In one of them, a ferromagnetic core is located between textolite discs 2 mm thick. On the outside of the discs the magnetizing coils are arranged in the form of copper plates with strip terminals for connection to the electrodes of the forming line (Fig. 1.11). The cores are isolated from each other by textolite discs 2–5 mm thick. For laying of strip forming lines, the outer diameter of these discs is larger than the diameter of the cores (Fig. 1.12). Studs tightening induction system are installed in special grooves made on the outer diameter of the discs. The leads of the magnetizing turns from opposite sides of the inductors are connected to each other, and electrodes of the strip forming lines are soldered to them. The location of the magnetizing coils of neighbouring inductors at a small distance from each other (through one isolating disc) ensures a minimum inductance of the coil (for more details, see Chapter 2). However, with such a design, the 'turn–turn' capacity of the neighbouring inductor increases.

Fig. 1.11. Appearance of inductors and insulating discs.

Fig. 1.12. Induction system of LIA.

Another variant of the induction system design allows to abandon the output insulator calculated for the total voltage of the injector section. In addition, this element of LIA undergoes intensive exposure to various factors from the cathode–anode gap (ultraviolet irradiation, bombardment by charged particles), which reduces its reliability. In the present variant, the insulator is distributed in the form of rings made of plexiglass installed between the ferromagnetic cores. The thickness of the rings is 16 mm. The rings have a developed surface with an electrical strength exceeding 80 kV/cm. The magnetizing coils of inductors are made of solid stainless sheet with a thickness of 1.5 mm. Together with the insulators they form the vacuum volume of the accelerator. Since the windings have an inner diameter smaller than the diameter of the insulators, they shield the surface of the insulators from the action of the above factors.

The equivalent circuit of the inductor. The equivalent circuit of the inductor should most fully reflect the picture of the ongoing physical processes, especially in the case of a single-turn winding. At the same time, a large number of reactive elements in the circuit result in the need for solving high-order differential equations and for complex transcendental dependences between currents and voltages. For example, the complete equivalent scheme for a pulsed transformer is described by a differential equation of the 9th order [14]. A sufficiently complete inductor equivalent scheme can be represented by six reactive and three ohmic elements (Fig. 1.13).

The equivalent scheme is selected by the method of estimating the energy contained in individual sections of the field, equating its energy stored in the corresponding reactivities, and calculating the energy losses at the relevant active 'consumer'. Nominals of the elements are determined by their geometric dimensions and position

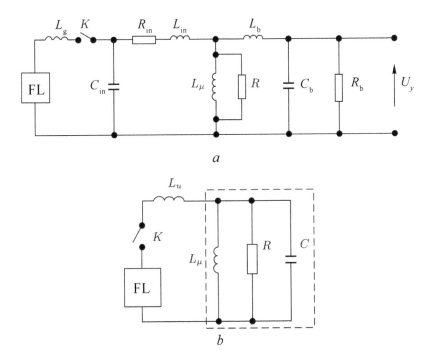

Fig. 1.13. Complete (*a*) and simplified (*b*) equivalent scheme of the inductor: L_g – inductance of the pulse generator (commutation circuits and supply lines); $C_{in} = C_c + C_{m.c}$ is the capacity of the inductor equal to the sum of the capacitances of the magnetizing coils relative to the core and the mutual capacitance between the inductors; L_{in} and L_b – leakage inductance of the magnetizing coils and inductance of the beam (load); L_μ – magnetization inductance; C_b – capacitance of the accelerating tract, dynamic capacitance of the beam (load); R_{in}, R and R_b – resistance equivalent magnetizing losses in coils and eddy currents and the magnetic viscosity losses in the beam (load).

in space relative to other elements. An example of calculating the nominals of the elements of the replacement circuit for the current of LIA 04/6 will be given in Ch. 2. In addition, a more detailed discussion for the case of the generator of microsecond pulses (GMP) is carried out in Ch. 5. The output characteristics of LIA and GMP were calculated on a computer using the Electronic Workbench software.

Let us analyze the effect of reactive elements on the parameters of the output pulse. To do this, we use a simplified equivalent circuit (Fig. 1.13 *b*), in which

$$L_u = L_{in} + L_g + L_b, \quad C = C_{in} + C_b, \quad \frac{1}{R} = \frac{1}{R_{in}} + \frac{1}{R_b}.$$

If the conditions

$$L_{in}\frac{di}{dt} \ll U; \quad \frac{dU}{dt} \ll \frac{U}{RC}, \tag{1.26}$$

are satisfied, then the shape of the peak of the accelerating voltage pulse is determined mainly by the law of variation of the wave impedance of the forming line and the parameters of the ferromagnetic cores of the inductor. In this case, the length of the pulse front depends primarily on the capacitance C, which shunts the load. Under the conditions (1.26) for the adopted equivalent inductor scheme, the front duration is defined as

$$\tau_{front} \approx 2\frac{L_{in} + R_b RC}{R_b + R}. \tag{1.27}$$

Here, the lengthening in the front due to the finite rapidity of action of the switch is not taken into account, and the equivalent resistance of the beam is assumed to be equal to the wave resistance of the line ρ. The pulse front duration is minimal when the wave impedance of the inductor ρ_{in}, formed by parasitic parameters, satisfies the condition

$$\rho_{in} = \rho = \sqrt{\frac{L_g + L_{in}}{C_b}}, \tag{1.28}$$

and the resistance of the beam (load) is a condition

$$R_b = \rho = \sqrt{\frac{L_g}{C_b}}. \tag{1.29}$$

Thus, as in any system with the lumped parameters, the system's wave impedance, formed by parasitic parameters, must be equal to the wave impedance of the current pulse generator.

By increasing the magnetizing current in the inductance L_μ, the voltage drop on the internal resistance of the pulse generator increases and, consequently, there is a slight decrease in the voltage at the load and the current flowing through it. Obviously, the voltage change on the load will be the greater, the longer the pulse, the smaller the magnetization inductance and the greater the internal resistance of the pulse generator R_g. The relative change in voltage during the pulse duration is

$$\frac{\Delta U}{U} = \frac{\tau R_g R}{(R_g + R)L_\mu}.$$

(1.30)

To increase the stability of the induced voltage, it is necessary to reduce the internal resistance of the generator or increase the load resistance (to reduce the number of accelerated particles) during the time of the pulse action or approximately in proportion to the value of tR/L_μ.

The need for an accurate calculation of the instantaneous values of voltage and currents in the inductor is dictated mainly by the fact that when the load is mismatched with the pulse generator, the secondary voltage can differ significantly from the exciting voltage, which will lead to an increase in the scatter of particles and, as a consequence, to fluctuations in the beam density along the length and its defocusing.

One should pay attention to the fact that the average power in the beam of the LIA reaches hundreds of kilowatts and in order to maintain a high efficiency of the induction system, it is necessary to strive to completely transfer the energy in the storage to it. For optimal energy transfer, either inhomogeneous artificial lines [18], whose theory was developed in [19], or homogeneous lines in combination with the correction system [20] are used.

Linear induction accelerators with ferrite induction systems. If we limit ourselves to a short duration of the accelerating voltage pulse, then ferrite can be used for the accelerator cores (for example, 300NN type, as it is done in the LIA SILUND-20). Studies have shown that ferrites of the manganese zinc group have magnetization reversal characteristics similar in shape to alloys with a rectangular hysteresis loop. The SILUND-20 has the following output parameters: electron energy 1.5–2 MeV; current 400 A; pulse duration 10–15 ns; pulse repetition frequency 20–50 Hz. Its induction system is also assembled from individual inductors, but since ferrite is a dielectric, there is no need for special insulation of the turns from the core.

Specially selected ferrites are also used for the sections of the ERA injector. However, their design is significantly different from the above. The section is one inductor with a voltage on the accelerating gap of 250 kV. Its internal cavity is filled with ferrite and transformer oil, which provides the necessary electrical strength. The section is connected directly with the forming line. With this design there is no need for a special accelerating tube and the impedance of the system is greatly increased, which facilitates the formation of a pulse. However, the difficulties caused by voltage increase, and, naturally, the reliability of the induction system drops. In addition a pulsed hydrogen thyratron can not be used here as a switching element of the pulse generator (in the developed thyratrons the anode voltage does not exceed 80 kV). As a result, it has to use a spark gap, limited to single impulses. The pulse generator is a double forming line with a spark gap as a switching element. The internal cavity of the forming line is filled with transformer oil. The bushing insulators between the air discharger (pressure 10 atm) and the oil forming line are made of an epoxy compound. Charging of the forming line is carried out from the PVG during 330 ns. In this case, the line voltage is close to the breakdown for the spark gap. If the spark gap is not ignited intentionally, then after 100 ns it turns on spontaneously. This mode of operation of the discharger provides a small spread in the response time (~1 ns). The direct joining of the forming line with the accelerating section and the use of an air discharger under pressure made it possible to obtain a sufficiently short duration of the voltage pulse front (~12 ns).

The most impressive advances in this area of accelerator technology include the Advanced Test Accelerator (ATA) manufactured at the Lawrence Livermore National Laboratory (LLNL, USA). It greatly exceeds existing high-current accelerators in terms of the parameters and operates in a pulse-periodic mode [21]. This accelerator consists of an injector for 2.5 MeV and 170 accelerating modules. The injector is the previously manufactured module ETA, which is upgraded to reduce the size, increase the repetition rate of pulses and has a lower electric field strength on insulators. The LIA ATA injection system contains a triode gun, which generates a current of 10 kA and accelerates electrons to 2.5 MeV. The electron source is a plasma cathode controlled by a grid, located at a distance of 2 cm from the cathode. The injector is made up of 10 induction modules designed for a voltage of 250–300 kV each. The acceleration rate reaches 12.5 MeV/m. Specially designed ferrite toroids are used in the primary circuit of the inductor of the ATA accelerator. Each inductor is powered by a pulse with an amplitude of 250 kV, a half-height duration of 70 ns and a 15 ns front, formed by 12-ohm double water-insulated forming lines. The electron beam passes through accelerating modules, which increase the electron energy to 47.5 MeV. The duration of the current pulse is 70 ns, the rise of the pulse between levels 0.1–0.9 occurs over 15 ns. At the length of the accelerating system of 85 m on the ATA installation, electron beams with a current of up to 10 kA, an electron energy of 47.5 MeV, a pulse repetition rate of 1 kHz were obtained with the formation of 10 pulses. Electrons are transported in a magnetic field with an induction of 0.3 T, created over the entire length of the accelerator. The ATA accelerator is designed to investigate the generation of radiation in free-electron lasers. The progress in the region of such high energies required the solution of a number of complex problems, the most important of which was the problem of ensuring the transverse stability of a high-current beam. The greatest success was achieved when the beam was transported in a laser-created ionized channel in benzol vapours

1.3.3. Forming lines

Short-pulse electron beams with fast current growth, are produced using single (SFL) or double (DFL) forming lines. These are lines with distributed parameters, which can have a wide variety of configurations (stripline, coaxial, radial). All of them are used in the LIA.

Single forming line. Consider one of the simple forming lines – strip forming line (Fig. 1.14). Several types of waves can propagate in such a line: TE, TM and TEM. The main mode here, obviously, is the lowest-frequency TEM-wave, which has a linear dependence $\omega = kc^*$, where $c^* = c/\sqrt{\varepsilon\mu}$ is the speed of propagation of the electromagnetic wave in the dielectric. The remaining modes in such a line can be excited only in the case $\omega > \pi c/d$. So, when a strip line is discharged to an active load a planar TEM-wave runs in it with the speed c^*. The ratio of the electric and magnetic fields in such a wave has the following form: $\sqrt{\varepsilon}E = \sqrt{\mu}H$, and their distributions along the section of the line can be considered uniform. In this case, the capacitance and inductance per unit length, as well as the line impedance, are

$$C_{SFL} = \frac{\varepsilon b}{4\pi d}, \quad L_{SFL} = \frac{4\pi\mu d}{b};$$

$$\rho_{SFL} = \frac{U_{FL}}{I} = \frac{4\pi Ed}{cbH} = \frac{4\pi d}{cb}\sqrt{\frac{\mu}{\varepsilon}}.$$

(1.31)

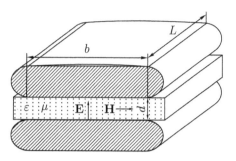

Fig. 1.14. The single strip line.

In accordance with the above formulas, the maximum energy accumulated in a line per unit length is

$$Q_{SFL} = \frac{\varepsilon E_m^2 bd}{8\pi},$$ (1.32)

where E_m is the maximum electric field inside the dielectric located between the strips in the line. When connecting a line pre-charged to a voltage $U_{SFL} = E_m d$, the power released at a load resistance ρ_{load} close in value to the wave impedance of the line ρ_{SFL}, is described by the expression

$$P = I^2 R = \frac{U_{SFL}^2 \rho_{load}}{\left(\rho_{load} + \rho_{SFL}\right)^2} = c\sqrt{\frac{\varepsilon}{\mu}} \frac{E_m^2 bd}{4\pi} \frac{\rho_{load}\rho_{SFL}}{\left(\rho_{load} + \rho_{SFL}\right)^2}.$$ (1.33)

Typically, the electrical length of the line is small compared to the time it is charged from the charge generator. In this case, the line operates as a lumped capacitance. The charging voltage of the line U_{SFL} is related to the output voltage of the generator of charge pulses U_{CPG} by the ratio

$$U_{SFL} = \frac{2U_{CPG}C_{CPG}}{C_{SFL} + C_{CPG}},$$ (1.34)

where C_{CPG} – impact capacitance of the CPG (for example, the capacitance of the capacitors of last stage of the magnetic pulse generator (for details, see Chapter 2).

During the discharge of the SFL, it is necessary to consider as a long line with a wave resistance $\rho_{SFL} = \sqrt{L_{SFL}/C_{SFL}}$. Let us study the process of discharge of a single forming line using the diagrams shown in Fig. 1.15. An open charged SFL with an electric length $\tau_{SFL} = \sqrt{L_{SFL}C_{SFL}}$ can be represented as a superposition of the forward and backward waves (Fig. 1.15 a). To satisfy the boundary conditions for the open line voltage polarity ofeach wave must be the same, and the polarity of the current must change when reflected from the open ends. Thus, the total current in the line is zero, and the voltage along it is constant and equal to U_{SFL}. When the discharger is switched on, a load with resistance ρ_{load} is connected to the output of the SFL (in the case of LIA, the load is an induction system). From the line, the energy begins to be ejected by a wave travelling in the positive direction (Fig. 1.15 b). If the wave impedances of the line and the load are equal, then the wave is not reflected at the output and all the energy stored in it is output in a time 2τ. Since the current flowing through the load is equal to $I_{SFL}/2$, the voltage at the load is equal to $U_{SFL}/2$, which is half the line charging voltage.

In the case of a coaxial forming line (Fig. 1.16), which is a doubly connected waveguide, the main mode is also the TEM-wave. The impedance of the line is

$$\rho'_{SFL} = \frac{60}{\sqrt{\varepsilon}} \ln \frac{R_2}{R_1},$$
(1.35)

and the pulse width is

$$\tau'_{SFL} = \frac{2L\sqrt{\varepsilon}}{c},$$
(1.36)

where ε is the permittivity of the insulation; L is the length of the coaxial line; R_1 and R_2 are the radii of the inner and outer electrodes. The electrode radii are chosen based on the required value of the wave impedance and providing electrical strength.

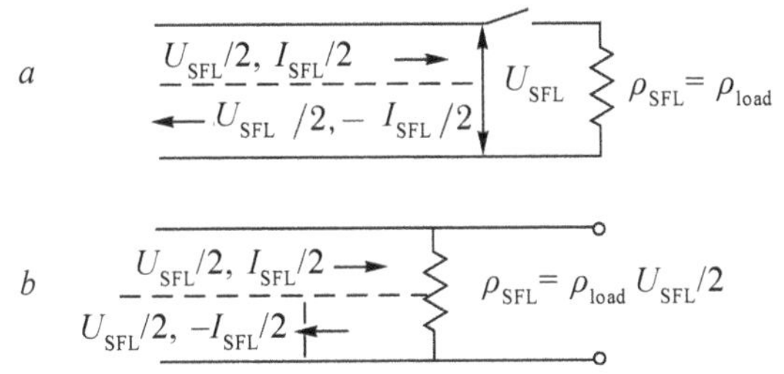

Fig. 1.15. Formation of a pulse in a charged line: *a*) until the discharger closes; *b*) after the discharger is closed.

The density of the energy stored in the line is directly proportional to εE_m^2, which requires the use of dielectrics with the highest value of the dielectric permittivity and the greatest electrical strength. At present, the maximum strength ($E_m \sim 0.5$ MV/cm) achieved by using paper–polymer–oil insulation ($\varepsilon \approx 2.4$–2.6), impregnated with transformer oil ($\varepsilon \approx 2.4$). However, this insulation is not restored and requires a complete replacement after the breakdown. Good characteristics were obtained using a mixture of purified water and ethylene glycol ($\varepsilon \sim 40$–80; $E_m \sim 0.25$ MV/cm). Such a dielectric has low conductivity at millisecond charging times. Typically, the water-insulated forming lines are used to create low-impedance LIAs because of the high dielectric permeability of water. At the same time, the forming lines with insulation based on solid dielectrics with transformer oil allow charging to a higher voltage with equal electrode sizes.

In contrast to the strip line, the distribution of electric and magnetic fields in the coaxial forming lines is non-uniform across the section and increases inversely with the radius. To obtain optimal parameters when switching such a line to the load, it is necessary to take into account the following circumstance. It is known from the experiment that the maximum achievable electric fields $E^{\pm}$ on positively and negatively charged electrodes immersed in a dielectric have different values. This property is characterized by the value $K = E^-/E^+$, which, as a rule, is greater than 1. Thus, for typical dielectrics used in the creation of powerful line-based devices, the value of K is: for water 2; for transformer oil 1.5. In view of this circumstance, the negative electrode in the line is usually chosen to be its most stressed internal electrode. The linear capacitance, inductance, and also the wave resistance of the coaxial FLs are determined by the following formulas:

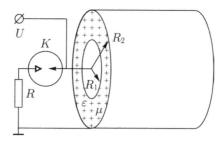

Fig. 1.16. The scheme of a coaxial single forming line.

$$C'_{SFL} = \frac{\varepsilon}{2\ln(R_2/R_1)}, \quad L'_{SFL} = 2\mu\ln\frac{R_2}{R_1};$$

$$\rho'_{SFL} = \sqrt{\frac{\mu}{\varepsilon}}\,\frac{2\mu\ln(R_2/R_1)}{c}. \tag{1.37}$$

Obviously, it is possible to accumulate the maximum energy in such a line only in the case of a negative charge on the internal electrode. Its value per unit length is given by

$$Q'_{SFL} = \frac{\varepsilon E_m^2 R_1^2 \ln(R_2/R_1)}{4}, \tag{1.38}$$

where E_m is the maximum electric field near the internal electrode of the line in the dielectric. Similarly, when a line pre-charged to a voltage $U = E_m R_1 \ln(R_2/R_1)$ is connected to a load resistance ρ_{load} close in value to the line impedance, the power released on the load has the form

$$P = I^2\rho_{load} = \frac{U_{SFL}^2\rho_{load}}{\left(\rho_{load}+\rho'_{SFL}\right)^2} = c\sqrt{\frac{\varepsilon}{\mu}}\,\frac{E_m^2\ln(R_2/R_1)}{2}\,\frac{\rho_{load}\rho'_{SFL}}{\left(\rho_{load}+\rho'_{SFL}\right)^2}. \tag{1.39}$$

If in addition to the active resistance ρ_{load} ($\rho_{load} \sim \rho'_{SFL}$) the line load also has a parasitic inductance L (Fig. 1.17), then the power generated at the resistance ρ_{load} when the line is switched is as follows:

$$P = \frac{U_{SFL}^2\rho_{load}}{\left(\rho_{load}+\rho'_{SFL}\right)^2}\left(1-e^{-t/\tau_L}\right)^2, \tag{1.40}$$

where $\tau_L = L/(\rho_{load}+\rho'_{SFL})$ is the characteristic time of the discharge circuit. The presence of inductance in the circuit results in a lengthening of the voltage front on the load with characteristic time $t = \tau_L$. If the line is charged with magnetic energy (as an inductive storage) from an external source of current of magnitude I_0 and the current at the output of the line flows through the breaker, then from the moment of a fast interruption of the current by the interrupter, the current in the load is described by expression

$$I = \frac{\rho'_{SFL}}{\rho_{load}+\rho'_{SFL}}I_0\left(1-e^{-t/\tau_C}\right), \tag{1.41}$$

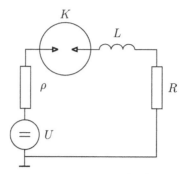

Fig. 1.17. The equivalent circuit of the inclusion of SFL.

where $\tau_C = \rho_{\text{load}}\,\rho'_{\text{SFL}}\,C/(\rho_{\text{load}} + \rho'_{\text{SFL}})$ is the characteristics time of shunting of the load resistance by the capacitance; C is the parasitic capacitance of the breaker and the load connected parallel to ρ_{load}. In this case, the power released by the line on the load is

$$P = \frac{I_0^2\,\rho_{\text{load}}\,\rho'^2_{\text{SFL}}}{\left(\rho_{\text{load}} + \rho'_{\text{SFL}}\right)^2}\left(1 - e^{-t/\tau_C}\right)^2. \tag{1.42}$$

Double forming line. The scheme of the coaxial double forming line (DFL) is shown in Fig. 1.18. At the initial time, the average electrode is charged from the generator. The potential at the outer electrode is zero. The potential of the central electrode, close to zero, provides during charging the inductance connecting the central electrode to the outer electrode. After the actuation of the left switch, the voltage on the external line is inverted, as a result of which the total voltage on the two lines is doubled. At this time, the right switch is switched on. It transfers the voltage to the load. Advantages of the DFL are undoubtedly the possibility of switching from the grounded end of the line, obtaining a doubled voltage in the open mode, and also the possibility of forming a voltage close to the charging voltage at the matched load.

The drawbacks of DFL when charging from one generator of charging pulses are the need to use inductance L, which, on the one hand, shunts the voltage front on the load, and on the other – forms a voltage pre-pulse, which must be eliminated by introducing an additional spark gap K (see Fig. 1.18, right), set to breakdown at a voltage significantly higher than the value of the pre-pulse. When charging DFL from CPG with positive and negative voltage polarities, the need for inductance L disappears.

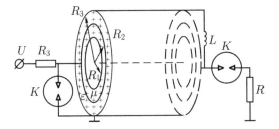

Fig. 1.18. The equivalent scheme of inclusion of the DFL.

The principle of the operation of the DFL is illustrated by the diagrams in Fig. 1.19. During the slow charging process of DFL (units and tens of microseconds), the charging inductance shorts the central and outer electrodes together, so that the voltage between them is zero (Fig. 1.19 *a*). When the discharger closes the external and intermediate electrodes, the polarity of the wave voltage propagating to the right is reversed (Fig. 1.19 *b*). After a time τ, the total voltage at the open end of the line turns out to be $2U_{DFL}$. This voltage is maintained for a time of 2τ (Fig. 1.19 *c*). (Charging inductance is not shown in the diagram, since when the line is discharged it is a high-resistance load.) If after a time τ after the discharger is shorted to an output of the DFL, a matched load with impedance ρ_{load} is connected, the line voltage drops to U_{DFL} and all energy from the line is outputted for a time of 2τ. Thus, when operating on a matched load, the charge and discharge voltages of the DFL are equal to each other.

During the charging of SFL and DFL from the pulse voltage generators, a prepulse is formed on the load. In the case of the SFL, the prepulse is due to the presence of a capacitive coupling through the output discharger, and in the case of the DFL it arises because of the presence of a charging inductance. As a rule, a pre-pulse has a negative effect, since it can create a particle beam, plasma at the cathode. There are several methods used in practice to reduce the amplitude of the prepulse. In the case of SFL, a resistor or additional inductance is used that connects the high-voltage electrode of the diode to ground, and in the case of the DFL, a prepulse discharger.

In the case of strip forming lines, unlike coaxial lines, the electric field strength at the edges of the electrodes sharply increases, which is illustrated by the formula for the ratio of the maximum electric field strength to the average strength:

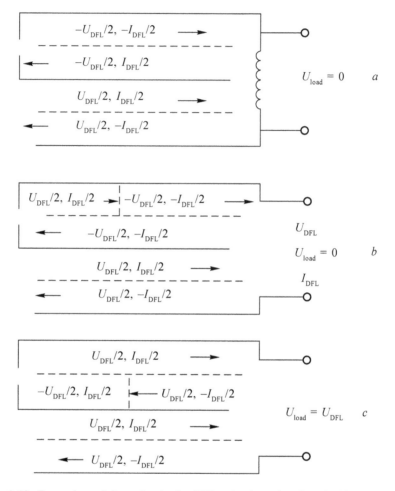

Fig. 1.19. Formation of the pulse in the DFL: *a*) when charging the line from the PVG; *b* and *c*) after closing the double line,

$$\frac{E_{max}}{E_{av}} \approx \frac{d_{ins}}{\Delta \ln\left[(\pi d_{ins} + \Delta)/\Delta\right]}, \qquad (1.43)$$

where d_{ins} is the insulation thickness (distance) between the electrodes of the strip forming line; Δ is the thickness of the strip electrode. For example, for commonly used copper strip electrodes DFL with a thickness $\Delta = 0.2$ mm at a distance $d_{ins} = 2$ mm, the amplification of the electric field strength at the edges reaches 3. As a result, without special measures, the use of the strip lines in high-voltage engineering is impossible. Methods for reducing the edge tension of strip forming lines will be set forth in Chap. 2.

Optimization of the diameters of the SFL and DFL electrodes. Let's consider the optimization of the geometry of coaxial SFL and DFL, provided that the value of K slightly exceeds 1. Optimization in the case $K > 1.5$ for the DFL is more complicated. It is considered in detail in [22].

Achieving the maximum voltage on the load ρ_{load} for SFL, which has an outer diameter R_2, it is necessary to find the extremum of the following expression for its variation with respect to the parameter $X = R_2/R_1$:

$$U_R = E_m R_1 \ln\left(\frac{R_2}{R_1}\right) \frac{\rho_{load}}{\rho_{load} + \rho'_{SFL}} = E_m R_2 \frac{\ln X}{X} \frac{\rho_{load}}{\rho_{load} + \rho'_{SFL}}. \quad (1.44)$$

It is seen that the maximum voltage on the load is achieved at a ratio of the radii of the line $X = e \approx 2.72$. The maximum power, determined by means of the expression

$$P_R = \frac{U^2_{SFL}\rho_{load}}{\left(\rho_{load} + \rho'_{SFL}\right)^2} = c\sqrt{\frac{\varepsilon}{\mu}} \frac{E_m^2 R_2^2 \ln X}{2X^2} \frac{\rho_{load}\,\rho'_{SFL}}{\left(\rho_{load} + \rho'_{SFL}\right)^2}, \quad (1.45)$$

For the DFL with a given size R_3, in the case of charging its internal and external lines from a single voltage source, we have the following equality:

$$U_1 = E_m R_1 \ln\frac{R_2}{R_1} = U_2 = E_m R_2 \ln\frac{R_3}{R_2} = E_m R_3 \frac{\ln X}{X} e^{-(1/X)\ln X}, \quad (1.46)$$

where, as before, $X = R_2/R_1$. Finding the extremum of expression (1.46) for the variation of X, it is easy to obtain that the maximum of the voltage on the lines is attained at $X = R_2/R_1 = e \approx 2.72$. The expression for the power in the case of a load of the DFL on the resistance ρ_{load}, which is close to the total wave resistance of such a line ($\rho_{DFL} = \rho_1 + \rho_2$, where ρ_1 and ρ_2 is the wave resistance of the internal and external lines, respectively), has the form

$$P_R = \frac{4U_{DFL}^2 \rho_{load}}{\left(\rho_{load} + \rho_{DFL}\right)^2} = c\sqrt{\frac{\varepsilon}{\mu}} \frac{4E_m^2 R_1^2 \ln^2\left(R_2/R_1\right)}{2\left[\ln\left(R_2/R_1\right) + \ln\left(R_3/R_2\right)\right]} \times$$

$$\times \frac{\rho_{load}\rho_{DFL}}{\left(\rho_{i\,\ddot{a}\ddot{a}} + \rho_{DFL}\right)^2} =$$

$$= c\sqrt{\frac{\varepsilon}{\mu}} \frac{2E_m^2 R_3^2 \ln X}{X^2 + X} e^{-(2/X)\ln X} \frac{\rho_{load}\rho_{DFL}}{\left(\rho_{load} + \rho_{DFL}\right)^2}.$$

(1.47)

When finding the maximum of this function with respect to X, it turns out that the maximum power on the load is outputted at $X = 1.52$. The ratio of wave impedances is $\rho_1/\rho_2 = 1.52$. Figure 1.20 shows the dependence on the parameter X of the voltage U and powers P_R for SFL and DFL under identical fields on the internal electrodes E_m, on the outer radii of the lines and the ratio ρ_{load}/ρ. Considering the ratio of voltages and powers for SFL and DFL, we can conclude that the advantage of DFL. However, it should be noted that the resistance of the DFL near the point at which the extremum of power is observed is 40% higher than the resistance of the SFL. As a result greater load is required in the case of DFL.

Radial (disc) forming lines. Radial forming lines (RFL) are used in the so-called 'non-iron' LIAs to produce powerful electron beams of short duration ($\tau \sim 10$–20 ns). For specified pulse durations, the use of ferromagnetic cores as inductors is inadvisable because of high energy losses in the steel during magnetization reversal. The scheme of such a LIA is shown in Fig. 1.21, and the principle of its operation will be described in detail below. The accelerator uses a pair of radial lines (*1* and *2*) connected at a large radius in a matched manner to avoid reflections. The property of disc lines is the constancy of the wave resistance for waves moving along the radius. If an angle α is specified between the inner flat and lateral conical electrodes, then the linear line capacitance per unit length along the radius is

$$C_{RFL} = \frac{\varepsilon \cdot 2\pi r\, dr}{4\pi r\, \mathrm{tg}(\alpha)\, dr} \frac{1}{} = \frac{\varepsilon}{2\, \mathrm{tg}\alpha},$$

(1.48)

and the linear inductance is determined from the magnetic energy stored per unit length of the line along the radius:

$$L_{RFL} \frac{I^2}{2c^2} dr = \frac{B_\theta^2}{8\pi\mu} 2\pi r\, dr\, r\, \mathrm{tg}\alpha,$$

(1.49)

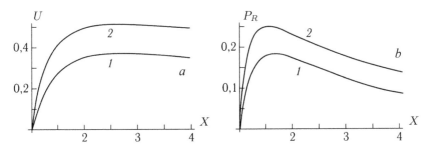

Fig. 1.20. Dependence on the parameter X of: *a*) the voltages created by the SFL (*1*) and DFL (*2*) on the load, with the same fields on the internal electrodes and the outer radii of the coaxial lines; *b*) the powers transferred to the load from the SFL (*1*) and DFL (*2*), with the same fields on the internal electrodes and the outer radii of the lines.

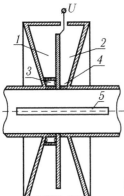

Fig. 1.21. Scheme of a "non-iron" LIA with RFL: *1, 2* – radial lines; *3* – annular discharger; *4* – insulator; *5* – electron beam.

where
$$L_{RFL} = 2\mu \, \mathrm{tg}\, \alpha. \tag{1.50}$$

Thus, the impedance of the radial line

$$Z_{RFL} = \sqrt{\frac{L_{RFL}}{C_{RFL}}} = 2\sqrt{\frac{\mu}{\varepsilon}}\mathrm{tg}\, \alpha. \tag{1.51}$$

If the charging voltage U_{RFL} is applied to the inner electrode of the double disk line and make the annular spark gap *3* is switched on, in the first line there will be a wave with amplitude $U = -U_{RFL}/2$ propagating along the radius which, having reached the outer radius of the line, goes in the opposite direction already in the second line. When the wave reaches the minimum radius at time $t = 0$, a reflected wave with a double amplitude appears in the open end of

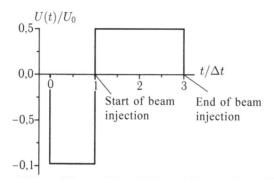

Fig. 1.22. Dependence of the voltage of the accelerating gap on time.

the line: $U(t) = -U_{RFL}$ (Fig. 1.22). It will support this voltage at the acceleration gap during the travel time of the wave back and forth, equal to $\Delta t = \dfrac{2(R_{max} - R_{min})}{(c / \sqrt{\varepsilon\mu})}$.

When the wave returns at the time $t = \Delta t$, its polarity reverses. At the same time, the injection of the beam begins, the current of which is chosen so that the line is matched, that is, $I_0 = U_{RFL}/(2Z_0)$. The voltage on the accelerating gap is halved and becomes equal to $U = U_{RFL}/2$. Then, for a time $t = 2\Delta t$ the beam accelerates in the gap. If we neglect all the losses, then the efficiency of energy transfer in such an accelerator can theoretically be close to 100%.

1.3.4. Switches of forming lines

The switches of the forming lines use gas spark gaps and pulsed hydrogen thyratrons. To quickly transfer the energy of the forming line to the load, it is necessary to have dischargers with low inductance and low ohmic losses. In addition, the spread of the response time should not exceed several nanoseconds. The spark gap must withstand a high operating voltage (50–100 kV) in the non-conductive state, have a wide range of operating voltages, a small amplitude of the starting pulse (several kilovolts), a minimum delay time (units of nanoseconds), a high current (up to 100 kA) provide a high frequency of repetition of pulses, have relatively small dimensions. Thus, the spark gap must satisfy numerous, often conflicting requirements.

The breakdown phenomena in the dischargers are quite complex, and their detailed theoretical description goes beyond the scope of this monograph. Here we confine ourselves to a brief description of physical processes and a review of the literature data.

It was experimentally established that the breakdown voltage of the gas is a function of the product of the gas pressure by the electrode spacing (Paschen's law) [23]. At very high values of pd_{el} (where p is the pressure of the gas in mm Hg, d_{el} is the electrode spacing), the dependence of the breakdown voltage is almost linear: $U_{break} = Apd_{el}$. For small pd_{el}

$$U_{break} = A_0 pd_{el} + B_0 \sqrt{pd_{el}}, \qquad (1.52)$$

where A, A_0 and B_0 are constants for a given gas. At relatively low gas pressures, the breakdown voltages in uniform fields for different electrode materials practically coincide. At pressures of the order of tens of atmospheres, the values of the breakdown voltages are largely determined by the state of the surface and the material of the electrodes. Experiments in which various metals have been used for the anode and the cathode have shown that the emission characteristics of the cathode have a greater effect on the breakdown voltage than the characteristics of the anode. The breakdown voltage is increased by changing the material of the cathode in the following sequence: sodium, aluminium, platinum, iron, stainless steel. With a thorough polishing of the surface, the breakdown voltage of the discharge gap increases. Numerous studies have shown that nitrogen is the most suitable gas for the dischargers. It has a sufficiently high electrical strength at elevated pressures in homogeneous fields and is a chemically inert gas.

We will present some time characteristics of an electric discharge in gases. The development of the discharge can be divided into three successive stages: the stage of discharge formation, the final stage and the arc stage. The stage of discharge formation begins when a breakdown voltage is reached on the discharge gap and at least one electron initiating the development of the discharge appears. The time between the moment of application of the breakdown voltage to the gap and the moment of appearance of the initiating electron is called the statistical time of the breakdown delay. The primary electrons arising in the gap acquire kinetic energy sufficient for shock ionization of the gas molecules. The secondary electrons, together with the primary electrons, repeat this process, forming an avalanche with an average number of electrons $n = n_0^{\alpha x}$ (where n_0 is the number of primary electrons, α is the first Townsend ionization coefficient, and x is the distance travelled by the avalanche). The empirical condition for the transition of an avalanche to a streamer has the form $\alpha x > 20$.

During the time from the appearance of one or several electrons before they grow into an avalanche (fast-developing streamer), the resistance of the discharge gap is still high and the voltage on it is practically equal to the applied voltage. Thus, the time of breakdown delay consists of the time of statistical delay and the time of discharge formation. The value of the statistical delay time depends on many conditions: the gas pressure (with increasing pressure it decreases), the gas volume between the electrodes, the intensity of preliminary ionization or irradiation of the discharge gap. At the final stage, the resistance of the discharge channel changes from a large value determined by the developed streamer to a small one, determined by a highly ionized plasma of breakdown. In accordance with the change in resistance, the voltage across the discharge gap also drops. The dependence of the voltage on the discharge gap on time is called the switching characteristic, and the time during which the voltage drops is the switching time. The switching characteristics depend both on the parameters of the discharge circuit and on the conditions in the gap. At the arc stage, the magnitude of the flowing current and the time of its flow are determined only by the parameters of the discharge circuit.

The dependence of the resistance of the spark channel on the value of the current in the circuit and the time of development of the discharge has the form

$$R_d = \sqrt{p d_{el}^2 \left(2a \int_0^t I^2 dt \right)^{-1}}, \qquad (1.53)$$

where I is the current; a is the constant for a given gas. This relation is valid under the condition that the discharge proceeds in such a short time that the energy losses as a result of the radiation and thermal conductivity processes are negligible. As a result of the calculation in the discharge circuit of the LIA, taking into account parasitic parameters and spark resistance, the following dependences of the pulse front duration on the conditions in the discharge gap and the circuit parameters were obtained:

$$t_f = 21\frac{pd_{el}^2}{aU_0^2} + 2.2\frac{L_{in}}{R}, \tag{1.54}$$

$$t_f = 26.3\frac{pd_{el}^2}{U_0^2} + 3.03\sqrt{L_{in}C}, \tag{1.55}$$

where the first terms represent the time characteristic of the discharge gap, and the second – the discharge circuit. Formula (1.54) refers to the case of a high current in the pulse when the influence of the inductance on the pulse front greatly exceeds the influence of the capacitance shunting the load. If the load resistance is high, then the influence of the capacitance can not be neglected and the process is described by formula (1.55). It follows from the above expressions that in order to reduce the duration of the pulse front, it is first necessary to reduce the parasitic parameters. With negligibly small parasitic parameters, the time constant of the discharge gap should be reduced. It is known that the switching time decreases with increasing gas pressure and electric field strength. In addition, it depends on the shape of the electrodes. In a homogeneous field ('sphere–sphere' electrodes), the commutation time is approximately half that in an inhomogeneous field ('tip–tip' electrodes).

At a spark current of the order of several kiloamperes or more, the expansion of the discharge channel is influenced by the process of increasing the conductivity of the spark. At high currents, it is very difficult to obtain a short switching time, since it increases with current in the circuit:

$$t_{sw} = 9 \cdot 10^{-9}\frac{d_{el}(Ip)^{1/3}}{U_0}. \tag{1.56}$$

As a rule, the LIAs use controlled gas dischargers. The action of the starting pulse on the discharge gap of the switch manifests itself either in increasing the electric field strength in the gap, or in weakening its electric strength. The first principle of operation is applied in the work of three-electrode and multi-electrode start dischargers, spark relays, dischargers with laser and with the electron beam. The most common are three-electrode dischargers and trigatrons.

The three-electrode discharger (Fig. 1.23 *a*) contains three electrodes. One high-voltage electrode (*1*) is connected to a high-voltage source, and the second (*2*) is grounded through a load.

A trigger pulse is applied to the control electrode (*3*) installed between the high voltage electrodes. The dischargers is adjusted in the following way: the length of the gap *1–3* is chosen such that it does not break through under the influence of the voltage U_0, and the length of the gap *2–3* is such that it does not break through the trigger pulse. When the starting pulse arrives at the control electrode, the field distribution is distorted, the gap *1–3* breaks and the control electrode takes the potential U_0. As a consequence, a full voltage is applied to the second gap and it also breaks through. The investigation of the operation of the three-electrode spark gap showed that in order to reduce the delay time of starting the discharger and to increase the stability, it is necessary to increase the amplitude and steepness of the front of the trigger pulse. With a steepness of the front of 40–50 kV/μs and pulsed amplitude of 50–70% of U_0 the scatter of the switching time of the discharger is of the order of 10^{-8} s. In general, it can be noted that three-electrode dischargers are characterized by a high amplitude of the starting pulse and a narrow range of operating voltages.

The trigatron (Fig. 1.23 *b*) is structurally different from the three-electrode discharger in that the control electrode (*3*) is located in the hole on the axis of the high-voltage electrode connected to the load (ground). There are two main ways to switch on a trigatron. The first is a longitudinal start-up, when the spark closes the discharge gap *1–3*, and a lateral start, when the initiating spark closes the gap *2–3*. The arc discharge distorts the field distribution near the electrode. In addition to the appeared electrons, it also generates photons that facilitate the formation of streamers and the closing of the discharger.

In addition to the controlled dischargers discussed above, some devices use dischargers controlled by ultraviolet, laser and soft X

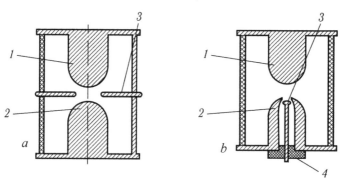

Fig. 1.23. Discharger with distortion of the field (*a*) and the trigatron (*b*): *1, 2* – high-voltage electrodes; *3* – control electrode; *4* – insulator.

radiation and an electron beam. In this case, less energy is required to ignite the discharge and the spread of the discharger response time can be reduced.

When the discharger operates at a repetition rate of pulses, the problem arises of reducing the deionization of the gas in the discharge gaps. The speed of recovery of the electrical strength of the gap is affected by the following factors.

1. When the electrodes are heated to 1600 K, the breakdown voltage of air at normal pressure is halved. In this case, the local temperature of that part of the surface of the electrodes to which the discharge arc was closed (and the arc column temperature can reach 20 000 K), decreases approximately ten times slower than required to provide the electric strength.

2. In the interelectrode gap, after the discharge, the heated gas remains, while the electric strength of the air when heated to 3000 K is halved.

3. If the interelectrode distance is small, the arc burns in a medium consisting of a mixture of heated gas particles and vapour of the electrode material. At arc currents above 1000 A, it burns practically only in the vapours of the electrode material. This leads to a decrease in the electrical strength of the gap, since the ionization potential of the metal atoms is much lower than that of the gases.

The first of these factors can be eliminated by cooling the electrodes, the second and third by blowing the gas through the discharge gaps.

To avoid the possibility of the appearance of locally overheated areas and to achieve uniform wear of the electrodes, the operation of the discharger with a high pulse repetition frequency is based on the principle of uniform distribution of many parallel channels of the discharge of small currents over a relatively large area of electrodes while simultaneously cooling them. Obtaining many parallel discharge channels in the gap is possible only if they are formed in a time much shorter than the switching time (otherwise the formation of one channel will lead to a decrease in the voltage at the electrodes, which will make it impossible to form other channels). If there are n channels that conduct current, then in the expressions (1.54)–(1.56) one must replace L by L/n and I by I/n. The design of the discharger, intended for commutation of the forming lines of the LIA at the Tomsk Polytechnic University, which maximally satisfies the above criteria, will be described in Ch. 2.

So, the development of a discharger that satisfies the numerous requirements imposed on the switches of the forming lines of the LIAs is a complicated technical task. The main disadvantage of the discharger is its limited service life. Usually dischargers lose their efficiency after 10^4–10^6 operating cycles. A system with a discharger can be used in laboratory installations operating in the single mode with a low repetition frequency.

Pulsed hydrogen thyratrons. Pulsed hydrogen thyratrons are a reliable, durable device, have a high stability of parameters, allow switching currents of 10–15 kA at a high repetition rate. The thyratron can withstand up to 10^{11} pulses, which allows it to successfully compete with other key elements of high-current electronics. The pulsed thyratrons withstand a large applied voltage at a zero grid potential. To cause a discharge between the cathode and the anode, it is first necessary to create a positive ignition pulse between the grid and the cathode, causing an auxiliary discharge in this gap. Then the discharge develops in the main gap.

One of the main trends in the development of the technology of the pulsed thyratrons is an increase in the rate of current rise and a decrease in the time of switching on the thyratron. For this purpose, complex grid electrodes, hollow anodes, etc., are used in the construction of thyratrons, and various methods for creating a plasma are employed. The delay time of the current pulse with respect to the grid pulse depends on the grid current parameters, the heating voltage, the anode voltage, and the repetition rate. To reduce its value and make it more stable, it is necessary to increase the steepness of the grid voltage front and the magnitude of the grid current pulse.

The merits of the pulsed thyratrons include the possibility of a parallel switching on of a large number of such devices. The disadvantages of the thyratrons are high inductance, long switching time (in comparison with gas dischargers), restrictions on switching current and voltage. It should be noted that with a decrease in the pulse duration up to 10^{-7} s, the amplitude of the switched current can increase several times in comparison with its certified value.

Thus, thyratrons are preferred in accelerators with a pulse duration of the order of 100 ns. As shown by experiments, with a pulse duration of tens of nanoseconds, the switching time of hydrogen thyratrons is 30–50 ns, as a result of which there is a need for elements making the pulse 'sharper'. One of the possible solutions to this problem is based on the use of the properties of shock electromagnetic waves [24]. Shock waves arise when electromagnetic waves propagate in a medium whose magnetic properties (in particular, its magnetic permeability) depend on the field strength H of the propagating wave. A similar medium is ferrite whose magnetic permeability decreases with increasing H. Consequently, the peak of the pulse propagates in it at a higher speed than the base of the pulse. In this way, the pulse front can be reduced to 10^{-9} s. At the same time, such short fronts are difficult to realize, since the presence of parasitic inductances and capacitances in the induction system (see Fig. 1.13) does not make it possible to obtain pulses shorter than 5–10 ns.

1.3.5. Electron guns

An important component of a linear induction accelerator is an electron gun. The design of the gun largely determines the magnitude of the current, the quality of the beam, and the reliability of the accelerator. Many LIAs use electron beams formed by thermionic cathodes with an optical system based on Piers optics.

The principle of operation of thermionic cathodes is based on the property of conductive materials to emit electrones when heated to a high temperature. The electrons inside the body are in the potential well. Escape from this well by tunnelling through the barrier can be realized only by the electrons with energy comparable to the work function Φ_w. The maximum density of the emission current increases with the temperature of the cathode T in accordance with the Richardson–Dashman formula:

$$j = AT^2 e^{-11600\Phi_w/T}, \tag{1.57}$$

where A is the constant characterizing the material $[A/(cm^2 \cdot K^2)]$; Φ_w is the work function, depending on the material [eV]; T is the temperature [K]. For the best cathodes the value of j reaches ~40 A/cm² at $T = 1400$ K. For example, at $T = 1400$ K, the current density from a barium-based cathode 411M, coated with a thin layer of ruthenium and osmium, having a work function of about $\Phi_w = 1.5$ eV and a constant $A = 350$ A/cm² · K², can reach 35 A/cm².

An important characteristic of the electron beam, which determines the measure of the intensity, is the perveance $P = I_0/U_0^{3/2}$ (the ratio of beam current I_0 to accelerating voltage U_0). In view of the smallness of the numerical value of the perveance, a more convenient value is usually used – the microperveance ($P_\mu = P \cdot 10^6$). Intense flows are those in which the perveance takes values greater than 10^{-8}–10^{-7} A/B$^{3/2}$. It should be noted that space charge forces play an important role in intense flows.

The problem of the formation of intense beams by electron guns is solved by two methods: the method of analysis and the synthesis method [25, 26]. In the first case, the configuration and potentials of the electrodes of the forming system are chosen approximately and a computer is used to calculate the electron trajectories taking into account the space charge. If the resulting beam does not satisfy the specified requirements, necessary changes are made in the shape and potentials of the electrodes and again the trajectories are again calculated. The process is continued until beam with the specified parameters is obtained. This method is very time-consuming and requires a high qualification of the developer.

More widely used is the synthesis method, in which the beam parameters (shape, perveance or energy and beam current) are given, and the electric and magnetic fields necessary for the formation of this beam are determined. In this method, two problems are solved – internal and external. The internal problem involves solving a system of equations describing the motion of electrons within the beam, and finding the relationships characterizing the electrical and geometric parameters of the beam. The external task is to find the electric fields created by a system of electrodes with certain potentials, and magnetic fields created by coils with current or permanent magnets. For the internal problem, the potential distribution in the beam is described by the Poisson equation, and for an external problem the potential distribution outside the beam is described by the Laplace equation.

The synthesis method is based on known solutions of internal problems for unbounded laminar flows between two parallel planes, two coaxial cylinders and two concentric spheres. The relation between the current I and voltage U in such flows is described by the 'three-halves power law' ($I = PU^{3/2}$). In this case, all trajectories are rectilinear and coincide with the lines of force of the electric field. The potential distribution along any trajectory satisfies the relationship $U(z) = A_p z^{4/3}$ (where A_p is a factor determined by the perveance; z is the coordinate measured along any trajectory). The straightness of the trajectories means that there is no force that distorts the trajectory, that is, the component of the electric field strength normal to the trajectory is equal to zero ($E_p = 0$).

The creation of a system for the formation of intense beams with the help of an electric field is reduced to 'excision' from unbounded flows, for which solutions of the internal problem, limited by the beams of the necessary configuration, are known. An indispensable condition here is the coincidence of the beam boundary with rectilinear trajectories. From an unlimited flow between two parallel planes, one can form a beam of any cross-section with boundaries perpendicular to the original planes. For example, it can be a beam in the form of a cylinder (a band beam). From the flux between two coaxial cylinders, it is possible to 'cut out' a tapered converging band beam, from the flux between two concentric spheres – a convergent conical axisymmetric beam.

A simple 'discarding' of the part of the flux remaining outside the cut out beam will lead to a change in the conditions on the beam boundary, in particular, the requirement $E_p = 0$ will not be satisfied. A stable bounded beam can be formed by creating an electric field outside it, equivalent to the space charge field of the discarded part of the flux. This field must be created by a system of electrodes located outside the beam. The shape and potential of the electrodes are determined from the solution of the Laplace equation with boundary conditions arising from the solution of the internal problem: the potential distribution along the beam boundary is given by the 'three-halves power law'; the normal to the beam boundary is the component $E_p = 0$ at any point on the surface of the beam. With sufficient accuracy for practical purposes, the external field forming the stable beam can be created by two electrodes – cathode (focusing), coinciding in shape with a zero equipotential surface, and anodic, coinciding in shape with an equipotential surface having the potential of an accelerating electrode (anode). Analytical solutions are used for for beams with rectilinear trajectories according to which a zero equipotential surface forms an angle of 67.5° with the boundary of the beam, and the remaining equipotentials (with $U > 0$) approach the beam boundary under the right angle.

The systems of formation of intense beams created on the basis of the considered principle are called Pierce guns. Such electron guns consist of an electron source – a cathode (usually a thermionic), a near-cathode (focusing) electrode and an anode with an aperture for the output of the beam (Fig. 1.24). The external field forming the beam must correspond quite accurately to the calculated field in the immediate vicinity of the beam boundary, which determines the configuration and potentials of the electrodes near the beam. Far

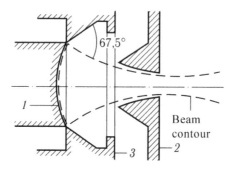

Fig. 1.24. Electrode system of the Pierce gun: *1* – cathode; *2* – anode; *3* – focusing electrode.

away from the beam, the shape of the electrodes is chosen taking into account the design and technological requirements.

Many electron guns must form beams with a high current density (up to tens and hundreds of A/cm^2). At the same time, the real thermionic cathodes have a limited emissive ability. The increase in current take-off dramatically reduces the life of cathodes. Therefore, electron guns with a high compression of the electron beam (the cross section of the formed beam at the exit from the anode aperture is tens and hundreds of times smaller than the area of the emitting surface of the cathode) are used. Most widely used are the Pierce guns, forming converging axisymmetric and band beams.

Since the conventional Pierce optics does not allow to produce the electron beams with the microperveance more than 3 A/V$^{3/2}$, grid or mesh anodes are used in the guns. In a three-electrode gun, the grid certainly works in a lighter mode than the mesh anode. At the same time, the introduction of the third electrode complicates the design of the gun and worsens the angular characteristics of the electron beam. Annular Pierce electron guns and magnetron guns provide electron beams with the microperveance more than 10 A/V$^{3/2}$. However, they have some drawbacks that make them difficult to use in accelerator technology. The most preferred of all the above is an electron gun designed for powerful klystrons [26]. In it, the magnetic and optical systems are chosen from the calculation of the coincidence of the direction and course of the electric and magnetic field lines. This improves the focusing ability of the gun optics and thereby provide a microperveance of about 6–7 A/V$^{3/2}$ without applying additional electrodes.

The injector sections of the LIAs containing electron guns use the same elements as in the accelerating sections, i.e. inductors. For example, in the Astron LIA the injector is a section similar in design to the main section but with a slightly larger diameter [3]. In the centre of the section there are located the gun electrodes, with the help of which not only the longitudinal but also the radial component of the electric field is created. Thus, compensation for the repulsive action of space charge forces is achieved. The required distribution of the field gradient is reached by choosing the distance between the electrodes and the number of inductors between consecutive electrodes. The accelerator gun has an oxide cathode in the form of a flat disc with a diameter of 17.8 cm. The anode aperture is covered with a mesh to eliminate the failure of the electric field. The gun gives a current of 1200 A at a voltage of 550 kV. This corresponds to a microperveance of 2.9 $A/V^{3/2}$. In the accelerator LIA-30/250, the section of the main accelerator is used as an injector with a metal rod passing along its axis. The cavity between the rod and the body of inductors is filled with transformer oil. The voltage of 300 kV is output through the bushing and fed to the cathode of the electron gun. The gun uses a special cathode with a high specific emission value with a 50 mm diameter. The current of the gun is more than 250 A.

High-current relativistic electron beams with a current of several kiloamperes or more are produced using explosive-emission cathodes [27]. According to electron microscope studies the surface of any material contains sharp microedges, the height of which is $\sim 10^{-4}$ cm, the average radius of less than 10^{-5} cm, and the apex radius is much smaller than the average radius. The concentration of microedges reaches 10^4 cm^{-2}. When a high voltage is applied to the cathode, the electric field at the ends of the microedges is amplified hundreds of times in comparison with the average value. This field leads to strong emission and can cause evaporation of microedges due to excessive ohmic heating. At current densities substantially exceeding the critical value for stable field emission, explosive evaporation of microedges and the formation of local emissions of cathode plasma (cathode flares) occur. Hydrodynamic expansion with a high velocity (of the order of several cm/μs) and fusion of cathode flares lead to a rapid formation of a plasma layer covering the entire surface of the cathode. The effective emitting surface sharply increases. Electrons are emitted from the surface of the expanding cathode plasma under the influence of a high electric field strength and their current is limited by the space charge of the electron cloud located in the cathode–anode gap. Since the cathode plasma can be regarded as a metal surface, the work function of which is on the average equal to zero, the source of electrons on the cathode has practically unlimited emissivity. In this case, the current flowing in the diode depends on the distribution of the potential in it, which varies under the action of the space charge of the electrons. In high-voltage diodes, the height of the potential barrier is negligibly small in comparison with the applied voltage. In this case, the minimum of the potential practically coincides with the surface of the cathode. Consequently, we can assume that the electric field vanishes at the cathode surface and that the initial velocity of the emitted electrons is zero.

In the case of a planar diode (Fig. 1.25 *a*) accurate evaluation of the current density may be carried out with the accuracy sufficient for practice using the well-known 'three-halves power law' up to the applied voltage of $U \leq 1.5$ MV convenient in practical calculations for

$$j = \frac{\sqrt{2}}{9\pi}\left(\frac{e}{m}\right)^{1/2}\frac{U^{3/2}}{d^2},$$ (1.58)

where *d* is the distance between the electrodes. Such diodes are used as electron sources for relativistic reflex triodes (see Chapter 3).

In practice, it is often necessary to remove the electron beam into the vacuum region, which is done by means of an external magnetic field. Since the beam is not output through a grounded anode, there are no problems associated with the destruction of the anode and the scattering of particles in the anode foil or grid. In addition, by using an external magnetic field in the diode, it is possible to control the transverse dimensions of the electron beam. In such diodes (Fig. 1.25 *b*) a coaxial (tubular) electron beam is formed, which in comparison with a uniform beam has a much higher current:

$$I_{in} = \frac{mc^3}{e}\frac{(\gamma - \gamma_b)\sqrt{1 - 1/\gamma_b^2}}{2\ln(R_{tube}/r_b)},$$ (1.59)

where $\gamma = 1 + eU/(mc^2)$ is the total electron energy; $\gamma_b = \sqrt{2\gamma + 0.25} - 0.5$ is the kinetic energy of electrons; R_{tube} is the drift tube (anode) radius; r_b is the average radius of the electron beam, approximately equal to the radius of the cathode ($r_b \approx r_c$). Such diodes are used to form electron beams in many devices of relativistic high-frequency electronics.

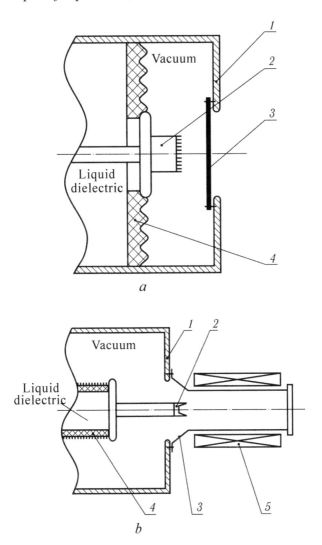

Fig. 1.25. The design of a planar diode (*a*) and a coaxial diode with magnetic insulation (*b*): *1* – case; *2* – cathode; *3* – anode; *4* – high-voltage insulator; *5* –

1.3.6. Magnetic systems for retaining the transverse beam dimensions

The problem of transporting powerful pulsed beams always arises when it is necessary to deliver an electron beam from the region of its formation to the site of application. Since a high-current electron beam has a high space charge, leaving the formation region under the

action of the electric field of this charge, it begins to expand rapidly
in radius and increases its size by a factor of *e* over the length

$$l \approx r_b \sqrt{\frac{m\gamma^3 v_z^3}{4eI}},$$ (1.60)

where r_b is the radius of the beam; v_z is its longitudinal velocity; $\gamma =$
$1 + eU/(mc^2)$; U is the applied voltage; I is the beam current [27].
For relativistic beams with currents of the scale 10 kA, the value of
l amounts to several transverse beam dimensions, which is clearly
not suitable for solving many applied tasks.

Without an effective solution to the problem of retaining the
radial size of the beam, reliable operation of LIA is not possible,
especially at a pulse repetition rate of several hertz and higher. At an
electron energy at the output of more than 3 MeV, the beam losses
can cause not only destruction of the accelerator sites adjacent to
the beam, but also inadmissible intensive activation. The formation
of a stable intense beam of a definite configuration is possible
only if the electric field of the electron beam is oppositely directed
by external (in relation to the beam) electric and magnetic fields.
Therefore, the electron gun must contain electrodes that create a
potential distribution near the beam boundary, which ensures that the
component of the electric field strength normal to the beam boundary
is equal to zero. In addition, to stabilize the beam, it is necessary
that when electrons move from its boundary to either side a force
arises that returns them to the beam boundary.

The expansion of the beam can be limited by using a longitudinal
magnetic field (homogeneous or decreasing in the direction of the
cathode) or a sequence of electron lenses (electrostatic or magnetic)
located along the beam. The electron guns forming beams with
parallel trajectories use a longitudinal uniform magnetic field whose
lines coincide with the trajectories of electrons, and near the cathode
– also with electric lines of force, which ensures the existence of an
extended stable beam. Another way is to compensate for the space
charge of the beam electrons by the ion background.

The presence of magnetic field inhomogeneities in the gaps of the
accelerating sections and at the junction between the sections causes
oscillations in the envelope of the beam. The inclination of the axis
of the beam relative to the axes of the focusing coils and accelerating
tubes, even by a fraction of a degree, causes the centre of gravity of
the beam to oscillate relative to the axis of the accelerating path. The
amplitude of the oscillations, as a rule, increases along the length of
the accelerator. Since the energy spectrum of the beam is nonuniform

along the pulse length due to the pulsed nature of the accelerating field, the phase of oscillations of the beam's centre of gravity and its envelope vary with time. Vibrational motion of the beam causes losses during its transportation.

An important question concerning the use of LIAs for the formation of high-current relativistic electron beams with their subsequent use is the study of the processes of passage of an electron beam in the accelerating path. To this end, special experiments were carried out at the Tomsk Polytechnic University, in which an accelerator consisting of an injector and an accelerating section was used [28]. Figure 1.26 schematically shows the injector section of the LIA and part of the accelerating tract of the accelerating section. They were made on the basis of the layout scheme developed at the Tomsk Polytechnic University (see Chapter 2). The general case *1* houses a ferromagnetic induction system of seven ferromagnetic cores *2*, each of which is covered by three magnetizing coils *3*. The turns are connected to three strip double forming lines *4* connected in parallel, consisting of four plates each and laid along an Archimedes spiral over the ferromagnetic cores. Potential electrodes of strip DFL are connected to the anodes of a multichannel spark discharger *5* of a trigatron type, operating in the field distortion mode with current division between the channels. The explosive-emission graphite cathode *6* is located opposite the grid *7*, which closes the entrance to the accelerating tube *8*, the inner surface of which is covered

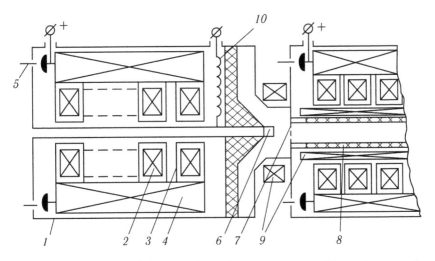

Fig. 1.26. Scheme of the injector and accelerating sections of LIA: *1* – case; *2* – cores; *3* – magnetizing coils; *4* – forming lines; *5* – discharger; *6* – cathode; *7* – grid; *8* – accelerating tube; *9* – magnetic field coils; *10* – demagnetizing winding.

with a weakly conducting layer. The coils *9* form a longitudinal magnetic field, the coil of the accelerating section in the form of a single-layered solenoid also fulfilling the functions of the voltage divider along the accelerating tube and the demagnetizing circuit of the cores. Demagnetization of the cores of the injection section is carried out along the winding *10* laid along the end of the insulator, which also improves the distribution of the electric field strength along the high-voltage insulator.

The main objective of this experimental study was to determine the relation between the energy of the injected beam and the magnitude of the current captured by the accelerating section. Figure 1.27 shows the accelerating path during operation of the cathode–anode gap in the regime of a planar cathode with an anode grid. The magnitude of the magnetic field *B* in the region of the accelerating chamber was adjusted to within the range 0.05 to 0.3 T.

Cathodes of different diameters at different cathode–grid distances, and varying the spatial distribution of the magnetic field in the region of the cathode (curves *1–4* in Fig. 1.27 *b*) were studied. The best results were obtained in the case when the side surface of the cathode was closed by a dielectric. It was established that for the electron energy from 200 to 420 keV increasing *B* from 0.1 to 0.15 T increases the beam current by 44%, and with further growth of *B* to 0.3 T by only 10%. The optimal magnetic field in the transition region for *B* = 0.1 T is a field that decreases along the length of

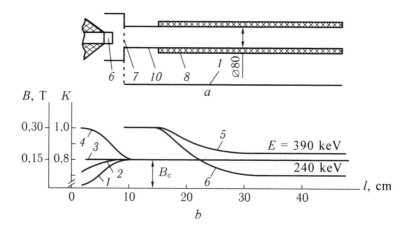

Fig. 1.27. Geometry of the cathode–anode gap (*a*): *1* – case; *6* – cathode; *7* – grid; *8* – accelerating tube; *10* – winding circuit. Distribution of the magnetic field (curves *1–4*) and the coefficient of beam current flow (curves *5, 6*) along the accelerating path (*b*).

the cathode towards the anode grid (curve *4*), and for $B = 0.3$ T – increasing (curve *2*).

The beam loss in the accelerating path was determined from the current measurements along the path with the accelerating section disconnected and the shorted turns of its ferromagnetic cores, since with the open turns, the beam energy is pumped into the forming line. The current passage coefficient K was defined as the ratio of the beam current along the accelerating path measured by the Faraday cylinder to the current at a distance of 2 cm from the anode grid (curves *5, 6*). In the experiments, graphite cathodes with a diameter of 10 to 30 mm were used at cathode–grid distances from 7 to 20 mm. The magnitude of the beam current, measured at distances $l = 10$–15 cm and $l > 30$ cm from the anode grid, coincides with the value of the limiting current of the uniform beam calculated for the geometric dimensions of the acceleration path, provided that in the first case the current is closed along the metal flange II, and in the second case it is closed along the accelerator case (Fig. 1.27 *a*).

As might be expected, in order to increase the current passage, other things being equal, it is necessary to increase the injection voltage and the beam diameter. When the cathode–grid distance was varied from 8 to 20 mm and the cathode diameter equal to 21 mm, the injection voltage varied from 200 to 390 kV, and the diode current amplitude decreased from 2350 to 1580 A. The beam current at the outlet of the section increased from 1070 to 1260 A, that is, the current passage coefficient increased from 0.47 to 0.82. With an increase in the diameter of the cathode up to 30 mm and an injection voltage up to 390 kV, the beam current at the output of the section was 1400 A ($K = 0.85$), and the switching on of the accelerating section and beam acceleration had practically no effect on the current passage coefficient. The accelerating section provided the energy gain in the beam to 550 keV.

As a result, it was concluded that when using the explosive-emission cathodes, a longitudinal magnetic field along the accelerating path at the level of 0.2–0.3 T is required.

In an electron gun with the compression of an electron beam of large radius to a beam of small radius, the limiting magnetic field decreases in the near-cathode region, which provides an approximate coincidence of electric and magnetic field lines. Such guns with a partially shielded cathode make it possible to form high-perveance beams. One of the variants of such a gun is described in [29]. It was made for the LIA ETA-II (10 MeV, 3 kA, 50 ns). The injector of the

LIA is an electron gun with quasi-Pierce optics with a voltage of 1 MV (Fig. 1.28). The cathode operates in the regime of space charge limitation with a current density of 15 A/cm². A beam diagnostic station is installed at the output of each of the ten accelerating gaps, A longitudinal magnetic field is produced by solenoids. The magnetic field is corrected using special coils.

The geometric dimensions of the LIA accelerator system are chosen based on the specified inhomogeneity of the particle energy at the output ($\Delta W/W$) under the assumption that the particles are accelerated without changing their radial coordinate:

$$\frac{R_{av}}{l_{tract}} < 4\left(\frac{R_{av}}{r_b}\right)^2 \frac{\Delta W}{W}, \tag{1.61}$$

where R_{av} is the average radius of the inductor; r_b is the radius of the electron beam; l_{tract} is the length of the accelerating tract.

In addition to space-charge forces, the beam is influenced by the Earth's scattering field (in the intervals between the elements of the induction system) and the scattering field. In some accelerators, the task of retaining the dimensions and position of the beam is solved by using short magnetic solenoids placed between the accelerating

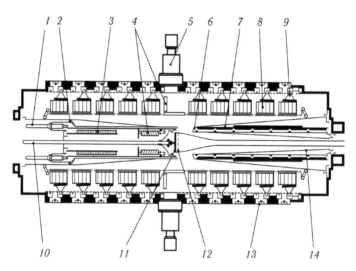

Fig. 1.28. Design of the injector of the ATA LIA: *1* – grid supply cables; *2* – oil-filled volume; *3* – ferrites; *4* – compensation coil; *5* – a branch pipe of a vacuum pump; *6* – an anode of an electron gun; *7* – focusing magnetic coils; *8* – ferrite inductors; *9* – leads of magnetizing turns; *10* – supply of the cathode heater; *11* – flange of the cathode; *12* – grid; *13* – the place of connection of the generator of charge pulses (250 kV, 70 ns); *14* – electron beam.

sections and having a relatively large aperture: from 15 to 50 cm (in combination with a correcting coil system). In such a system of the Astron LIA the most efficient beam passage was 90%, whereas good passage was considered to be 75%.

When reconstructing the accelerator in order to bring the current of the beam to 1000 A, the length of the accelerating sections was reduced by half, i.e., up to 50 cm, and 29 focusing solenoids and 9 correcting magnets were installed along the beam path. The location of the solenoids and correcting magnets was determined by computer calculation. It was assumed that along the acceleration path there exists a magnetic field with an induction of 0.45 T. However, when starting such an accelerator, it became clear that, because of the instability, the beam current can not exceed 500 A. The authors suggested that this instability is due to the generation of a high-frequency electromagnetic field, as well as the interaction of the beam with charges and image currents.

Intensive electron beams excite an electromagnetic field that has a magnetic field transverse to the axis. Under its action, the beam deviates from the axis of the accelerator and hits the walls of the chamber. To evaluate the deflecting force, the theory of interaction of a high-current electron beam with charges and image currents has been developed. The image force is

$$F_{im} = q\frac{2I_b}{\beta^2 r_1}r,$$

where $q = 1 - (a/r_1)^2\beta^2$; a is the inner radius of the tract; r_1 is the inner radius of the inductor core. Taking into account the image force in the calculation of the dynamics of a high-current beam leads to additional conditions imposed on the distance between the solenoids, and also on the relationship between the dimensions of the beam and the induction system. In particular, the aperture of the tract should be 4–8 times larger than the beam dimensions. The task of retaining beam dimensions is facilitated by using a continuous axisymmetric magnetic field. Indeed, under the influence of a magnetic field perpendicular to the axis, the beam deviates by an amount Δr_1, defined by expression

$$\Delta r_1 = \frac{BL_{dev}^2}{m_0 c\gamma}, \tag{1.62}$$

wherein L_{dev} is the distance over which the beam deviation is determined. In the case of beam propagation in a continuous longitudinal magnetic field, the presence of a field component perpendicular to the axis leads to the inclination of the axis and the deviation of the beam from the axis of the accelerator is given by

$$\Delta r_2 = \frac{BL_{dev}}{B_\phi},$$ (1.63)

where B_ϕ is the field needed to focus the beam [30]:

$$B_\phi = \frac{m_0}{r_b}\left(\frac{2I}{\pi\varepsilon_0 ep}\right)^{1/2},$$ (1.64)

and p is the impulse of the beam electrons.

From (1.62)–(1.64) we obtain

$$\frac{\Delta r_2}{\Delta r_1} \approx 46\frac{r_b}{L}\frac{\gamma^{3/2}}{I^{3/2}}.$$ (1.65)

Analysis (1.65) shows that in the initial part of the acceleration, the most problematic from the point of view of retaining the beam dimensions, the limiting current in a system with a continuous field is 5–10 times greater than when using short solenoids. According to estimates, in the case of a longitudinal magnetic field, the beam deflection decreases under the action of charge and image current. The equation determining the shape of the accelerated axisymmetric beam in a longitudinal magnetic field during the follow-up of the pulse has the form [31]

$$p\left(m_0^2c^2 + p^2\right)\frac{d^2r}{dp^2} + p^2\frac{dr}{dt} + \frac{r}{2}\left(\frac{B}{E}\right)^2 p = \frac{I_b m_0^2 c^2}{2\pi\varepsilon_0 eE^2 r}.$$ (1.66)

In deriving this equation, it was assumed that: 1) the potential due to the space charge of the beam does not affect the longitudinal motion of the particles; 2) the accelerating field has only a longitudinal component; 3) the magnetic field at the cathode is zero (Brillouin flux); 4) the beam is laminar, the conditions of paraxiality are fulfilled, and the effects occurring at the front and the trailing edge of the pulse are ignored. If the field at the cathode is not zero, then the term due to the initial magnetization of the beam enters into (1.66). In the LIA, the magnetic field at the cathode is usually zero. The term on the right-hand side of (1.66) corresponds to the repulsive force, and the last term on the left-hand side is the force that contracts the beam to the axis. Since the force causing spreading of the beam is inversely proportional to its radius, and the force attracting the beam is directly proportional to it, for each $p(z)$ there must be an equilibrium value $r = r_1(p)$, at which the force acting on the peripheral particle of the beam is zero. From (1.66) we have the following equation for such an equilibrium trajectory:

$$r_1(p) = \frac{m_0}{B_\phi} \left(\frac{2I_b}{\pi \varepsilon_0 eEp} \right)^{1/2}.$$
(1.67)

If the peripheral electron is located on an equilibrium trajectory, then the corresponding radius decreases monotonically in inverse proportion to the square root of p. For small deviations of the initial conditions from equilibrium, the beam undergoes pulsations the period of which increases with increasing energy:

$$2 \left(\frac{B}{E} \right)^2 \ll 1.$$
(1.68)

The following results were obtained in [12] by numerical solution of equation (1.67) for the case of an accelerating field homogeneous in z in LIA. If the beam enters the region of the magnetic field under equilibrium conditions, that is, if the beam radius and the quantity $r_1(p)$ determined by (1.67) are equal, as well as with the coincidence of the angular divergence of the beam with its equilibrium value, the beam radius decreases monotonically with respect to the measure of acceleration of electrons (curve *1* in Fig. 1.29). However, if the initial conditions do not correspond to the equilibrium conditions, then the envelope of the beam experiences oscillations that increase as it moves (curve *2* in Fig. 1.29). In the same work, the problem of choosing the conditions necessary to realize the equilibrium of the electron beam in the course of its motion inside the accelerator path of the LIA is considered in detail for the cases of laminar and non-laminar beams, as well as shielded and non-shielded cathodes. It is shown numerically that in a system with a longitudinal homogeneous magnetic field, the stability of the beam equilibrium both during acceleration and drift in the case of deviation of the initial conditions from equilibrium is ensured when the beam current $\Delta I/I \approx 20\%$ and the magnetic field inhomogeneity value is less than 5%. In

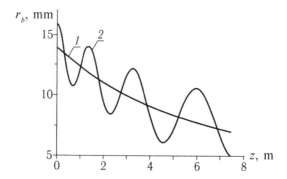

Fig. 1.29. The envelopes of a laminar electron beam at a current of 200 A, electron energies of 0.55 MeV and $E_z = 3$ kV/cm: *1* – the case of equilibrium values of the radius and angular divergence of the beam; *2* – the radius of the beam exceeds the equilibrium radius by 2 mm with zero divergence of the beam.

addition, in the case of a periodic dependence of the magnetic field on the coordinate *z*, due to the finite extension of the segments of the solenoid located between the accelerating sections, only near a certain optimal value of the magnetic field does the beam pass the entire accelerating path without losses. A deviation from this value results in the beam falling to the walls of the accelerator.

Since the repulsive force of the space charge decreases as I/γ^2, at a beam energy of 3–5 MeV it is possible to abandon the use of a continuous magnetic field, and to retain the radial dimensions of the beam with the help of short magnetic solenoids located between sections of the induction system. The location of the solenoids is determined by numerically solving equation (1.66) on the sections between two adjacent solenoids for different beam convergence values. With this calculation, a trajectory with a value of r_{max}/r_{min}, corresponding to a stable motion of the beam with respect to the deviations of its initial parameters from the equilibrium ones is selected. Analysis of the behaviour of the beam with deviations of the parameters of the focusing system from the nominal shows that the tolerances for the deviation of the parameters are not rigid. As noted above, the dynamics of the beam is analyzed assuming the laminarity of the electron beam and the fulfillment of the paraxiality conditions. This assumption is sufficiently valid for a uniform distribution of the charge density over the beam cross section. An experimental study of the parameters of the electron beam accelerated in the LIA revealed that the distribution of the charge density over the cross section can be uneven. The forces acting on the electrons of the space charge repulsion and the total force in this case are essentially non-linear. Such a non-linearity leads to the intersection of trajectories and to the formation of a non-laminar (multistep) structure of the electron beam. An analysis of the non-laminar electron flux shows that the distribution of the charge density over the beam cross section does not remain constant in different cross sections, and the maximum charge density can occur at the beam boundary. A possible way to reduce the effect of the space charge of the beam accelerated in the LIA is to compensate it with the ions of a residual gas. At residual gas pressures less than 10^{-3} Torr, the compensation time exceeds the pulse duration of the accelerating voltage, which leads to the necessity of preliminary ionization of the residual gas, for example by an electron beam or a high-frequency field. Investigations [32, 33] have shown that beam–plasma interactions arise during acceleration, which lead to a loss of a third of the beam energy and to the expansion of the energy spectrum of accelerated electrons. At the same time, it is attractive that the current reaches 1000 A, even at a small beam energy (40 keV). The use of a plasma with a decrease in its density along the length of the accelerator leads to a sharp decrease in the efficiency of the beam–plasma interaction. The beam loss decreases, and the energy spectrum of the electrons becomes narrower.

The technique of producing a high-current electron beam for the ATA LIA (47.5 MeV, 10 kA, 70 ns) is described in [21] using a Kr–F laser with an energy of 0.4 J in the pulse. The accelerating tube is filled with benzol vapours at a pressure of 10^{-8} Pa which are ionized the laser and form a narrow plasma channel at the accelerator axis. The required quality of the transport of the electron beam through the plasma channel requires careful adjustment of the profile of the longitudinal magnetic field and the beam cross section in accordance with the transverse configuration of benzol vapours.

The generalizing work [34] determines the amplitude of the beam envelope oscillations arising due to the instability of the feeding systems of the elements of the focusing–accelerating channel. The appearance of oscillations of the envelope or a change in its size may be due to the following reasons: deviation from the calculated values of the parameters of the focusing channel (magnetic field strength) and the beam (current, initial radius, energy); deviation from the calculated value of the magnetic field at the cathode. The dependence of the envelope radius increment $(\Delta R/R)$ on deviations from the calculated values of particle energy, initial beam radius, magnetic field intensity and beam current can be written as follows:

$$\left(\frac{\Delta R}{R}\right)_{\gamma} = -\frac{1}{2(1+K)}\left(\frac{\gamma^2 K}{\gamma - 1} - 2\right)\frac{\Delta\gamma}{\gamma};$$
(1.69)

$$\left(\frac{\Delta R}{R}\right)_{r_b} = -\frac{K}{1+K}\left(\frac{\Delta R}{R}\right)_H;$$
(1.70)

$$\left(\frac{\Delta R}{R}\right)_H = -\frac{\Delta H}{H};$$
(1.71)

$$\left(\frac{\Delta R}{R}\right)_{I_b} = -\frac{1}{2(1+K)}\frac{\Delta I_b}{I_b},$$
(1.72)

where

$$K = \frac{21R^2}{I_b \beta \gamma V_z^2}.$$
(1.73)

When passing through the focusing–accelerating channel, the beam meets in its path a series of focusing (n_{foc}) and accelerating (n_{acc}) elements. It is possible to add the effects of the above errors at which the error of the envelope radius will increase. Let us take the root of the dispersion $\langle \Delta R/R \rangle$ as a measure of its increase. Then the root-mean-square value of the envelope radius in the absence of discrete places of the beam loss will be

$$\left\langle\frac{\Delta R}{R}\right\rangle^2 = \frac{n_{\text{acc}}}{4(1+K)^2}\left(\frac{\gamma^2 K}{\gamma^2 - 1} - 2\right)^2\left\langle\frac{\Delta\gamma}{\gamma}\right\rangle^2 + \frac{K^2}{(1+K)^2}\left\langle\frac{\Delta R}{R}\right\rangle_H^2 +$$

$$+ n_{\text{foc}}\left\langle\frac{\Delta H}{H}\right\rangle^2 + \frac{K^2}{4(1+K)^2}\left\langle\frac{\Delta I_b}{I_b}\right\rangle^2.$$
(1.74)

The error in the magnitude of the relativistic factor in the n-th accelerating element is determined by an acceleration voltage error $\Delta U/U$ both of the given element and all previously considered elements. As before, assuming the errors to be independent, we can write

$$\left\langle \frac{\Delta\gamma}{\gamma} \right\rangle^2 = n\frac{(\gamma-1)^2}{\gamma^2}\left\langle \frac{\Delta U}{U} \right\rangle^2. \qquad (1.75)$$

The presence of a magnetic field H_c on the cathode also leads to an increase in the envelope radius, whose amplitude is equal to

$$\left\langle \frac{\Delta R}{R} \right\rangle_{H_c} = \frac{1+K}{8}\left(\frac{\delta H_c}{H} \right)^2. \qquad (1.76)$$

In a LIA with focusing by a longitudinal magnetic field, the beam can oscillate relative to the axis of the focusing channel [35]. These oscillations are caused by the presence at the centre of gravity of the beam of the initial deflection and the initial angle of inclination of the beam axis relative to the channel axis, as well as by the passage of the beam through sections with radial components of the electric and magnetic fields when the beam axis does not coincide with the axes of these sections. The effect of the azimuthal components of the electric and magnetic fields is similar, but they can be considered absent because of the sufficient simplicity in providing the axial symmetry of the components of the focusing–accelerating channel.

Accelerating gaps of electron guns and accelerating sections can be considered as a set of thin electric lenses with a radial field located at the entrance and exit of the gap (unless there is no mesh or foil only at the input) and an inner section with a longitudinal electric field. In this case, thin edge lenses have symmetry axes, and the section with a longitudinal field does not. Similarly, at the entrance and exit of the focusing elements there are thin lenses with a radial magnetic field with symmetry axes and inside them are sections with a longitudinal field that do not have a distinguished axis. Thus, the entire focusing–accelerating path is divided into a large number of sections, whose axes can be displaced relative to each other by position and angle.

The conducting chamber reduces the frequency of coherent oscillations of the beam in comparison with the case of beam motion in free space, but increases their amplitude. To reduce the amplitude of coherent oscillations of the beam, its radius should be made smaller than the radius of the chamber, that is, 'tear' the beam from the walls. In addition, the oscillation frequency can be increased by choosing the intensity of the focusing magnetic field above the agreed value. However, in this case, oscillations of the envelope of the beam arise, i.e., one cause of the loss of the beam particles is replaced by the other. Nevertheless, for a certain excess of the field strength over the agreed value, a minimum of particle losses is possible.

It should be taken into account that the permissible amplitude of the envelope oscillation depends on the purpose of the accelerator. If the accelerator is simply a source of ionizing radiation and only the integral effect of irradiation is important, the amplitude of the oscillations must be such that the beam particles do not fall on the wall of the vacuum chamber. In this case, finding within the tolerance limits for the instability of the focusing magnetic field the amplitude of the pulses of the accelerating voltage, the initial beam radius, the beam current, and the magnetic field at the cathode is relatively easy to provide. The situation is completely different in the case when the beam from the LIA must be injected into some other device or used to power the microwave device, i.e., when correct beam passage is necessary. The tolerances for the output parameters and the position of the beam in the LIA become rather rigid. Ensuring compliance with them is a difficult task (especially the stabilization of accelerating voltage is particularly difficult).

For clarity, see below Table 1.3 [36].

An analysis of the data in the table shows that if the LIA is intended to be used as an injector for another installation and the correct passage of the beam in it is essential, the instability of the accelerating voltage and the intensity of the focusing magnetic fields in the accelerator for an energy of several MeV should not exceed 1%. The radius of the beam should be much less than the radius of the channel (their ratio should be no more than 0.3). With a ratio of these radii close to unity and a matched magnetic field, the frequency of the dipole coherent oscillations of the beam is small, which leads to a large amplitude of these oscillations and to particle losses. The amplitude of the dipole coherent oscillations essentially depends on the initial displacement and the initial angle of the beam. With correct installation, the allowable beam offset at the output of the accelerating gap of the gun is ~1 mm, and the angle of its inclination is not more than a third of a degree. The greatest contribution to the increase in the amplitude of the coherent oscillations of the beam is made by the slope of the magnetic axes of the focusing elements. If the beam is correctly routed to the LIA with an energy of several MeV, the tolerance for the deviation of any end of the magnetic axis of the focusing element is approximately 0.1 mm. The tolerance for the error in the lateral alignment of the accelerating sections and long

Table 1.3. Admissible deviations of the parameters of the LIA

No.	Deviation of the parameter	Measurement units	Tolerances Transport lossless	Tolerances Correct transport
1.	The initial beam deflection	mm	2	0,7
2.	The initial angle of the beam	deg	1	0,3
3.	Accuracy of alignment of accelerating sections	mm	0,2	0,07
4.	Offset of the axis of the vacuum chamber in the gap	mm	2	0,7
5.	Acceleration voltage deviation		0.075	0,01
6.	Deviation of the focusing magnetic field		0.05	0,007
7.	Deviation of the beam current at the input		0.2	0,03
8.	Deviation of the beam radius at the input	mm	0.1	0,015
9.	Magnetic field on the cathode	Gauss	50	20

sections of the vacuum chamber has the same order. If the vector of the magnetic field does not coincide with the axis of the conducting chamber at an angle between the field vector and the axis much less than unity and the length of the chamber much shorter than the wavelength, the beam oscillates around the axis of the chamber. This effect is used when passing a beam through a long focusing channel with a longitudinal field, unshielded from external transverse fields, for example, from Earth's magnetic field.

The assessment of the current for correct passage $I_{c.w.}$ for the LIA 5/5000 and ATA accelerators is given in [37]. In the first case, the transportation channel area $S = 50$ cm², and $I_{c.w.} = 0.6$ kA at $\gamma = 1.75$, in the second case, $S = 320$ cm², and $I_{c.w.} = 25$ kA at $\gamma = 5$.

1.4. 'Non-iron' LIAs

In the former USSR and later in Russia researchers under the leadership of A.I. Pavlovsky and V.S. Bosamykin developed another class of linear induction accelerators in which ferromagnetic cores are absent. They are called the 'non-iron' LIA [38].

The principle of operation of such accelerators is based on excitation of an electric field between two sections of the internal surface of a united closed conducting cavity, which simultaneously plays the role of an electromagnetic screen (Fig. 1.30). The voltage at the ends of the line connecting by volume any two points belonging to the surface of the screen is

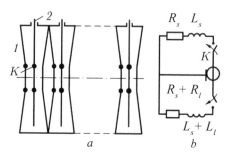

Fig. 1.30. Inductors on lines with distributed parameters (*a*) and their equivalent circuit (*b*).

$$U = \oint Edl = -\frac{d}{dt}\sum_{i=1}^{N}\varphi_i, \qquad (1.77)$$

where $\varphi_i = \int_S BdS$ is the value of one of the magnetic fluxes penetrating the area between this line and the screen. The total voltage in similar LIAs consisting of the same type of inductor elements as in accelerators with ferromagnetic cores, acts only along the accelerating path. The 'non-iron' LIA allow to master the energy region above 10^7 eV while maintaining the high currents inherent in nanosecond accelerators (HCEA). The first 'non-iron' accelerator provided an electron beam current of up to 2 kA at an energy of 2 MeV and an accelerating voltage frequency of 6.2 MHz [39].

Transients in the inductor (neglecting the inhomogeneity of the line) are similar to the processes in a cable charged up to a voltage U_0 with characteristic impedance ρ_0, which is closed at one end via a switch K with the inductance L_s to resistance R_s. Simultaneously from the other end the cable starts discharging through a series-connected load with parameters L_l and R_l and a switch with parameters L_s and R_s. To accelerate charged particles, a voltage pulse of one polarity is used. Therefore, no more than half of the stored energy is transmitted to the beam. Of course, one can use the second part of the pulse to accelerate, but the first pulse of the doubled voltage amplitude will act on the accelerating tube in this case (see Fig. 1.22). There are circuits of inductors that make it possible to form voltage pulses of the same polarity, but the voltage in them varies according to the cosine law.

The change in the magnetic flux in the short-circuited line is due to the propagation of the current wave:

$$\int_0^\tau U\,dt = \Psi = -I_m L_{\text{lin}} v\tau,\tag{1.78}$$

where I_m is the amplitude value of the magnetization current of the line equal to the difference between the total line current (switch current) and the load current; L_{lin} is the inductance per unit length; v is the wave propagation velocity; τ is the wave duration. In view of the smallness of the inductance the magnetizing current reaches high values, increasing linearly with increasing duration of the formation of the accelerating voltage. Therefore, the maximum reasonable acceleration time for a 'non-iron' linear induction accelerator does not exceed two to three dozen nanoseconds.

1.4.1. Inductors based on toroidal circuits

Below we discuss in more detail the possible schemes, as well as the electrodynamic and structural features of the 'non-iron' LIA. The material is written on the basis of review articles [38, 40]. The simplest 'non-iron' inductor is a single-turn transformer with a primary circuit in the form of a conducting torus which is excited by an external AC source. The secondary circuit, also having a toroidal configuration, concentrates the induction emf in the accelerating gap. To reduce the losses, the thickness of the conducting walls of the tori must be greater than the depth of the skin layer at the operating frequency. If the power source is represented by one or more long lines connected in parallel with impedance ρ, the energy w_{rz}, scattered in the load resistance during the time t, depends on the coupling coefficient, the inductance of the primary circuit L_1, and also on the active resistances of the primary and secondary circuits. Analysis of the case of an ideal coupling (without taking into account the energy losses in the line and in the switch), when

$$W_{R_2} = \frac{U_{\text{RFL}}^2}{2\rho}\frac{L_1}{\rho + R_2}\left\{1 - \exp\left[-\frac{2\rho R_2}{L_1(\rho + R_2)}t\right]\right\},\tag{1.79}$$

shows that the efficiency of the inductor tends to unity for a sufficiently short duration of a unipolar pulse ($t \ll L_1/p$) and a matched load.

To increase the intensity of the accelerating electric field, it is advantageous to combine all the elements forming the primary circuit of the pulsed 'non-iron' transformer (connecting circuits, capacitors and the switch) into a single-turn toroidal oscillatory circuit. One of the variants of such an inductor is shown in Fig. 1.31 *a*. An alternating magnetic flux is excited in the discharge of the annular capacitor C on the inductance L_t of a torus, in which the azimuthal gap is an annular switch P. The ground plate of the primary circuit is simultaneously a part of the secondary circuit.

When the switch is activated, the load current I_2 flows along the path *AEFGDA*, which does not cover the magnetic flux $\varphi = L_t I_1$ in the primary circuit. Capacitor C (charged gap capacitance *DE* is connected in parallel to C) is directly discharged through the load R_2. The equivalent circuit of an inductor with partially joined circuits is given in Fig. 1.31 *b*. If in the azimuthal rupture of the torus, for example in a switch, there is a lumped leakage inductance L_p with a non-toroidal magnetic flux or an active resistance R_1, the direction of the electric field due to the flow of the total current   (*I* $= I_1 + I_2$) is opposite to the field propagating from the capacitor. Accordingly, the accelerating voltage decreases. The same results are obtained when considering transient processes by means of mutual inductance of the circuits ($M = L_t$). The accelerating system can operate in an oscillatory or aperiodic mode, realized at a high primary inductance L_t. In both cases, neglecting the inductance of

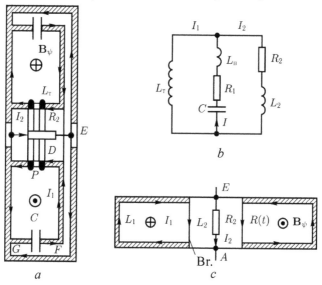

Fig. 1.31. Constructive (*a*) and equivalent (*b*) schemes of a 'non-iron' inductor, as well as a toroidal inductive storage with a break in the current circuit (*c*): Br – breaking unit (foil); *P* – switch.

the secondary loop L_2 and the parasitic inductance L_p the expression for the efficiency has the form

$$\eta = \frac{\rho_1^2 R_2}{\left(\rho_1^2 + R_1 R_2\right)\left(R_1 + R_2\right)},$$ (1.80)

where $\rho_1 = \sqrt{L_t/C}$. If $R_1 = 0$, all the stored energy is dissipated in the load. If $R_1 > 0$, the efficiency is maximal at $R_2 = \rho_1$. The frequency range of the inductor in the oscillatory mode is

$$\left[\left(L_p + L_t\right)C\right]^{-1/2} \leq \omega \leq \left[\left(L_p + \frac{L_t L_2}{L_t + L_2}\right)C\right]^{-1/2}.$$ (1.81)

The inductor can be formed by parallel connected synchronized generators of pulse voltages according to the Arkadiev–Marks scheme. The high voltage inside the primary circuit relative to the grounded *BCDG* screen (Fig. 1.32 *a*) occurs during the successive operation of spark gaps. It is then balanced by the self-induction emf. Figure 1.32 *b* shows the inductor circuit based on a generator with recharging of one of the ring capacitors, C_2, through the toroidal inductance L_2. The ring switch P is an integral part of the active (L_2-C_2) and passive (L_1-C_1) circuits. If the capacitance of the capacitors is the same, the inductance L_1 must be significantly higher than the inductance L_2. Such circuits allow increasing the output voltage of each inductor at the same charging voltage.

For inductors with the combined elements of the primary circuit elements there is a region of changing parameters ($d_{in} \ll \lambda_{osc}$, where d_{in} is the characteristic dimension of the inductor and λ_{osc} is the natural oscillation wavelength; $\rho_c \ll \rho_1$ where ρ_c is the wave impedance of the capacitor; $R_2 \gg \rho_c$), beyond which the constructional design under discussion loses its advantages. For $d_{in} \approx \lambda_{osc}$ and $\rho_1 \gg \rho_c$, the circuit can be excited at its own frequency, so the use of switches becomes unnecessary. Such cases as $\rho_1 \gg \rho_c$, $R_2 \approx \rho_c$ or $R_2 \approx \rho_c$, $\rho_1 \approx \rho_c$, also require other constructive solutions.

With a high level of stored energy (more than 1 MJ), it is advisable to use energy-intensive power supplies – magnetocumulative generators (MCGs) along with capacitive storage devices to obtain single electron current pulses. The resulting problems are associated with the formation of short ($\leq 10^{-6}$ s) voltage pulses. One way to solve them is to use the MCG as a charger for fast capacitive storage devices. From the point of view of increasing the specific energy

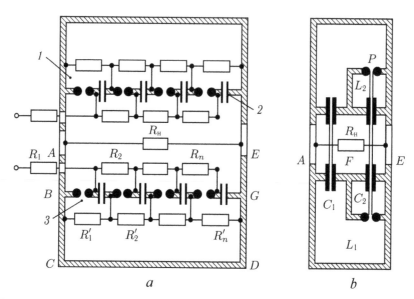

Fig. 1.32. Inductors based on Arkadiev–Marks pulse voltage generators (*a*) and generators with charge exchange of annular capacitors (*b*): *1* – pulse voltage generators; *2* – capacitors; *3* – spark gaps; $R_1,...,R_n$ and $R'_1,...,R'_n$ are charging resistors.

capacity it is promising to use liquid dielectrics subjected to a pulsed action of pressure.

For the formation of high-voltage pulses of duration $\sim 10^{-7}$ s inductive storage devices also use a method based on the phenomenon of electrical exploding of conductors [41]. The magnetic energy in the final stage of the MCG operation is stored in a closed toroidal circuit with a foil or parallel wires being an integral part (see Fig. 1.31 *b*). The duration of the voltage pulse depends on the inductance of the circuit $L_1 = L_t$, the phase transition time t_p and the electrical resistance of the breaking unit, and the amplitude increases according to the growth of the active resistance: $U = IR(t)$. When the current is transferred to the secondary circuit (L_2, R_2), a significant fraction of the energy is lost in the break. In this case, the energetically favourable mode is the one in which the break time is much shorter than the current decay time: $t_{br} \ll (L_1 + L_2)/R_2$. With a simultaneous current break in several connected toroidal circuits, the voltage along their axis is summed up (as in other linear induction systems). It is obvious that the greatest efficiency of the accelerator can be realized if the breaker itself is used as a source of electrons ($L_2 = 0$; see Fig. 1.31 *b*), and the circuit geometry, with other things being equal, ensures the maximum power of energy release. Dissipation in the breaking unit of magnetic energy W accumulated over time t is accompanied by the appearance of the voltage amplitude U at an average electric field strength at break E, which in the actual case is not more than some permissible value. The expression for W has the form

$$W = \frac{\varphi^2}{2L_1} \approx \frac{\pi}{\mu_0} \frac{1}{\ln(r_2/r_1)} EUt^2. \tag{1.82}$$

Formula (1.82) allows one to estimate the upper limit of the possible energy content of the electron beam when using the energy of inductive storage devices. Note that if an abnormal resistance determined by the cumulation of a cylindrical plasma layer that is formed during the explosion of thin conductors in a toroidal circuit is used to form a short voltage pulse, the device acquires features characteristic of installations with a 'plasma focus'.

1.4.2. Inductors on lines with distributed parameters

The minimum duration of a pulse in an oscillatory circuit is determined by the time of propagation of an electromagnetic wave along the dielectric of a capacitive storage device. In this case, when considering the transient processes in the inductor, it is necessary to take into account the features peculiar to long lines. Figure 1.33 *a* shows the schemes of inductors on lines with distributed parameters ($\rho_1 \approx \rho_c$), which are formed by a high-voltage plate F and a grounded shield. When the ring switch BF is closed, the line FG is connected to the accelerating gap R_l. If we ignore the heterogeneity of the toroidal line, then the transient processes in the inductor are analogous to those in a cable charged to U_0 with a wave resistance ρ which is closed at one end by means of a switch with inductance L_s and resistance R, and from the other end it simultaneously starts to be discharged through the series-connected load (L_l, R_l) and switch (L_s, R_s) (Fig. 1.33 *b*). The reflection coefficients in such a scheme, neglecting the inductance of the switch and the load, have the form

$$K_s = \frac{\rho - R_s}{\rho + R_s} (R_s < \rho); \quad K_l = \frac{R - \rho}{R + \rho}. \tag{1.83}$$

Figure 1.33 *c* shows the idealized time dependences of voltage on the load for various $R = R_s + R_l$. Taking into account the heterogeneity of the line, the leakage inductance and the transition resistance of the switch leads to smoothing of jumplike changes in voltage and current. The change in the magnetic flux in the short-circuited long line is due to the propagation of the current wave

$$U_0 = -\varphi = -IvL_{\text{lin}},$$

where L_{lin} is the linear inductance. The alternating magnetic field in the inductor is not connected with the external space, since it is shielded by metal plates and separated from the open end of the line by the length of its undischarged part. Thus, the distribution of the electric field outside the inductor with an ideal ring switch is determined by the configuration of the open end of the *FG* line, and the load characteristics correspond to the discharge of the radial line on the impedance of the accelerating gap. Assuming that the inductor's active resistance is concentrated in the load and the switch, one can obtain an expression for the efficiency:

$$\eta = \frac{W_{R_l}}{W_0} = \frac{2R_l\rho}{(R+\rho)^2}\frac{1+K_s^2}{1-K_s^2K_l^2}; \qquad (1.84)$$

where $W_0 = CU_{\text{RFL}}^2$.

The schematic diagram of the inductor in which $\rho_1 \gg \rho_c = \rho$ is shown in Fig. 1.33 *a*. The line formed by the high-voltage plate *F* and the grounded screen, with the help of the switch *P*, is simultaneously connected to the lumped inductance of the torus L_t and the resistive load R_l. The peculiarity of this scheme is the possibility of forming single rectangular voltage pulses. The coefficient of efficiency of the inductor is

$$\eta = \frac{4R_l\rho}{(R+\rho)^2}\frac{1+K_l^2}{1-K_l^2}. \qquad (1.85)$$

At $R_s = 0$, during the unipolar voltage pulse in the matched load, all the energy accumulated by the capacitance of the line is dissipated. It is obvious that the limiting transition to the aperiodic regime in the inductor (see Fig. 1.31 *a*) occurs at $\rho_1 \gg R_l \gg \rho$. When placing the ring switch at a certain distance from one of the ends of the line, in particular equal to a quarter of its length (Fig. 1.34 *b*), the accelerating voltage of the inductor is doubled.

The electric field in the acceleration region is maximal if the inductor consists of only two radial lines, and the switch *P* is located on the average radius of the inductor. When a current wave propagating along a short *PE* line reaches the accelerating gap *AE*, a voltage pulse is formed on the matched resistance the amplitude of which is equal to the charging voltage. During the first pulse, approximately half of the energy stored in the inductor is transferred to the load. A unipolar voltage pulse can be formed on a matched load by placing the switch at equal distances from the ends of the radial line, increasing the recharge time of one of its halves (*PF*) by means of a high separating inductance L_t (Fig. 1.34 c).

In addition to the above, schematics of inductors with radial transmission lines are possible which are formed by increasing the radial dimensions of the grounded plates *AB* and *EG* between the load and the ring switch *BF* (see Fig. 1.33 a). When the switch operates, a cylindrical converging electromagnetic wave propagates along the

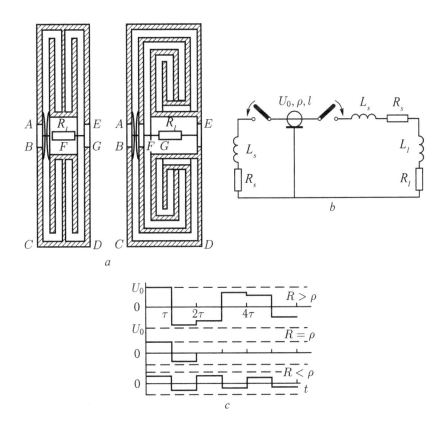

Fig. 1.33. Inductors on lines with distributed parameters ($\rho_1 \approx \rho_c$): *a*) schematic diagrams of inductors; *b*) the electrical circuit; *c*) the active component of the voltage U_R of the load ($R = R_l + R_s$).

transmission line. In circuits with charging voltage of different signs
the number of switches is reduced by half.

The radial line occupies an intermediate position between a
homogeneous long line and a toroidal oscillating circuit with lumped
parameters. Its wave resistance varies with the radius as

$$\rho_{\text{RFL}}\left(r\right) = \frac{1}{2\pi}\sqrt{\frac{\mu_0}{\varepsilon\varepsilon_0}\frac{\Delta}{r}}, \tag{1.86}$$

where Δ is the length of the high-voltage gap; ε is the relative
permittivity (the speed of propagation of electromagnetic waves
remains constant). If the charged radial line is closed along the axis
by a conducting channel of radius r_1 then in the process of wave
motion the initial voltage jump decreases in amplitude so that the
voltage at the front is

$$U\left(r\right) \approx U_{\text{RFL}}\left(1 - \sqrt{\frac{r_1}{r}}\right). \tag{1.87}$$

On the outer radius of the line r_2, the voltage changes the sign after
a time

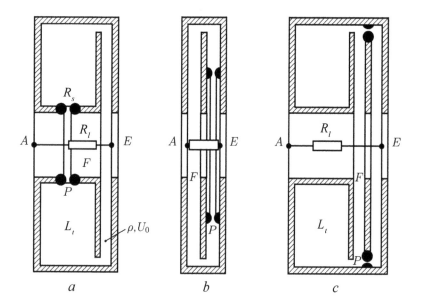

Fig. 1.34. Inductors: *a*) on a long line with a lumped inductance of the primary
circuit L_t ($\rho_1 \gg \rho_c$); *b*) on a double radial line (switch *P* is removed from the end of
the line by a quarter of its length); *c*) on a double line with a dividing inductance L_t.

$$\tau_p = k\sqrt{L_p C}, \tag{1.88}$$

where $L_p = \dfrac{\mu_0}{2\pi}\Delta\ln\dfrac{r_2}{r_1}$ is the total inductance of the circuit. Coefficient $1 \le k \le \pi$ depends of the degree of non-uniformity. The value of $k = 1$ refers to the case of a homogeneous line ($r_2 \approx r_1$, $\Delta \ll r_1$) with a wave resistance $\rho_p(r) = \rho_p(r_1)$, when the pulses have a rectangular shape, and the recharge time at the outer radius corresponds to the range of the waves: $\tau = \dfrac{r_2 - r_1}{v}$. Experiments with a highly non-uniform radial line (insulation – water; $r_2 = 0.5$ m; $r_1 = (10^{-4}–10^{-2})$ m; $\Delta = 3\cdot10^{-2}$ m) show that in this case, the output voltage $U(r_2)$ varies in time practically in accordance with the law of cosine with small distortions of the amplitude, arising from repeated reflections of the initial voltage jump. For $r_1/r_2 = 50–1000$, the coefficient k increases from 2.5 to π. The non-uniformity of the radial line can be used to transform nanosecond pulses. At the same time, it should be noted that the line remains uniform if its high-voltage gap varies in proportion to the radius.

1.4.3. Energy capabilities of 'non-iron' LIAs

The limiting energy of particles accelerated in linear induction accelerators is determined by the cost of facilities and the content of tasks solved with their help.

   The effect of the time on the acceleration energy, especially important for heavy particles and for electrons at an energy of $>10^7$ eV, is eliminated by introducing the delay of activation of each subsequent inductor at a time $\tau_3 \approx h/u$ (where u is the particle velocity, h is the longitudinal dimension of the inductor). For accelerating electrons to energies of tens of MeV or more when the beam currents $>10^5$ A and the pulse duration $\sim10^{-7}$ s the most promising are systems on lines with distributed parameters. The contribution of each inductor to the total acceleration energy is limited by the permissible electric field strength E_0 in the dielectric and the condition for obtaining voltage pulses whose shape is close to rectangular: $U_{RFL} = E_0\Delta \ll E_0 l$ (where l is the length of the line that determines the pulse duration t; $\Delta \le h$ is the size of the high-voltage gap). Nanosecond generators on lines of coaxial or planar geometry, usually loaded on a cylindrical beam of a small diameter in comparison with the radial dimensions, are also used as power sources for high-current accelerators. The transfer of electromagnetic energy into a load with a high specific energy intensity is again determined by the section representing the radial line. Distracting from the non-uniformity of the lines, we estimate the energy capabilities of the unit element of the accelerator. Using the transmitting radial line over which the voltage wave propagates with amplitude $U = E\Delta$ during time t into the load (hollow cylinder of radius r_1) the following energy can be transferred

$$W \leq \varepsilon\varepsilon_0 E^2 \pi \left(r_2^2 - r_1^2 \right) \Delta = \frac{\pi}{\mu_0} E U t^2 \left(1 + 2\frac{r_1\sqrt{\varepsilon}}{ct} \right). \qquad (1.89)$$

At the same time, the energy transferred from the forming line charged to $U_{RFL} = 2U$ and connected directly to the load

$$W \leq \frac{\pi}{\mu_0} U E_0 t^2 \left(1 + 4\frac{r_1\sqrt{\varepsilon}}{ct} \right), \quad \text{where } E_0 = \frac{U_{RFL}}{2\Delta}, \qquad (1.90)$$

is half the energy transferred at small r_1.

The above expressions determine the upper limit of energy release also for the set of lines forming the power source of the 'non-iron' LIA, or the nanosecond accelerator with voltage NU. It seems natural that they resemble the expression (1.82) for an inductive storage device with a break in the current circuit. At a given electric field, which depends on the electrical strength of the structure, and the selected voltage, the power of the accelerators on lines with distributed parameters can be increased only by increasing the radial dimensions of the lines and the load. Thus, the electrical strength of the currently known dielectrics limits the possibility of energy concentration by means of such high-current accelerators, which is especially important if short electron beams are required ($\leq 10^{-8}$ s).

In conclusion, we note that the additional voltage concentration by metal electrodes widely used in accelerator tubes seems to be applicable for systems with the transverse size of the acceleration region comparable to the high-voltage gap of the power source. When the length of the acceleration region is much larger than its transverse dimension, which is noted in LIAs at energies $>10^7$ eV, it is more advantageous to take energy from the power source by means of a load with the active resistance distributed along the entire length. This load, in particular, is also the electron beam.

1.4.4. Design and parameters of 'non-iron' LIAs

The technology of 'non-iron' LIAs is quite complicated. This can be seen by considering their construction. The development of this accelerator technology required solving a number of electrophysical problems, among which we can note: 1) the creation of a multi-element accelerating system using hundreds of switches connected with a given program with subnanosecond accuracy, commutation of low-impedance ($\rho \approx 1$ Ohm) forming lines; 2) development of methods for the formation and transportation over significant (about 10 m) distances of high-current (with a current of tens of kiloamperes) electron beams.

In 1967, the first 'non-iron' LIA-2, was created, which provided an electron beam current of up to 2 kA at 2 MeV and a pulse duration at half-height of 40 ns. A longitudinal magnetic field was used to transport the beam along a path 2.5 m long. Ring storage capacitors were represented by K-15-4 ceramic capacitors, charged in a pulse and connected in parallel, with an operating voltage of 50 kV. Switching was carried out by four-channel dischargers with 'distortion of the field'.

The successful solution of the above tasks ended with the commissioning in 1977 of LIA-10 [42]. The LIA-10 accelerating system consists of 16 series-connected modules, the contribution of each of which to the electron energy is about 1 MeV. A separate accelerating module is a set of three functionally connected units – a block of inductors, a pulse voltage generator PVG-500, which provides impulse charging of the capacitance of the inductor block, and a generator that forms the triggers of inductor switches. The modules are autonomous both in design and in electrical circuit.

The most important part of the LIA is the inductor block (Figure 1.35). It consists of three series-connected sections on radial lines with insulation by deionized high-resistance water, connected in parallel on the charging circuit with one PVG-500. A separate inductor is formed by a toroidal grounded body, open on the inside diameter, with an annular high-voltage electrode mounted inside it. The radial line of the inductor closes around the perimeter of one of the output gaps on its internal radius with the aid of a multichannel ring switch. In this case, the other output gap through the same switch is connected to the load – the electron beam. The ring switch is formed by ten (in other modifications by eight) separate gas-filled dischargers of trigatron type at 500 kV. The spread of the switch-on time does not exceed 1 ns.

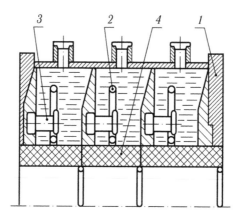

Fig. 1.35. The block of inductors LIA-10: *1* – case; *2* – annular high-voltage electrode; *3* – switch; *4* – accelerating tube.

In the near-axis area of the inductor block there is an accelerating tube made of polyethylene separating the vacuum cavity of the path from the cavity of the inductor filled with deionized water. The inner diameter of the tube is 380 or 280 mm in various versions of the inductor blocks.

A non-loaded inductor with uniform lines should ideally generate rectangular pulses of an accelerating voltage with an amplitude close to the value of the charging voltage. The duration of the first pulse is 20 ns, the second and subsequent pulses of alternating polarity 40 ns. The non-uniformity of the lines, the effect of the inductances of the dischargers and the line with a finite electrical length between the inductor proper and the load lead to some distortions of the ideal pulse shape (Fig. 1.36).

Parameters of the inductor block: charging voltage 500 kV; the duration of the charge pulse is 0.8 μs; the maximum amplitude of the accelerating voltage is 1.4 MV; short circuit current 180 kA. Dimensions: outer diameter 1 m; longitudinal dimension 0.45 m.

The capacitance of the inductor block is charged from the PVG-500 generator, made according to the modified Arkadiev–Marks

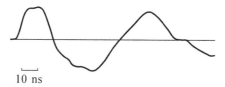

10 ns

Fig. 1.36. Shape of voltage pulse.

scheme. Five stages of PVG-500 are formed by the IK-100-0.25 capacitors, gas-filled trigatron type switches for 100 kV, liquid resistors and current-carrying buses. Isolation is carried out with transformer oil. All dischargers, except for the discharger of the first cascade, are triggered from the pulses generated when the discharger of the previous cascade is triggered by resistive circuits. The spread of the turn-on time of the PVG discharger at a constant voltage of 100 kV is ±3 ns with a reserve of electric strength of the gas gap equal to two. The electrical circuit of the PVG is shielded by a grounded metal housing that is connected to the inductor block by a coaxial high-voltage input of 500 kV containing a filter to suppress the back pulses propagating through the input when switching inductor lines. The energy reserve of one generator is 6.25 kJ, the impulse current is 50 kA, the charging voltage is 100 kV. The dimensions of the PVG are $1.3 \times 0.75 \times 0.75$ m^3.

The PVG is triggered by a pulse with an amplitude of 50 kV, a duration of 30 ns and with a 10 ns front. To start multichannel switches of the inductor unit, 30 (in another version 24) simultaneous positive-polarity pulses with a front up to 5 ns, duration of at least 15 ns, amplitude of about 65 kV are necessary. The generator of start pulses is a double forming line with glycerin insulation, activated when a single-channel discharger is triggered. To sharpen the front up to 2 ns, a discharger with ring electrodes is used, operating in a multi-spark mode. The impedance of the DFL is 3.3 ohms, the pulse charging voltage is 100 kV, the charging time is 0.8 μs. The generator of start pulses is switched on by a single pulse with the same parameters as those realized at its output. In this case, the spread of the response time of each generator is not worse than 1 ns.

Based on the accelerator modules, electron beam generators were constructed in which the accelerating systems contained 2, 3 and 4 inductor blocks. The total voltage of the series-connected inductor units is fed to an auto-emission coaxial vacuum diode with magnetic insulation by a longitudinal magnetic field.

The basic concept of the diode system of these units is the maximum possible reduction of the inductance of the diode due to the choice of the shape of the electrodes and the provision of the required electrical strength by means of the magnetic insulation of the system as a whole. For this purpose, the diode electrodes are placed in a longitudinal magnetic field formed by a solenoid located on the surface of the accelerating tube (Fig. 1.37). The inductance is minimized by making the cathode holder and the outer surface of the anode conical. It was experimentally established that to prevent current leakage, the strength of electric fields on the electrode surface should not exceed 150 kV/cm. A coaxial electron beam is formed in the diode system and its transverse dimensions are determined by the diode geometry and the insulating magnetic field.

Generation by the inductor of a train of pulses of different polarity makes it possible to operate the installations based on accelerating modules both on the first and on the second half-wave of the voltage. Figures 1.38 and 1.39 show oscillograms of the current in the diode and then of the electron beam in the region of drift in a three-block accelerator when operating on the first and second half-waves of the voltage. The amplitude of the output current reaches 35–40 kA with pulse duration at the base of 22 and 45 ns, respectively. The possibility of changing the program for starting inductors in this accelerator makes it possible to vary both the shape of the current pulses and their amplitudes.

The considered method of magnetic insulation leads to effective suppression of radial leakage currents from the cathode. However, the

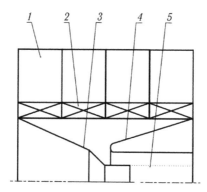

Fig. 1.37. Coaxial diode with magnetic insulation of accelerator LIA-10: *1* – block of inductors; *2* – solenoid; *3* – cathode; *4* – anode; *5* – electron beam.

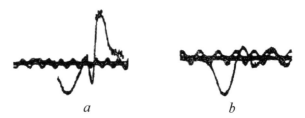

a   *b*

Fig. 1.38. The oscillograms of the current in the three-block accelerator when operating on the first half-wave of the voltage: *a*) the current is in the diode; *b*) the current of the electron beam. Markers 100 MHz.

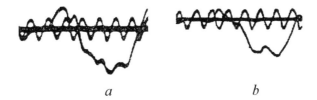

a   *b*

Fig. 1.39. Oscillograms of the current in a three-block accelerator when operating on the second half-wave of the voltage: *a*) the current in the diode; *b*) the current of the electron beam. Markers 100 MHz.

presence of lines of force that simultaneously cross the surface of the cathode holder and the anode makes it possible to form axial leakage currents, which can somewhat reduce the output characteristics of the injector. In order to prevent this process, a system is used that provides complete magnetic insulation of the cathode electrode. Due to the introduction into the internal cavity of the cathode holder of a conducting body with a specially calculated shape, it is possible to form a magnetic field with one of the lines of force (surfaces of equal flux) of this field coinciding with the surface of the cathode holder and cathode (Fig. 1.40). This circumstance, as well as the absence of lines of force, simultaneously crossing the anode and cathode electrodes, leads to more than halving of the leakage currents. Two-module and three-module variants of accelerators have been constructed. Note that in the three-module version with full magnetic insulation of the cathode electrode, the output current reaches 80% of the current in the diode.

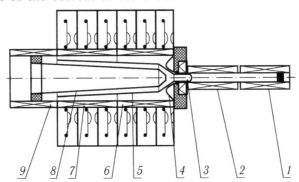

Fig. 1.40. Scheme of a two-block accelerator with magnetic insulation of the entire cathode electrode and with a magnetic field deflector (when operating on the second half-wave voltage): *1, 2* – solenoids of the beam transport path; *3* – cathode; *4* – anode; *5* – cathode holder; *6, 7* – magnetic field solenoids in the diode; *8* – magnetic field deflector; *9* – additional solenoid.

In the accelerators based on inductors on lines with distributed parameters, the following characteristics of electron currents are realized:

– in the two-module: beam current 30 kA; electron energy is 1.5 MeV;

– in the three-module: beam current 40 kA; electron energy is 2.1 MeV;

– in the four-module: beam current 45 kA; electron energy is 2.6 MeV.

The duration of the current pulse in all installations is ≈20 ns when operating on the first and ≈40 ns when operating on the second half-wave of the accelerating voltage.

The four-module accelerator was used as an injector of the modernized installation LIA-10 which consisted of 4 injector and 12 accelerating units. At the output of LIA-10 an electron beam was produced with a current of the order of 35 kA and an electron energy of 14 MeV for a pulse duration of 20 and 40 ns when operating at the first and second half-waves of the accelerating voltage, respectively. The beam was transported in the acceleration path by the magnetic field with an amplitude of 0.5 T. In 1994, the LIA-10M accelerator was built on the basis of water-insulated lines with a step change in the wave resistance. A beam of electrons with an energy of up to 25 MeV, a pulsed current of up to 50 kA, a current duration of 20 ns, a bremsstrahlung dose of 1 m from the target to 7.5 Gy was obtained.

One of the methods for increasing the voltage generated by a group of inductively connected in series inductors of the LIA is to mount to their output a transmission line with a large wave resistance. This method, used in high-voltage generators of unipolar pulses [43], is more effective in the case of generators of bipolar pulses to which the inductors of the 'non-iron' LIAs belong. When the load (electron beam) is switched on at the moment when the transmission line with an electric length equal to τ arrives at the output of the second accelerating voltage pulse with a duration of 2τ, the voltage on the load is determined by the relation

$$U = \frac{4U_0 RW^2}{\left(W_0 + W\right)^2 \left(R + W\right)}, \tag{1.91}$$

where

$$W = \frac{1}{2}W_0 \left(1 + \sqrt{1 + \frac{8R}{W_0}}\right) \tag{1.92}$$

is the optimum wave impedance of the transmission line for given total wave resistance of the inductor group W_0 and the load resistance R; U_0 is the effective charge voltage of a group of inductors connected in series. At $R/W_0 = 10$, the output voltage is $1.85U_0$. According to calculations, when using two LIA or modules and a two-stage transmission line with the length and impedance of the first and second stages equal to 0.8 m, 15 Ω and 80 Ω respectively, the ratio U/U_0 will reach a value of 2.6 at a beam current of 20 kA.

The VNIIEF (The All-Russian Scientific Research Unstitute of Experimental Physics) produced an I-3000 accelerator, based on two LIA modules and a coaxial vacuum transmission line ($W = 43$ ohm) 2 m long (Fig. 1.41). The line is formed by a grounded electrode with an internal diameter of 510 mm and a cantilever high-voltage electrode with a diameter of 240 mm. Dimensions of the accelerator: length 3.5 m; width 3.7 m; height 2.3 m. The weight of the installation is 2.3 t. The output characteristics of the I-3000: electron energy up to 3 MeV; beam current 10–20 kA; the duration of the pulse at half-height 16 ns.

The LIA-30 commissioned in 1988 was created on the basis of inductors proposed for the first time in the VNIIEF with water-insulated radial lines switched by multichannel dischargers at 500 kV. It has an electron acceleration energy of 40 MeV and provides an electron beam with a current of up to 100 kA in a pulse of 25 ns duration and a pulsed dose of bremsstrahlung up to 100 Gy per 1 m from the target.

In 2004, the compression of a relativistic electron beam in a solenoid of nine sections with an increase in the longitudinal magnetic field from 0.5 to 2.0 T at a length of 1.1 m was mastered to expand the possibilities of radiation studies at the LIA-30 accelerator. The beam compression provided about four times increase in the dose and dose rate of bremsstrahlung (up to 240 krad and $1.5 \cdot 10^{13}$ rad/s, respectively).

A radiation–irradiation complex PULSAR with a BR-1 pulsed nuclear reactor and a number of other electrophysical units was constructed on the basis of the LIA-30. It is recognized as unique all over the world. The complex is designed to investigate the separate and combined effects of pulses of bremsstrahlung and gamma-neutron radiation. A less powerful irradiation complex LIA-10M-GIR2 was created based on the accelerator LIA-10M with the nuclear reactor GIR-2.

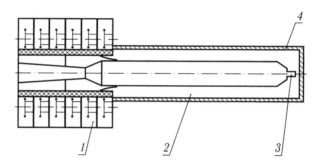

Fig. 1.41. Scheme LIA I-3000: *1* – block inductors; *2* – high voltage transmission line; *3* – cathode; *4* – anode.

The main problems of creating such accelerators are associated with the electrical strength of high-voltage elements and insulation, with the synchronization of a large number of spark gaps, and with the transportation of a beam having a current of tens of kiloamperes inside the acceleration path over long distances.

To power relativistic microwave devices the VNIIEF uses a 'Corvet' 'non-iron' LIA, previously used as an injector of the LIA-10. The accelerator consists of four modules, each of which is a set of three functionally connected units: a block of inductors, a pulse voltage generator PVG-500, which provides impulse charging of the capacitance of the inductor block, and a generator that generates the triggering pulses of inductor switches. Modules are autonomous as for design and the electric scheme. The inductor block consists of three series-connected sections on radial lines with insulation from deionized high-resistance water connected in parallel along the charging circuit with one PVG-500. A separate switch is formed by a toroidal grounded housing, open at the inner diameter, and an annular high-voltage electrode mounted inside it. The radial line of the inductor along the perimeter of one of the output gaps is connected on its internal radius by means of a multichannel ring switch. In this case, the other output gap through the same switch is connected to the load. The ring switch is formed by ten separate gas-filled trigatron at 500 kV. The spread of the response time of dischargers does not exceed 2 ns. On the axis of the inductor block there is an accelerating tube made of polyethylene separating the vacuum cavity of the accelerator path from the inductor cavity filled with deionized water.

A non-loaded inductor with uniform lines should ideally generate rectangular pulses of accelerating voltage with an amplitude close to the value of the charging voltage. The duration of the first pulse at the base is 20 ns, the second and subsequent pulses of alternating polarity 40 ns. The parameters of the inductor block: charging voltage 500 kV; the charging pulse duration 520 ns; the maximum amplitude of accelerating voltage 1.5 MV; short circuit current 180 kA.

The data presented indicate a wide range of possibilities for varying the parameters of electron beams formed by LIAs with inductors on radial lines.

1.5. Linear ferrite accelerators

In the 1970s, a number of ideas for the creation of a vortex electric field on the axis of a system containing ferromagnetic cores appeared [44, 45]. Thus, it was proposed to create vortex electric fields by exciting a free, uniform precession of the magnetic moment in the material of the ferromagnetic core (mainly in ferrites). The principle of operation of such accelerators, known as 'Liferus' (Russian abbreviation of the linear ferrite accelerator), is based on the fact that if there is an angle between the directions of the magnetic field strength **H** and the magnetic moment **M**, then the magnetic moment precesses around the direction **H** according to the Landau–Lifshitz equation:

$$\frac{dM}{dt} = -\gamma_m [\mathbf{M} \cdot \mathbf{H}], \tag{1.93}$$

where γ_m is the gyromagnetic ratio of the inductor material. During the precession, the projection of the vector **M** varies in time, which leads to the appearance of vortex electric fields, which can be used to accelerate charged particles.

There are several possible versions of the accelerator, whose operation is based on the principle considered above.

In the first variant the set of ferromagnetic inductors is magnetized by the longitudinal magnetic field H_z, created by a solenoid mounted outside the cores. Then a beam is introduced along the axis of the cores, which creates a pulsed magnetic field H_ϕ in the inductor core. The change in the total component of the magnetic field **H** causes a precession of the magnetic moment to occur, and the vortex electric field on the axis of the system has the form

$$U_z = \mu_0 k \gamma \frac{H_\phi H_z}{\sqrt{H_\phi^2 + H_z^2}} \mu_s \sin\left(\gamma_m \sqrt{H_\phi^2 + H_z^2}\, t\right), \tag{1.94}$$

where μ_s is the initial magnetization of the ferrite, approximately equal to the saturation magnetization; k is a coefficient that depends on the geometric parameters of the system. Part of the beam is accelerated. Thus, the acceleration is due to a redistribution of the initial energy of the particles.

In the second variant, a beam is passed along the axis of the cores and its current magnetizes the ferrite to saturation in the radial direction. Then an impulse field H_z is created. In this case the accelerating voltage is proportional to the value H_ϕ.

It is possible to create a vortex emf on the axis of the system when precession is excited in a saturated ferrite field H_z by interrupting the beam current previously introduced into the accelerator path. In this case, the tail of the beam will be accelerated.

In addition, pulse fronts are required that create a magnetic field H_z or H_ϕ in nanosecond fractions, which makes this method, with the available switching technique and generation of high power pulses, very difficult for the task in hand.

References

1.   Bowersy A., Elektrische Hochspannungen. – Berlin, 1939.
2.   Christofilos N.S., et al., Rev. Scient. Instr. 1964. V. 35, No. 7. P. 886.
3.   Beal J.W., et al., IEEE Trans. Nucl. Soc. 1969. V. NS-16, No. 3. Part 1. P. 294.
4.   Veksler V.I., Atomnaya energiya. 1957, No. 5. P. 427–430.
5.   Sarantsev V.P., Perelstein E.A., Development of collective ion acceleration methods at JINR, in: Orbits of Cooperation. – Dubna, 1987. P. 116–123.
6.   Furman E.G., et al., Prib. Tekh. Eksper. 1993, No. 6. P. 45–55.
7.   Didenko A.N., Linear induction accelerators for relativistic electronics, in: Relativistic high-frequency electronics. Issue. 3. – Gorky: IPF of the USSR Academy of Sciences, 1981. P. 22–35.
8.   Pavlovsky A.P., et al., Linear induction accelerators for microwave generators, in: Relativistic high-frequency electronics. Issue. 7. – Gorky: IPF of the USSR Academy of Sciences, 1992. P. 81–103.
9.   Peterson G.M.,et al., in: Proceedings of the 2nd All-Union Conference on Accelerators of Charged Particles. V. 1. – Moscow: Nauka, 1972. P. 206–213.
10.   Mondelli A., Roberson C.W., IEEE Trans. on Nuclear Science. 1983. V. NS-30, No. 4. P. 3212–3214.
11.   Bystritsky V.M., et al., in: Physics of Elementary Particles and the Atomic Nucleus. 1992. Vol. 23. Issue. 1. P. 19–57.
12.   Vakhrushin Yu.P., Anatsky A.I., Linear induction accelerators. – Moscow: Atomizdat, 1978.
13.   Itzhoki Ya.S., Pulsed transformers. – Moscow: Sov. Radio, 1950.
14.   Vdovin S.S., Designing of pulse transformers. – Moscow: Energia, 1991.
15.   Vintizenko I.I., Furman E.G., Izv. VUZ, Fizika. 1998, No. 4. Appendix. P. 111–119.
16.   Vintizenko I.I., Izv. VUZ, Fizika, 2007, No. 10/2. P. 136–141.
17.   Leiss J.E., et al., The design and performance of a long-pulse high-current linear induction accelerator at the National Bureau of Standarts, in: Particle Accelerators.

USA. 1980. V. 10. P. 224–234.

18. Anatsky A.I., et al., in: Proceedings of the All-Union Conference on Accelerators of Charged Particles. V. 2. – Moscow: Atomizdat, 1970. P. 231–236.

19. Litvinenko O.N., Soshnikov V.I., The theory of non-uniform lines and their application in radio engineering. – Moscow: Sov. Radio, 1964.

20. Lamb W.A.S., Nucl. Science. IRE Trans. 1962. V. 9, No. 2. P. 53–56.

21. Prono D.S., IEEE Trans. Nucl. Sc. 1985, No. 5. P. 12.

22. Rudakov L.I.,et al., Generation and focusing of high-current relativistic electron beams. – Moscow: Energoatomizdat, 1990.

23. Mick D., Krags D. Electrical breakdown in gases. Moscow: Izd-vo Inostr. Lit., 1960.

24. Kataev I.G., Shock electromagnetic waves. Moscow: Sov. Radio, 1963.

25. Alyamovskii I.V. Electronic beams and electronic guns. Moscow: Sov. Radio, 1966.

26. Molokovsky S.I., Sushkov A.D., Intensive electron and ion beams. – Leningrad: Energia, 1972.

27. Bugaev S.P., et al., The phenomenon of explosive electron emission. – Invention. Diploma No. 176, Otkr. izobr. prom. obraztsy, tov. znaky. 1976, No. 41. C. 3.

28. Vasilyev V.V., Furman E.G., Vopr. at. nauki tekh. Ser. Tekh. fiz. eksper. 1986. Issue. 1 (27). P. 35–38.

29. Prono D.S., et al., Proc. IEEE Part. Conf. Sci. And Technology. – NY, 1989. V. 3. P. 1441–1443.

30. Lebedev A.N., Khlestkob V.S., High-current intense beams of charged particles. – Moscow: MIFI Publishing House, 1983. – 74 p.

31. Vakhrushin Yu.P., Kuznetsov V.S., Zh. Teor. Fiz. 1969. V. 34, No. 3. P. 506–512.

32. Lutsenko E.I., et al., Zh. Ekper. Teor Fiz. 1969. P. 57, No. 11. P. 1575–1584.

33. Lutsenko E.I., et al., ibid, 1970. V. 40, No. 3. P. 529–534.

34. Plotnikov V.K., Admissible errors in the parameters of the focusing channel of a linear induction accelerator with an intense electron beam. 1. Conditions for aligning the beam with the focusing channel and oscillation of the envelope. Preprint No. 62. – Moscow: ITEF, 1986.

35. Plotnikov V.K., Admissible errors in the parameters of the focusing channel of a linear induction accelerator with an intense electron beam. 2. Oscillations of the beam axis, Preprint No. 63. – Moscow: ITEF, 1986.

36. Plotnikov V.K., Admissible errors in the parameters of the focusing channel of a linear induction accelerator with an intense electron beam. 3. Estimation of tolerances on the example of the accelerator LIA 5/5000, Preprint No. 64. – Moscow: ITEF, 1986.

37. Martynov S.V., et al., Experimental study of the effect of increasing the phase volume of an intense electron beam, Preprint No. 84–107. Moscow: ITEF, 1984.

38. Pavlovskii A.I. Bosamykin V.S., Atomnaya Energiya. 1974. Vol. 37. No. 3. P. 228–233.

39. Pavlovsky A.I., et al., ibid. 1970. Vol. 28. No. 5. P. 432–434.

40. Pavlovskii A.I., et al., Linear induction accelerators for microwave generators. Relativistic high-frequency electronics. Proc. Issue. 7. Nizhnyi Novgorod: IPF AN, 1992.

41. Bosamykin V.S., et al., Otkr. izobr. prom. obraztsy, tov. znaky. 1970, No. 31. P. 216.

42. Pavlovskii A.I., et al., DAN SSSR. 1980. Vol. 250, No. 5. P. 1118–1122.

43. Maisonnier Ch., et al., Rev. Scient. Instrum. 1966. V. 37, No. 10. P. 1380–1384.

44. Rakitskii A.A., Shenderovich A.M., in: Problems of Atomic Science and Technology. Ser. FVEYA. – Kharkov, 1973. Issue. 4 (6). P. 35–37.

45. Rakitskii A.A., Shneiderovich A.M., in: Voprosy atomnoi nauki i tekhniki. Ser. Lin. uskoriteli. – Kharkov, 1975. Issue 1(1). P. 37, 38.

2

Linear induction accelerators of the Tomsk Polytechnic University

Introduction

Linear induction accelerators are developed in various scientific centers of several countries, such as Russia, USA, Japan and France (see Table 1.1). A major contribution to the development of accelerator technology was made by specialists from the Tomsk Polytechnic University. To create these accelerators, the original layout scheme and element base were used: low-impedance strip forming lines [1]; multichannel spark dischargers with current divided between channels [2]; non-linear saturation chokes [3]. The accelerators produced in the TPU are used for the formation of electron beams, the power supply to relativistic microwave devices (undulators, klystrons, magnetrons, reflex triodes), pumping of gas lasers, etc.

2.1. Layout scheme of the LIA

The general layout scheme was used as a basis for both LIA injector modules designed to generate a high-voltage pulse at the cathode (anode) of the electron gun, and for accelerating modules in which the electron beam are accelerated. The block diagram of the LIA with a multichannel discharger is presented in Fig. 1.3. Figure 2.1 shows the design of the injector modules of the LIA, combining in a common case a ferromagnetic induction system, low impedance forming lines, a switch, and a core demagnetization system. The accelerating module differs from the injector by the presence of a vacuum path. Acceleration tracts are made of a dielectric tube with a weakly conducting coating, which serves to drain the charge of electrons that enter the inner surface and equalize the potential along the length of the path. A solenoid is laid on top of the tube, one terminal of which is connected to the case of the LIA, the other to the current source. The solenoid current demagnetizes the cores of the induction system and creates a longitudinal focusing magnetic field to transport the electron beam.

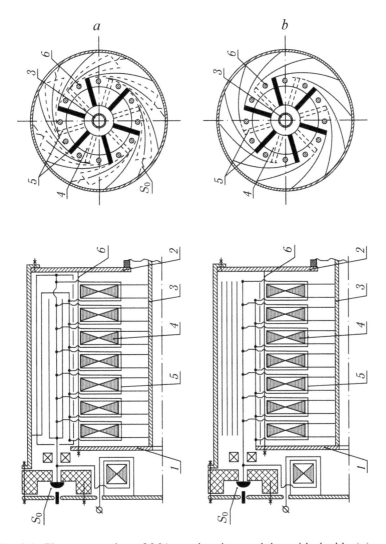

Fig. 2.1. The construction of LIA accelerating modules with double (*a*) and with single (*b*) forming lines: *1, 2* – flanges of the induction system; *3* – high-voltage electrode; *4* – cores of the induction system; *5* – magnetizing turns; *6* – electrodes of strip forming lines; S_0 – multichannel spark discharger.

Ferromagnetic cores *4* are covered by magnetizing coils *5*, which are connected to the terminals of the electrodes *6* of the forming line (strip double or single forming line). The double forming line consists of four electrodes laid along the Archimedes› spiral around the cores. Figure 2.1 *a* shows a system of sic parallel-switched double forming lines (DFLs). Potential electrodes of each line are connected to anodes S_0 of a multichannel discharger. The ground and free electrodes are connected to magnetizing turns of the cores, and the ground electrodes are contacted with the other end to the accelerators case, i.e., connected to the cathode *2* of the discharger. Depending on the required value of the internal wave resistance of the LIA, 4 to 12 parallel-switched DFLs are used.

In the case of the single forming line (SFL), the system consists of several pairs of potential and ground electrodes (Fig. 2.1 *b*).

The pulse duration τ, the stored energy Q, the wave impedance Z of the strip DFL and the beam current I_b for N_c of the cores at the maximum charge voltage U_0 of the DFL are related to the line sizes and the insulation thickness d_{ins} between the electrodes by the following relationships [4]

$$\tau = \frac{2l\sqrt{\varepsilon}}{c}; \tag{2.1}$$

$$Q = \frac{2\varepsilon_0 \varepsilon U_0^2 mhl}{d_{ins}}; \tag{2.2}$$

$$I_b = \frac{K_{en}Q}{U_0 N_c \tau}; \tag{2.3}$$

$$Z = \frac{377 N_c^2 d_{ins}}{K_{en} mh\sqrt{\varepsilon}}, \tag{2.4}$$

where ε_0 is the absolute permittivity of vacuum; ε is the relative permittivity of insulation; *m* is is the number of lines connected in parallel; K_{en} is the coefficient of transformation of the energy of the DFL into the kinetic energy of the electrons; *h* and *l* are the width and length of the electrodes.

The efficiency of the LIA is mainly determined by the ratio of the magnetization current of the cores I_μ and the beam current:

$$\eta_{LIA} \approx \frac{I_b}{I_b + I_\mu}. \tag{2.5}$$

The linear size of the section depends on the number of cores N_c. In this case, for the cross section of the steel of one core with one magnetizing turn, the following condition should be satisfied

$$S \geqslant \frac{U_0 \tau}{\Delta B}. \tag{2.6}$$

The outer diameter of the winding of the DFL is defined as

$$D_m = \left[D_c^2 + 16 \left(\delta_{el} + d_{ins} \right) \frac{ml}{\pi} \right]^{1/2}, \tag{2.7}$$

where D_c is the outer diameter of the induction system; δ_{el} is the thickness of the electrodes of the DFL (or SFL).

The relations (2.1)–(2.7) can be used to estimate the output parameters (U, I_b and τ) and the geometric dimensions of the LIA.

2.2. Low-impedance strip forming lines

Comparison of strip and coaxial forming lines used to excite the induction system of a linear induction accelerator reveals the preferences of the first ones due to the convenience of installation and much smaller dimensions for equal line impedances. However, the strip lines have a significant drawback – they show an increase in the electric field strength at the edge of the electrodes, as briefly mentioned in the first chapter.

Figure 2.2 *a* shows the equivalent of the DFL strip circuit for the case of four coupled parallel lines, and Fig. 2.2 *b* shows the execution of weakly conducting layers located at the edges of the electrodes with a width of πd (where d is the thickness of the insulation between the plates). The maximum value of the electric field strength in the direction of the x-axis (Fig. 2.2 *b*) can be estimated by considering the edge of the electrode as a sector of a coaxial cable with the centre of coordinates at the point 0 in the electrode case at a distance $\Delta/2$ from its edge. Then

$$E_x \approx \frac{0.5 U_0}{x \ln \left[(\pi d + \Delta) / \Delta \right]}, \quad \text{at} \quad \frac{\Delta}{2} < x < \frac{\pi d + \Delta}{2}; \tag{2.8}$$

$$E_x \approx \frac{E_{0.5\pi d}(\pi d + \Delta)}{x}, \text{ at } x > \frac{\pi d + \Delta}{2}, \tag{2.9}$$

where $E_{0.5\pi d}$ is the electric field strength at $x = (\pi d + \Delta)/2$.

The potential distribution along the x axis for $\Delta/2 < x < (\pi d + \Delta)/2$ has the form

$$U_x \approx 0.5 U_0 \frac{\ln(2x/\Delta)}{\ln\left[(\pi d + \Delta)/\Delta\right]}, \tag{2.10}$$

and for $x > (\pi d + \Delta)/2$, the potential varies slowly from $0.5 U_0$ to U_0 for potential electrodes and from $0.5 U_0$ to 0 for the grounded electrodes, depending on the distance to the case 5. It should be borne in mind that the strength along the x-axis of the electric field of grounded electrodes *1* and *3* on the equipotential boundary $0.5 U_0$ changes sign. An exact solution for the potential distribution along the x-axis for the case of two electrodes of a flat capacitor is given in [5]. From the point of view of reducing the weight and dimensions of the LIA, the thickness of the electrodes should be taken as low as possible. However, the edge effect plays an increasingly noticeable role, which is clearly demonstrated by the ratio of the maximum strength of the electric field to the average strength:

$$\frac{E_m}{E_0} \approx \frac{d}{\Delta \ln\left[(\pi d + \Delta)/\Delta\right]}. \tag{2.11}$$

The influence of the edge effect can be reduced by placing all the electrodes along the contour of the weakly conducting layers with a volume resistivity of the layer material ρ of 10^4–10^5 times smaller than the isolation line (Fig. 2.2 *b*), while applying a pulsed charge of the capacitance of the strip FL. The electric field strength in this case will be determined by the voltage drop along the layer caused by the charge current of the interlayer capacitances. Changing the potential and the electric field strength in the low conductive layer (Fig. 2.2 *b*) is determined by solving the diffusion equation (thermal conductivity) [6]. The material of the weakly conducting layer in this case aliphatic epoxy oligomer DEG-1 cured with polyethylene polyamine ($\varepsilon = 4.5$), as the insulation between the electrodes – syntoflex ($\varepsilon = 2.5$). The author of [1] performed measurements of the potential distribution using measuring electrodes filled in the body

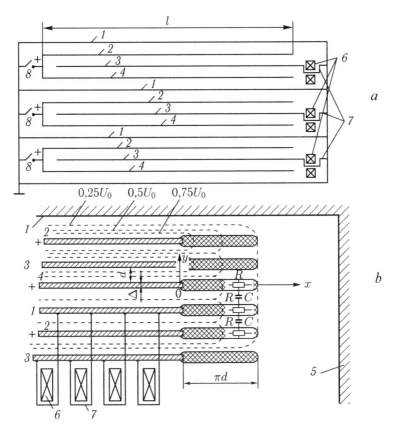

Fig. 2.2. The equivalent circuit (*a*) and the transverse section of the DFL with weakly conducting layers at the edge of the electrodes (*b*). The dashed lines show the equipotentials of the electric field (0.25 U_0, 0.5 U_0, 0.75 U_0) : *1* – grounded electrode; *2, 4* – potential electrode; *3* – free electrode; *5* – accelerator case; *6* – the core; *7* – magnetizing turns; *8* – discharger.

of a weakly conducting layer at different distances from the edge of the electrode. When pulses of a rectangular shape were applied to the line, the pulses from the measuring electrodes were recorded in an oscilloscope and the distribution of the electric field strength was estimated. It has been experimentally shown that the use of the weakly conducting layers makes it possible to get the electric field strength at the edges of the electrodes below its average value between the electrodes under the condition of a pulsed charge of the forming line in a time equal to 5–15 constants of the self-discharge of the layer material. Application of this method to reduce the influence of the edge effect in the strip DFL of the accelerator made it possible to raise the electric field strength by a factor of 1.5 with decreasing the thickness of the plates to 0.1 mm without changing the service life of the DFL insulation.

The research results were used to formulate the following conditions for choosing the parameters of the line, material, and geometry of the layer. A weakly conducting layer must be applied along the contour of all electrodes, regardless of whether they are grounded or not. The width of the grounded and non-grounded electrodes must be the same, and the layers must be placed one above the other. In this case, the lowest electric field strength occurs at the edge of the electrodes and at the edge of the layer. The displacement of the electrodes in the transverse direction leads to an increase in the loss in the material of the layer lying above or below the metal of the electrode, and to a decrease in losses in the layers projecting relative to the others. It is advisable to choose the width of a weakly conducting layer not exceeding the depth of penetration into the layer of the edge electric field of the electrodes. With an increase in the width of the layer, the losses in the layer material increase proportionally to it, and the electric field strength increases and the efficiency decreases at the contact point (the energy stored in the interlayer capacitance is not transferred to the load). As the material of the weakly conducting layer one should use dielectrics with a volume resistance less than the resistance of the basic insulation. Resistance should differ by no less than 4 orders of magnitude. Thus, when using a paper-oil insulation line with $\rho \geq 10^{12}$ ohm·m the resistance of the layer should be $\leq 10^8$ ohm·m. Therefore, the permissible electric field of the material layer in the order of magnitude must correspond to dielectric strength. It is advisable to use the layer materials having an inversely proportional dependence of ρ on the electric field and temperature. In this case, the layer material will be more heavily loaded in the central part, and the intensity of the electric field at the place of contact between the layer and the electrode will decrease. The said requirements are satisfied by the layer material layer DEG resin-1 having a volume resistivity of 10^6 ohms · m at $t = 20°C$. In the cured state, this is an elastic formation showing good adhesion to copper. Particular attention should be given to maintaining the matched operating mode of the line and, if possible, eliminating idling and short circuit modes in the operation of the strip line, resulting in a multiple increase in losses in the material of the layer at the point of contact with the edge of the electrode. In addition, it is necessary to exclude regimes in which the maximum voltage acts unjustifiably for a long time, i.e., provide a charge time not exceeding 5–15 self-discharge constants of the layer material.

In conclusion, it should be noted that this method of reducing the edge effect can be used in the case of any isolation of the capacitor type operating in a pulsed 'charge–discharge' mode.

2.3. Switches of forming lines – multichannel dischargers

One of the most important elements of the LIA is the switch of strip forming lines, which largely determines the rate of current rise, the amplitude and duration of output pulses on the load, the operating life and the limiting pulse repetition frequency. The switch should be designed for a current of 60–420 kA with a duration of 10^{-7} s and have an inductance of no more than 10^{-8} H. In the case of the pulse-periodic regime, this can be achieved in two ways: by using multichannel spark switching or magnetic commutation based on saturation chokes. Figure 2.3 shows the diagram of connection of the discharger and a magnetic commutator for the forming line. Located at the end of the LIA section, the switches have almost the same dimensions and successfully fit into its design, forming a low-inductive connection with the line. This section deals with the design and operation of multichannel spark dischargers. The magnetic commutators will be described in section 2.7.

The kiloampere electron beams can be produced in the LIA in the pulse repetition rate regime with wave resistance of the lines in fractions of ohms only when a multichannel discharger is used when all channels are synchronously switched on and the current is divided evenly between the channels. Indeed, the pulse rise time in switching of the FL is determined by the switching time of the device t_{sw}, the inductance L_{dis} of the discharge circuit and the characteristic impedance of the line ρ_{li}:

$$t_{fr} = t_{sw} + 2.2 \frac{L_{dis}}{\rho_{li}} \qquad (2.12)$$

In the case of a multi-channel switching both the value t_{sw} (by reducing the current flowing through the channel) and the value L_{dis} (due to the parallel operation of many spark gaps) are reduced. When choosing the number of channels it must be assumed that one charge switched by the channel must not exceed $8 \cdot 10^{-4}$ C. This value can be justified, proceeding from the streamer mechanism of breakdown of the gas gap during operation of the discharger in the

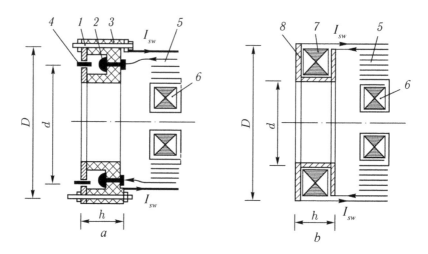

Fig. 2.3. Schemes for connecting a multi-channel spark discharger (*a*) and a magnetic commutator (*b*) to the electrodes of the forming lines: *1* – ground electrode of the discharger; *2* – anodes; *3* – isolator; *4* – control electrode of the discharger; *5* – electrodes of forming lines; *6* – cores of the induction system; *7* – magnetic commutator; *8* – magnetizing turn of the magnetic commutator (I_{sw} – current through the switches).

trigger mode by distortion of the field. The gas ions formed during the development of the streamer, because of low mobility, do not have time to depart from the near-anode region and completely compensate for the space charge of electrons near the anode. Since after the breakdown of the spark gap the multiplication of charge carriers sharply decreases, as the current flows through the spark channel, the near-anode potential drop begins to increase due to ion drift. This leads to acceleration of the electrons moving to the anode, the heating, melting and evaporation of its surface and subsequent compensation of the space charge in the near-anode region by ions of the anode material. As a result, the life of the discharger is sharply reduced and the maximum possible frequency of its operation is limited. As a rule, the switched on discharger simultaneously shorts the electrodes of the forming line and the charging power source. Therefore, to exclude the flow of additional current through spark gaps, it is necessary to carry out switching at the end of the line charge. These requirements are met by low-inductance ring multichannel dischargers developed at the Tomsk Polytechnic University. They have an upper limit on the frequency of operation on the order of 50 Hz in continuous mode and up to 200 Hz in the regime of short packets of pulses (the latter is determined by the time of restoring the electrical strength of the gap) [7].

Figure 2.4 shows a detailed functional diagram of a multichannel discharger. Electrode *1* and cathode *2* are connected to the strip DFL. Spark gaps formed by the anodes *4* and control electrodes *3* with a common cathode *2* are placed in a toroidal dielectric chamber. The electrodes *3* are introduced into a spark gap of length *h* to a depth Δ. The anodes *4* are covered by ferrite rings *5* with a shorted turn *6*. Each discharger is structurally made by a single block in the form of a ring and is located on the end part of the case of each section [2].

When the pulsed voltage is applied to the turns of the IT transformer (Fig. 2.4, time moment $t = 0$), the forming line is charged. At the same time, the starting capacitors C are charged along the R, C circuit. The voltage drop at the resistors R maintains a positive potential U_r on the control electrodes which reduces the electric field strength in the control electrode–anode gap so that the breakdown voltage of the spark gap channel is determined by the electrical strength of the gap h between the cathode and the anode in the same way as in a plane-parallel electrode system.

At time t_1 (Fig. 2.5), when the charging current in the secondary winding of the pulse transformer passes through the zero value, the triple-wound saturation throttle Th_s remagnetized and generates a

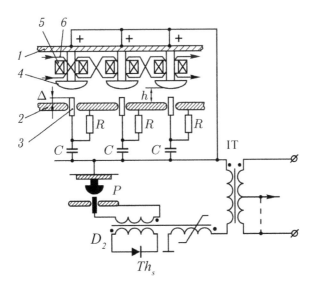

Fig. 2.4. Design of a multichannel discharger: *1* – high-voltage electrode; *2* – cathode; *3* – control electrodes; *4* – anodes; *5* – ferromagnetic rings; *6* – turn.

high-voltage pulse, starting a control trigger discharger *P*. To exclude the pulse from the saturation throttle, the one-turn winding shorted by the diode D_2 is used. With the passage of the current through the zero value, the transfer of energy from the primary storage device C_s to the forming lines ends. With an agreed capacity ratio: $C_s \approx K_{IT}^2 C_{FL}$ (where K_{IT} is the IT transformation ratio), the multi-channel surge discharger shorts the almost completely discharged primary capacitive storage through the transformer. This eliminates the flow through the current channels of the power source, including the pulse current transformer remaining in the primary winding, which, for a relatively small amount, but of considerable duration, can heat up the anode of a separate channel. The closure of the magnetizing current through the discharger is prevented by shunting part of the primary winding of the pulse transformer with a diode (not shown in Fig. 2.4).

The discharger is built according to the scheme of spark gaps of the trigatron type, but it is controlled as dischargers with distortion of the field. When the start-up discharger *P* is turned on and the voltage is inverted on the resistors *R*, and hence on the control electrodes, the discharge begins to develop at the spark gaps of the anode *4*– control electrode *3*. Then, when the voltage at the control electrodes

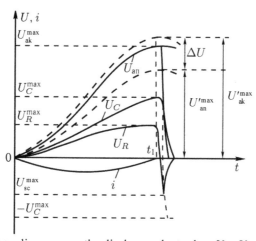

Fig. 2.5. Voltage diagrams on the discharge electrodes; U_{ak}, U_C, U_R – the voltage at the anode of the discharger, the starting capacitor *C* and the bias resistor *R*; $U'_{ak\,max}$, $U'_{an\,max}$ – breakdown voltage of the interelectrode gap in the presence and absence of potential at the control electrode; ΔU – zone of multi-channel triggering of the discharger; U_{ac}^{max} – maximum voltage between anode and cathode, U_C^{max}; U_R^{max} – maximum voltage on capacitance *C* and resistor *R* at the moment the discharger is switched on; U_{sc}^{max} is the breakdown voltage of the control electrode–cathode gap; *i* – current of the secondary winding of the pulse transformer.

becomes greater than the breakdown voltage of the gap, the control electrode *3* − cathode *2* (U_{sc}^{max}; Fig. 2.5), there is a breakdown of the latter. Since the time of formation of the streamer is proportional to the length of the spark gap, providing an appropriate rate of voltage inverting on the control electrodes, it is possible to achieve simultaneous completion of the development of spark channels in the anode *4* − the control electrode *3* and the control electrode *3* −cathode *2* gaps. Such a trigger mode of the spark channel ensures the maximum speed of the discharger.

As is well known, the breakdown of even the same spark gaps is of a statistical nature, and with the advancing switching on of the channel, the current in it will be much greater in comparison with the currents in the channels that joined later. In order to equalize the currents in the channels and to reduce the statistical spread of the time of their activation in the structure under consideration, forced current division in the spark channels is applied by an anode divider, made according to the scheme with a common short-circuited turn. To reduce the inductance, the total coil is divided into two turns *6* (see Fig. 2.4) in each of which a current flows in the steady-state equal to half of the current switched by one spark channel, minus the magnetization reversal current of the ferromagnetic cores *5*. The opposite direction of the current in the turns provides a small intrinsic inductance of the divider. In advance of the inclusion of a number of spark gaps the current therein is limited to the magnetization reversal current of the ferromagnetic cores *5*. At the same time, emf is induced in sections of short-circuited turns that cover the switched channels. This leads to an additional charge of interelectrode capacitances of non-switched channels. We note that even if only one spark channel was not switched on and the ferromagnetic cores of the anode divider are not saturated, the current through the spark gap does not exceed the sum of the magnetization reversal currents of the cores. This current is much less than the discharge current of the forming line, and the voltage of the interelectrode capacitance of the non-switched spark gap tends to a value many times greater than the original voltage.

The required value of the cross section of ferromagnetic cores can be estimated proceeding from the provision of the condition for the switching on of all channels and the division of the current between them during the time of the current pulse in the FL:

$$S \approx \frac{U_{ak}^{max} \Delta t + \Delta U_d \tau}{\Delta B},$$ (2.13)

where Δt is the maximum spread in the time of switching on of individual channels; τ is the duration of the current pulse in the FL; ΔU_d is the difference of voltage on spark channels; ΔB is the induction rate in the core. The quantity ΔU_d is proportional to the product of the difference of the lengths of the spark gaps and the average voltage drop at the spark channel during the current flow. Obviously, the choice for determining the cross section S in the expression (2.13) is a sum of the first term that depends largely on $d(U_{as} - U_R)/dt$ (i.e., the rate of change of voltage on the control electrodes in the breakdown of the starting discharger P). The use of a trigatron type design makes it possible to reduce the values of the interelectrode capacitance of the control electrode to ~3–5 pF. The use of standard RF 50-ohm cables provides $d(U_{as} - U_R)/dt$ values up to 10^{13}–10^{14} V/s. In this case, the spread in the time of channel switching can be brought to ~0.5 ns.

The considered principle of operation of a trigatron-type multichannel discharger with control due to field distortion is the basis for the construction of the blocks of dischargers of linear induction accelerators developed at the Tomsk Polytechnic University [2]. One of them was used to carry out an experimental study of such a discharger [8]. The eighteen-channel discharger operated in an atmosphere of compressed nitrogen at a voltage of 50 kV, which corresponded to 0.9–0.95 of the self-breakdown voltage in the nominal mode. The discharger switched the strip DFL with a wave resistance of 0.25 ohms. The current amplitude through the discharger was ≈180 kA for a base duration of 80 ns. The characteristic dimensions of the spark gaps (see Fig. 2.4): $h = 9$ mm; $\Delta = 2.2$ mm; diameter of the control electrode 5 mm; holes in the cathode of 10 mm. The anode diameter corresponded to the diameters of ferrite cores 400NN K40 × 24 × 16, covered by two short-circuited turns from a wire with a diameter of 1.2 mm in fluoroplastic insulation. The profile of the electrodes forming the spark gap was chosen so as to eliminate the inhomogeneity of the electric field at their surface. For the starting circuit, the capacitors 470 pF (two in series) were chosen as C, and the resistors 22 kOhm (two in series) as R. With a charge duration of the DFL of ~110 µs, they made it possible to provide, by the time of switching, the potentials $U_{ak}^{max} = 50$ kV and $U_R^{max} = 8$ kV (see Fig. 2.5). At the same time, the voltage at the capacitor C was 20 kV, and the potential at the anode of the starting discharger P −28 kV.

The discharger worked with a frequency of 50 Hz at a relatively small ($\approx$2 l/s) pumping rate there through of gas (nitrogen) at a pressure of 2.1 · 10^6 Pa. The operating life was ~10^6 pulses before the next inspection, consisting in cleaning the inner surface of the enclosures of the dischargers. The measured current of one channel of the discharger was $\approx$10±3 kA and the switching charge did not exceed 8 · 10^{-4} C. At the time of ~10^7 pulses, there was no significant erosion of the surfaces of the electrodes.

Information on the operation of individual channels was obtained by oscillographing signals from the photocathodes to which light from the spark gap was fed along the optical fibres. Light guides were inserted into carefully polished cylindrical grooves made in the discharger case opposite each channel. Processing of ~500 oscillograms (20–30 per channel) showed that the rms spread in the time of channel switching was ≤2.1 ns.

Figure 2.6 shows histograms of the current distribution in the presence of an anode current divider and without it. As can be seen from the histograms, at $\Delta U_j/U_k \leq 0.4$, forced division allows one to halve the spread of currents in the channels. It should be noted that in

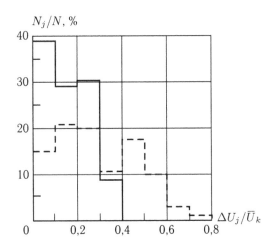

Fig. 2.6. Histograms illustrating the unevenness of current division along the channels of a discharger with a current divider (solid line) and without it (dashed line): N is the total number of activations in all channels; $\Delta U_j = U_{jk} - U_k$, where U_{jk} is the voltage amplitude from the photocathode of the k-th channel for j-th activation; U_k is the average voltage value from the k-th channel in a series of 30–40 activations; N_j is the number of j-th channel activations.

the presence of a divider of cases, no spark gaps were observed, while in its absence the number of failures reached 15%.

The experience of development and operation of multi-channel dischargers with the number of channels 9, 18 and 24 in a common unit, as well as synchronization for the joint operation of several such dischargers (several sections of one accelerator) allows us to draw the following main conclusions. It is advisable to switch the low-impedance forming line in the pulse–periodic regime using multi-channel dischargers, while ensuring the required inductance of the switched circuit and not exceeding the limiting value of the commutated charge by one channel. For multichannel triggering in dischargers, it is necessary to ensure a faster development of the breakdown of the control electrode–anode gap (larger gap), controlling the potential of the control electrode, which has the smallest capacitance relative to the other electrodes. To expand the zone of multi-channel operation, it is necessary to supply a part of the anode potential to the control electrode introduced into the spark gap, ensuring that the discharger is started by distorting the field. It is advisable to use anode current dividers according to the scheme with a common turn, divided into two sets of closed short circuits. Thus, a guaranteed division of the current between the discharger channels is achieved. To increase the service life of the discharger, it is necessary to pump operating gas to cool the spark gap electrodes and to clean it, primarily from nitrogen oxides, as well as from other compounds. Particular attention should be given to the operating conditions of the start-up device in terms of the maximum commutated charge. For the starting circuits considered above, when using a starting discharger, up to four blocks of eighteen-channel dischargers can be synchronized. With a larger number of blocks of dischargers or channels, either a multichannel design of the start-up discharger or several start-up dischargers synchronized with an additional start-up discharger should be used. In this case, the durability of the starting discharger is determined only by the energy that is switched by it and stored in the starting circuits.

A computer simulation program was created to optimize the values of h and Δ, the values of the starting circuit elements (R and C), the construction of the triple wound saturation throttle and the electrode profiles that determine the operation of the discharger. The experience of operating accelerators based on optimized multi-channel dischargers has demonstrated high reliability and long service life ($>10^6$ pulses) of dischargers with the number of channels from 12 to 24.

2.4. Schemes of power supply for LIA with multichannel dischargers

To charge the LIA forming lines with spark dischargers simple and reliable power circuits have been developed that can simultaneously generate current in the magnetic system of the transport channel or the magnetic system of the microwave device. Power circuits use capacitive storage devices based on impulse capacitors, a primary switch–ignitron or a block of high-speed thyristors, a pulse

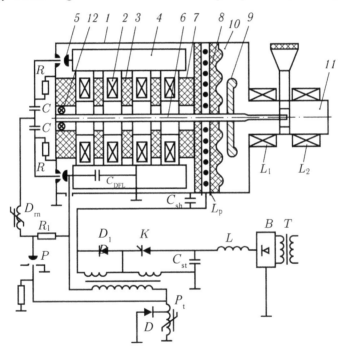

Fig. 2.7. Layout scheme of the LIA and the circuit diagram of power supply of the LIA for the pulse-periodic relativistic magnetron: *1* – case of the LIA; *2* – ferromagnetic core; *3* – magnetizing turn; *4* – DFL; *5* – anode of a multichannel discharger; *6* – cathode holder; *7* – insulator between the cores; *8* – high-voltage insulator; *9* – protective screen; *10* – demagnetization spiral; *11* – relativistic magnetron; *12* – Rogowski's coil; L_1, L_2 – coils of the magnetic system; C_{DFL} – forming line; C_{sh} – shunting capacitor at the output of a demagnetizing spiral; *P* – starting discharger; *R*, R_1, *C*, Dr_n – elements of a start circuit of the multichannel discharger; *T* – mains transformer; *B* – rectifier; *L* – charge inductance; C_{st} – storage capacitance; *K* – thyristor switch; P_t – pulse transformer; D_1, L_d – diode and inductance of demagnetization circuit; *D* – diode of the circuit of the starting discharger.

transformer installed in a separate housing or located inside a section, as well as a charge circuit of the storage (charge inductance, pulse transformer, rectifier).

Figure 2.7 shows the most commonly used power supply scheme and shows the voltage diagrams on the elements and demagnetization current. After charging the primary energy storage C_{st} the control circuit send a triggering signal to activate the thyristor K to discharge the storage to the primary winding of the high-voltage pulse transformer P_t. At the same time, the voltage of the storage C_{st} is applied to the demagnetizing circuit L_p (inductance of the spiral coil), C_{sh} (capacitance of the shunt capacitor) and the demagnetizing current flows along in L_p. At the same time, the voltage induced in the secondary winding I_p is applied to the capacitance of the DFL (C_{DFL}) and charges it to the maximum value U_0. The parameters of the elements are chosen to obtain the specified values of the voltages on the starting discharger and on the control electrodes of the multichannel discharger of the LIA. When the charging current in the secondary winding P_t passes through zero, the saturation choke coil Dr_n is remagnetized and generates a voltage pulse, triggering the start discharger P. To exclude the choke coil voltage pulse Dr_n at the start of charging of the DFL a diode, shorting the part of the secondary winding of the choke coil, is used. After triggering the start discharger P the spark channels of a multichannel discharger are triggered and the switching of the discharger leads to the discharge of the DFL through the induction system.

The voltage on the power supply elements when $C_{st} = 200$ µF and $U_{C_{st}} = 1.9$ kV (with a coefficient of transformation of P_t equal to 38) at the time of switching the DFL (at 60 µs charging time) are as follows: the charging voltage $U_{DFL} = 60$ kV; voltage on the starting discharger $U_p = 32$ kV; voltage on the control electrodes of the multichannel discharger $U_R = 9$ kV. In this case, the maximum demagnetization current $I_d = 600$ A.

The injector was investigated at loading by an electron beam with a charging voltage of the DFL of 60 kV. In this case graphite cathodes with a diameter of 50 mm with rounded edges and a diameter of 20 mm with a flat end surface were used. The anode was in the form of aluminium foils 70 mm in diameter of various thicknesses. Analysis of oscillograms of the beam current I_b, registered for a foil thickness of 0.12 mm in the case of cathodes of two diameters, showed that for a length of the accelerating gap of 15 mm (cathode diameter 50 mm) and 13 mm (cathode diameter 20 mm) beam current is greater than for other lengths ranging from 0 to 20 mm. Obviously, such parameters of the accelerating gap correspond to the matched load of the LIA. Estimation of the voltage applied to the cathode was carried out according to the three-halves power law. It showed that for the cathode 50 mm in diameter the voltage is 530 kV at a total current of $I_0 = 7.8$ kA and beam current after the foil $I_b = 3.2$ kA and for the cathode 20 mm diameter – with $I_0 = 7.3$ kA and $I_b = 4.4$ kA. The pulse duration at half-height, estimated from the oscillograms of the total current of the injector, was 40 ns.

The power supply scheme of the LIA with a common capacitive storage. In 1988–1989 the Tomsk Polytechnic University developed the LIA 04/7, designed to supply a pulse-periodic relativistic magnetron. The accelerator was placed on a mobile platform. The layout of the LIA 04/7 was distinguished by the absence of a solid insulator separating the cathode holder from the case (see Fig. 2.8). As shown by the first experiments with the relativistic magnetron, this element is the least reliable, since it undergoes intensive action from the cathode–anode gap (ultraviolet irradiation, bombardment by charged particles). In this case, the insulator was distributed in the form of plexiglass rings installed between ferromagnetic cores. The thickness of the rings was 16 mm; the rings had a developed surface; their electrical strength exceeded 80 kV/cm. The magnetizing coils of the inductors were made solid (the 'capacitive' inductor option, see Chapter 1). Together with the insulators they formed the vacuum volume of the accelerator. Since the windings had an inner diameter smaller than the diameter of the insulators, they shielded the surface of the insulators from the action of the above factors.

In the design, special conditions were imposed on the power supply scheme of the installation from the on-board network with a power consumption of not more than 40 kW. The power supply of the LIA should provide: 1) charging the forming lines; 2) formation of a demagnetizing field of inductors; 3) formation of a pulsed magnetic field; 4) synchronous activation of the channels of a multichannel discharger. When choosing and developing the scheme, it was taken into account that its most energy-intensive part is the power supply system of the magnetic field coils. It was based on an economical unipolar circuit with energy recovery. A schematic circuit diagram is shown in Fig. 2.8 [9, 10]. Its main parts are: 1) a power source consisting of the charging inductance L, the mains transformer T and the rectifier B; 2) capacitive storage of the magnetic field (C_1, C_2) and the charge of the DFL (C_{st}); 3) switching block of thyristors (T_1)

and ignitron *I*; 4) pulse charge transformer of the DFL (P_t); 5) diode D_1 and resistor R_2, performing protective functions and excluding oscillations in the circuit capacitive storage (C_1, C_2) – inductivities of coils of the Helmholtz pair (L_1, L_2); 6) diode D_4, connected to the demagnetization inductance of the induction system (L_d); 7) starting device of the multi-channel discharger, including multi-channel discharger bias circuit (R, R_1, C) and start-up discharger *P*.

The power scheme works as follows. Energy is supplied from the AC mains through the transformer *T*, the rectifier *B* and inductance *L* to capacitive storage C_1 and C_2 of the former of the magnetic

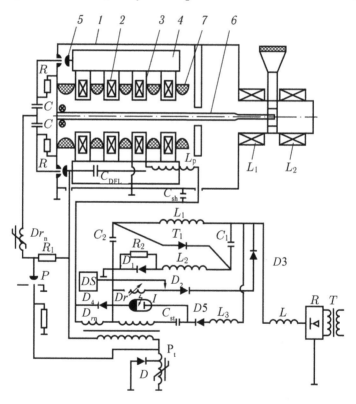

Fig. 2.8. The layout of the LIA and the principal electrical circuit for power supply of the LIA and the magnetic system for the pulsed-periodic relativistic magnetron on a mobile platform: *1* – LIA case; *2* – ferromagnetic core; *3* – magnetizing turn; *4* – DFL; *5* – anode of a multichannel discharger; *6* – cathode holder; *7* – insulator; L_1, L_2 – coils of the magnetic system; C_{DFL} – forming line; C_{sh} – shunting capacitor of the demagnetization spiral; *P* – starting discharger; *R*, R_1, *C*, D_{rn} – elements of the start circuit of the multichannel discharger; *T* – mains transformer; *B* – rectifier; *L* – charge inductance; C_{st} – storage capacity; *I* – an ignitron; P_t – pulse transformer; D_1 and L_d – diode and inductance of the demagnetization circuit; *D* – diode of the starting circuit of the starting discharger.

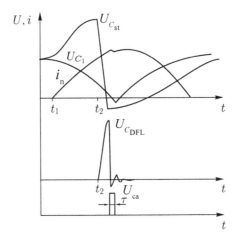

Fig. 2.9. Voltage diagrams on the elements of the power supply circuit of the LIA: U_{C_1}, $U_{C_{st}}$, $U_{C_{DFL}}$ – voltage diagrams on capacitive storage devices C_1, C_{st}, DFL; i_n – current in the magnetic system; U_{ca} – is the voltage at the cathode–anode gap of the RM of duration τ.

field and storage C_{st} of the charge of DFL. When turning on at time t_1 (Fig. 2.9) the unit of thyristors T_1 current starts to flow through the turns of the coils of the magnetic field L_1, L_2 and through the circuit L_3, L_5, and storage unit C_{st} is recharged to a voltage close to $U_{C_1}+U_{C_2}$. Upon reaching the required voltage on C_{st} (it is defined by the installation control system) ignitron I is switched on. The storage C_{st} begins to be discharged through the primary winding of the pulse transformer, charging the DFL and simultaneously forming the demagnetization current of the cores of the induction system by means of an additional winding of the transformer (inductance L_d). To reduce the amplitude of the induced voltage on the diode D_4, the demagnetization spiral is shunted by the capacitor C_{sh}. At time t_2, the DFL is charged up to a predetermined voltage. The charging current of the line goes through zero value, the ignitron turns off and the starting device forms the trigger pulse of the multichannel discharger. Switching off of the thyristor block occurs under the action of reverse voltage with a partial overcharge of the C_1 and C_2 storage units. The value of the recharge (the duration of the reverse voltage) is determined by the magnitude of the flux-linkaxe choke coil Dr, previously demagnetized from the external source DS. When the thyristor unit is turned off, the current of the magnetic field coils passes along the circuit D_1, D_2, D_3 into the capacitive storage units C_1 and C_2. The demagnetization current is closed through the diode

D_4. The magnitude of the magnetic field induction is regulated by the capacitance of the C_1 and C_2 storage units.

The ignitron switched on at angles $60° \leq \omega_c t \leq 80°$, where $t = t_2 - t_1$; ω_c – is the eigenfrequency of the discharge circuit (C_1, C_2– L_1, L_2). The magnetic field varies according to the law

$$B \approx \frac{U_{C_1}^{max} K_1}{\rho_c} \sin(\omega_c t), \qquad (2.14)$$

where ρ_c is the wave impedance of the circuit; K_1 is the coefficient determined by the geometry of the coils; $U_{C_1}^{max}$ is the amplitude of the charging voltage of the storage unit C_1. At time t_2, the DFL is charged to the specified voltage:

$$U_{C_{DFL}} \approx 2U_{C_{st}}^{max} K_t \cos(\omega_c t), \qquad (2.15)$$

where K_t is the transformer ratio of the pulse transformer.

The pulse of the output voltage U_i is proportional to $U_{C_{DFL}}$ with a proportionality coefficient N_c (the number of cores of the induction system). Thus, in the described scheme with the chosen parameters K_1, K_t, N_c and ρ_c, the ratio

$$\frac{U_i}{B} = \frac{\rho_c K_t N_c}{K_1} \mathrm{ctg}(\omega_c t) \qquad (2.16)$$

does not depend on the voltage of the storage units and is determined only by the moment of switching on of the ignitron. In the interval $60° \leq \omega_c t \leq 80°$, the output voltage of the LIA for small changes in the magnitude of the magnetic field (5–10%) changes by a factor of 2–3. The use of such a power scheme makes it easy to adjust the voltage and magnetic field to fulfill the condition for synchronizing the microwave wave and the electron beam in a relativistic magnetron.

The pulse repetition rate is limited by the duration of the current pulse of the magnetic system. In comparison with the traditional scheme without energy recovery (10 Hz), it is increased to 20 Hz. The decrease in the duration of the current pulse is

$$\Delta t = \frac{\pi(\Omega - 2\omega_c) - 2\Omega \omega_c t}{2\Omega \omega_c} + \frac{2\pi R}{\rho_c \omega_c}, \qquad (2.17)$$

where Ω is the circular frequency of the contour storage C_{st} -DFL;

R is the equivalent loss resistance in the LIA supply circuit. In proportion to the reduction in duration, the thermal losses in the coils of the magnetic system are also reduced.

The developed power circuit of the LIA made it possible to generate output voltage pulses with an amplitude up to 300 kV at a current of ~3.3 kA with a duration of ~70 ns at a repetition frequency of 20 Hz. The magnetic system created a field with induction up to 0.4 T. In general, the power scheme was located in a frame with dimensions of $1.2 \times 0.9 \times 1 \ \text{m}^3$. The output parameters of the relativistic magnetron reached 200 MW, which corresponds to an electron efficiency of ~20%. In addition, the pulse–periodic relativistic magnetron was equipped with a harmonic filter for non-linear radar detection of radioelectronic devices and a high-directivity pyramidal antenna of microwave radiation.

The described installations were used for research on the effect of powerful electromagnetic radiation on various semiconductor elements, radioelectronic devices and biological objects. In the most perfect of models of the with LIA with a pulsed power of an electron beam ~4 GW, the dimensions of the installation were: diameter 700 mm; length 900 mm; weight 1000 kg [11, 12].

In conclusion of the representation of this type of LIA we will describe the results of experiments using a peaker. The scheme of the sharpening of the generated pulse mounted on the cathode holder in series with the diode gap [9] was investigated at the injector section of the LIA 0.5/7. The design of the accelerator is shown in Fig. 2.10.

The breakdown voltage of the peaker 7 was regulated by changing the length d of the cathode–anode gap and the gas pressure P (nitrogen) in the discharger. Typical oscillograms of the beam current at a charging voltage on the DFL $U_{DFL}^{max} = 46$ kV for a planar diode system are shown in Fig. 2.11. Curve *1* corresponds to the values $d_d = 0$ and $P = 0$ (there is no peaking), and curve *2* – $d_d = 9$ mm and $P = 6$ atm.

Figure 2.12 shows the dependence of the beam current amplitude I, the pulse energy Q and its duration t_p at a level of $0.1-0.9$ on the parameters of the peaker. The measurements were carried out as follows. When studying the effects of the interelectrode gap d_d of the discharger under nitrogen pressure therein $P = 0$ atm on the beam parameters the obtained optimum value was $d_d = 10$ mm, corresponding to the maximum beam current $I = 3.1$ kA and energy $Q = 40$ J. Other things being equal, the further increase in the self-breakdown voltage of the peaker (by increasing the gap $d_d > 10$ mm)

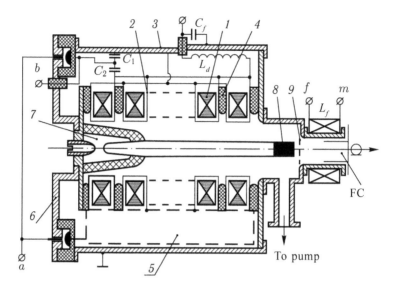

Fig. 2.10. LIA 0.5/7 accelerator circuit: *1* – inductors; *2* – magnetizing turns of inductors; *3* – case; *4* – insulators; *5* – DFL; *6* – multichannel discharger; *7* – peaker; *8* – cathode; *9* – anode (titanium foil); FC – Faraday's cylinder; L_d – demagnetization circuit inductance; L_f – the solenoid; C_1, C_2 – capacitances of the arms of the DFL; C_f – shunting capacitor.

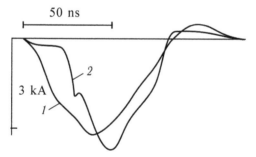

Fig. 2.11. Typical oscillograms of beam current: *1* – without a peaker; *2* – with a peaker.

did not lead to an improvement in the beam parameters, which is due to a proportional increase in the inductance of the loop closing the beam. The pressure control turned out to be more effective (Fig. 2.12 *b*), which was carried out at the optimal gap d_d = 10 mm. Here, too, a local maximum beam energy, Q = 48 J, was found at a pressure P = 2 atm and, correspondingly, the maximum efficiency of the accelerator. With further increase of the gap d_d the beam current and the pulse power of the accelerator increased monotonically, but at

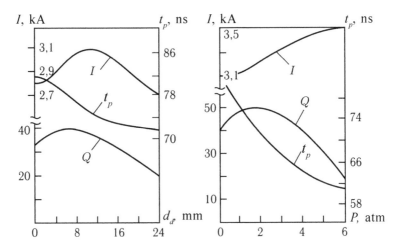

Fig. 2.12. Dependences of the beam parameters on the length of the interelectrode gap at $P = 0$ atm (a) and on the gas pressure in the peaker at $d_d = 10$ cm (b).

the same time the beam energy and the efficiency of the accelerator decreased. The latter is explained by the increase in losses for the magnetization reversal of the inductor cores, since before the moment of breakdown of the peaker the inductors operate in the idling mode, being under doubled voltage. A further increase in the gas pressure was limited by the mechanical strength of the discharger case, which, with a chosen charging voltage at the DFL, $U_{FL}^{max} = 46$ kV, did not allow reaching the maximum possible current and impulse power of the beam, which are estimated to be 3.8–4.0 kA and 1.9–2.2 GW, respectively.

More than a dozen linear induction accelerators were fabricated on the basis of multichannel spark dischargers and handed over to the customers of the Tomsk Polytechnic University. Similar accelerators were used in experiments on the generation of microwave radiation by relativistic magnetrons, which for the first time demonstrated the possibility of a pulse–periodic operation of such a generator. One of the main disadvantages of LIA with spark dischargers is the limited pulse repetition frequency: not more than 50 Hz in the continuous mode and 200 Hz in the burst mode with a limited number of pulses (it is determined by the time of restoration of the electrical strength of the gas gap in the environment of traditionally used gases).

2.5. Linear induction accelerators on magnetic elements

A switch, capable of switching a current of hundreds of kiloamperes with a virtually unlimited resource, with a frequency of one kilohertz, is a magnetic commutator (MC), which is a saturation choke. To implement such a switch with minimal dimensions and, accordingly, with a minimum inductance, it is required to charge the forming lines in a time of hundreds of nanoseconds from magnetic pulse generators (MPG) [13−15]. A magnetic pulse generator is a sequence of the stages (links) of the *LC*-circuits with increasing natural frequency. The circuit is formed by a capacitor and saturation choke coil with a core of ferromagnetic material. Such systems are characterized by reliability, high efficiency of energy compression, and the possibility of forming pulses of high power. The MPGs were used as a basis for a significant number of LIAs, primarily in the USSR, Russia (including the Tomsk Polytechnic University) [16−18] and in the USA [15, 19, 20].

A block diagram of such installations is shown in Fig. 2.13.

The block diagram of the LIA contains the following main elements: the power source PS for charging the capacitors of the magnetic pulse generator MPG (MPG implements the energy compression for charging the FL); magnetic commutator MC, which connects the FL to the induction system IS, increasing the voltage from low (LV) to high (HV) values. Note that the induction system can be structurally integrated with the electron gun EG. The MPG, MC and IS of the accelerator use ferromagnetic cores.

Non-linear magnetic elements (elements of ferromagnetic material with a rectangular hysteresis loop) are widely used in various fields of technology (automation elements, power plants, pulse forming elements in radio engineering, etc.). Such characteristics of devices based on magnetic elements, such as reliability, efficiency, mass, size, power consumption, are largely determined by the properties of magnetic materials. The cores of pulse transformers and saturation chokes are produced from materials with low coercive force and

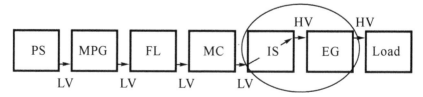

Fig. 2.13. Block diagram of a linear induction accelerator on magnetic elements.

high induction in the unsaturated region, minimum permeability in the saturation state and maximum in the unsaturated state (i.e., with a rectangular hysteresis loop). Permalloy satisfies most of the listed properties. An important advantage of permalloy is that, as a result of rolling into thin sheets and subsequent processing, it acquires a predominant orientation in the direction of the external magnetic field. Therefore, its magnetization, even in relatively small fields, is close to the saturation magnetization. The consequence of this is not only a high residual induction, but also a small magnetic permeability of saturation. At values of the electric field strength H hundreds and thousands of times higher than the coercive force, the value of μ_n approaches 1.

Permalloy cores are made of a strip from 0.1 mm to several microns thick. With a decrease in the thickness of rolled metal, the fraction of the volume occupied by the non-ferromagnetic insulating medium increases, i.e., the filling factor of the ferromagnetic material decreases. In addition, the thickness of rolled steel largely determines the magnetic properties of the cores. The coercive force depends most strongly on the thickness of the rolled steel: it decreases with decreasing thickness. Saturation induction, on the other hand, is practically independent of the thickness of the rolled product. When choosing the thickness of the tape, it is necessary to be guided by the fact that the energy losses due to eddy currents should constitute a small part of the total losses in the core. At the same time, one should not allow an unjustified understatement of the thickness of rolled products, since this can lead to a deterioration in other indicators and an excessive increase in the cost of the core. In addition, as the thickness of the strip decreases, its looseness increases. The latter causes difficulties in the production of large cores.

In addition to its main function, transformers and saturation chokes are used as magnetic commutators. Therefore, it is desirable that in the unsaturated state their total resistance to the pulsed magnetizing current is maximal, and in the saturated state the voltage on their windings is minimal. According to the law of electromagnetic induction, the voltage u_1 on the transformer winding (saturation choke) is equal to

$$u_1 = 10^{-8} \omega_1 S \frac{dB_{av}}{dt}, \tag{2.18}$$

where ω_1 is the number of turns of the primary winding; S is the steel cross section of ferromagnetic cores [cm^2]; B_{av} is the average value of the magnetic induction in the core steel. The magnetization characteristic of the core is non-linear and ambiguous, that is, the magnetization process depends both on the current values of B and H, and on their previous values. The problem is complicated by the fact that the course of the curve $B-H$ corresponds to a static hysteresis loop only if B or H are changed quite slowly. The dynamic magnetization curve (see Fig. 1.9) differs significantly from the static curve and is determined not only by the values of B and H, but also by their time derivatives.

A typical construction of a toroidal core of the rectangular cross-section wound by a thin strip of a ferromagnetic alloy is shown in Fig. 2.14. Saturation chokes and pulse transformers are made of several cores *1* with insulation *2* between them. Toroidal cores of the rectangular cross-section (D_{outer}, d_{inner}, l – outer and inner diameters and width of the core) allow maximum use of the properties of magnetic materials. The adjacent turns of the strip are isolated from each other by a special insulating layer several microns thick. To protect against mechanical damage, the core is located in a frame of a non-magnetic material and impregnated with an epoxy compound. The winding sections *3* (D_i, d_i, l_i – outer, inner diameters and winding width) are wound around the frame *2* made from the insulating material. It is preferable to use single-layer windings. At the same time, it is necessary to ensure the greatest possible uniformity in the distribution of current along the surface of the toroid. For this, the

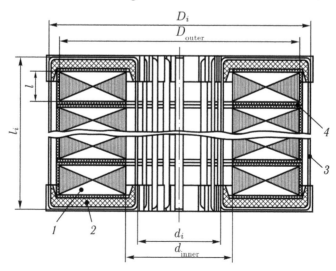

Fig. 2.14. Magnetic element of LIA: *1* – ferromagnetic core; *2* – frame; *3* – winding; *4* – insulation between the cores.

windings must be fixed opposite each other along the entire length of the frame. For better distribution over the surface of the frame with a small number of turns, it is advisable to wind the winding with several wires connected in parallel.

Pulse transformers used in the LIA on magnetic elements, increase the voltage to the necessary level, providing matching between the MPG and the primary storage device. They all have a similar design. In some MPG schemes (in particular, used in the Tomsk Polytechnic University), the capacitor of the first compression stage of the MPG is recharged through the secondary winding of the pulse transformer when the core is saturated. Thus, the first link in the compression of energy is organized. It is advisable to use for the cores of the pulse transformer the same materials as for the saturation chokes. This is convenient both from the point of view of unification of parts and materials, and also because the area of the hysteresis loop allows obtaining high transformation ratios.

In order to obtain transformers and saturation chokes of minimum dimensions, it is necessary to provide the maximum induction in the cores. In other words, it is desirable that the magnetization reversal takes place along the maximum hysteresis loop. Moreover, this will reduce the leakage inductance and the capacitance of the compression *LC*-circuits. In the constructed MPGs the pulsed transformer and choke cores become saturated in the direction opposite to the direction of the operating magnetization during the reverse magnetization reversal and remain in this state until the next operating impulse arrives. However, the energy left in the capacitors of the compression links can cause currents to flow through the windings of the magnetic elements and premature magnetization of their cores. Therefore, the magnetization is reversed by direct current. The direction of the current must be such that the saturation of the core occurs in the direction opposite to the operating magnetization.

2.6. Pulse–periodic LIA 4/2 acelerator

In this section, we describe a four-module linear induction accelerator operating in the packet–pulse regime [3]. It was manufactured at the Tomsk Polytechnic University in the early 90s of the last century. Unlike the LIA with multichannel dischargers, in the LIA 4/2 to ensure high pulse repetition rate in the packet, the switching of the forming lines of modules is performed by magnetic elements – commutators in the form of saturation chokes. A feature of this LIA is the use of a magnetic pulse generator for successive effective compression of the pulse energy in time and a controlled cathode based on a dielectric emitter to form pulses of beam current with a short front. This accelerator served as the first experiment of the Tomsk Polytechnic University in the direction of the development of a new class of accelerators – LIA on magnetic elements.

The LIA 4/2 was developed on the basis of a traditional layout scheme that constructively and electrically unites all the main components of the accelerator in a single module package: a ferromagnetic induction system forming a line with a commutator, a cathode unit (or a beam transport path), demagnetization of inductor cores and forming of the focusing magnetic field. The magnetic commutator of the forming lines was designed with the smallest dimensions and, accordingly, with a small inductance, comparable with the inductance of a multichannel discharger. To create a switch with such characteristics it is required to charge the forming lines for of hundreds of nanoseconds from the magnetic pulse generators [13].The 'precise' nanosecond synchronization of parallel modules and sharpening of the front of the beam current being formed were carried out using a cathode based on a controlled dielectric emitter (see section 3.2.3).

The appearance of LIA 4/2 is shown in Fig. 2.15. The accelerator consists of four modules, two of which are accelerating, and two (cathode and anode) form an injector section. The modules are installed on a common frame with the elements of pulse power circuits placed in it. Next to the frame there are two magnetic pulse generators. A ferromagnetic induction system consisting of twenty-one cores 4 with dimensions $380 \times 150 \times 25$ mm^3 made of permalloy 50 NP, a strip 0.01 mm thick and 25 mm wide, is coaxially located in the case 1 (Fig. 2.16) of each module between the flanges 2 and 3, as well as the magnetic commutator L_1, which is a single-turn saturation choke made of two of the same cores.

Above the cores there are strip single forming lines, conventionally shown in the figure with the capacitances C_1 and C_2. The line $C_1 = 0.08$ µF with a wave resistance $\rho = 0.3$ Ohm and electric length $\tau_{FL} = 20$ ns is connected in parallel with the magnetizing turns 5 of all cores 4. The cathode holder 6, terminating in a profiled screen 7, is installed along the axis of the induction system of the cathode module. Between it and the flange 2 there is an insulator 8, on top of which is laid a single-layer helix L_{p1}, performing the functions of a voltage divider and a demagnetizing circuit element. For high-frequency isolation from the voltage induced by the module, the

Fig. 2.15. Appearance of LIA 4/2.

inductance output L_{p1} is shunted to a case with a capacitor C_7. The cathode assembly consists of a focusing electrode *9* and a controlled emitter *10* with an emitting surface of 80 cm^2 made on the basis of a plate of barium titanate (BaTiO$_3$) 5 mm thick. The emitter lining *11* through the central electrode *12*, the blocking capacitor C_8 and the thyratron *T* is connected to the accelerator case (flange *3*). The cathode holder *6* and the electrode *12* form a coaxial transmission line with a wave resistance of ~10 ohms.

On the axis of the induction system of the anode module there is a metal tube *13* with a diameter of 64 mm of the beam transport path, over which the focusing solenoid L_{c1} is laid. The pipe *13* ends with an anode nozzle *14*. As in the cathode module, there is an isolator *8* with the inductance L_{p1}, which performs the same functions. The inductances L_{p1} are connected to the demagnetization impulse system. Between the cathode and anode modules is a vacuum volume *15*. The elements *9*, *10*, *14* of the cathode–anode space are chosen close to the Pierce optics to form a non-magnetized beam with zero field on the cathode slice, which is then transported and accelerated in an increasing magnetic field.

In accelerating modules, the beam is accelerated in dielectric vacuum paths *16* with an internal diameter of 78 mm over which the focusing solenoids L_{c2} are placed. All other elements in all modules are of the same type and perform the same functions. The transition regions between the modules focusing the magnetic field are formed by the coils L_{f1} and L_{f2} connected with solenoids L_{s1} and L_{s2} to a pulse power system. The total current and accelerator beam current

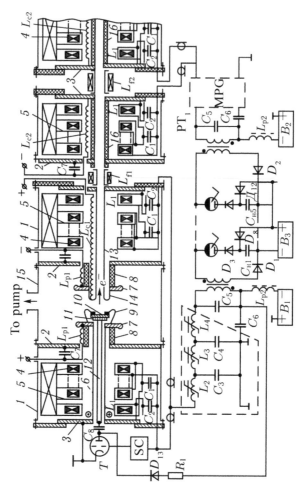

Fig. 2.16. Functional diagram of the accelerator: SC − startup circuit; L_1 − magnetic commutators; L_{p1} − single-layer demagnetization spirals; L_{s1}, L_{f1}, L_{s2}, L_{f2} − focusing coils; L_{p2} − ballast inductance; PT$_1$ − W_1 : W_2 = 2 : 26 on three cores K380 × 150 × 25; 50NP − 0.02 mm; L_2 − one turn on 7 cores K250 × 110 × 25; 50NP − 0.01 mm; L_3 − 3 turns on 7 cores K250 × 110 × 25; 50NP − 0.01 mm; L_4 − 9 turns on 8 cores K250 × 110 × 25; 50NP − 0.02 mm; B_1, B_2 − direct current sources 0−10 A; B_3 − constant voltage source 1−3 kV; T − thyratron; I_{p1}−I_{p2} − ignitron; D_1 − D_{13} − diodes; R_1 − resistor; C_1 − C_4 − SFL (C_1 = 0.08 µF, C_2 = 0.1 µF, C_3 = 0.21 µF, C_4 = 0.25 µF); C_5 − capacitor; C_6 − capacitor; C_7 − capacitor 33 nF; C_{n1} − C_{n5} − capacitors 300 µF.

at the output of the injection part are measured with Rogowski coils D_τ, and the beam current at the output with the Faraday cylinder.

Charging of the forming lines of the modules is of the two-channel type. The cathode and anode modules in the first channel, as well as the two accelerating modules in the second channel, are connected to the corresponding magnetic pulse generators in the form of separate blocks using feeders (sets of coaxial cables). Capacitors C_5, C_6 and saturable pulse transformer PT_1 form the first compression stage. Demagnetization of the cores L_1–L_4 and PT_1 is carried out by direct current from regulated rectifiers B_1, B_2 through ballast inductance L_{p2}. The primary windings of pulse transformers of both channels are connected through the decoupling diodes D_1, D_2 to capacitive storage devices C_{st1}–C_{st5} of the pulse modulators on ignitrons I_{p1}–I_{p5} (according to the number of pulses in the packet). The storage units are charged from the rectifier B_3 through the diodes D_8–D_{12}.

Let us consider the process of forming an accelerating voltage with the help of the one shown in Fig. 2.17 of the equivalent circuit for the injector module of the accelerator. In the initial state of the currents I_{p1}, I_{p2} flow through the inductors L_{p1}, L_{p2}, demagnetizing to negative saturation the cores of the induction systems of the modules, the saturation chokes L_2–L_4, the magnetic commutator L_1 and the pulse transformer PT_1. The current flowing through the coil of the solenoid L_s excites in the drift tube *13* a longitudinal magnetic field of the required magnitude.

When the ignitron I_{p1} is switched on at time t_1, the accumulator C_{n1} starts to charge the capacitances C_5, C_6 through the pulse transformer PT_1. In this case, the capacitance C_6 is charged through the windings of the saturated cores of the induction systems and chokes, L_1–L_4, which become saturated even deeper since the directions of the C_6 charge and demagnetization of L_{p1}, L_{p2} coincide. Choke L_4 of the input link of the MPG does not receive any voltage at this time, since the capacitors C_5, C_6 are charged against it in relation to it. A part of the voltage of the capacitor C_6 is supplied through the blocking capacitance C_8 and the diode D_{13} to the dielectric emitter plate *11* to accumulate the required charge on its emitting surface *10*:

$$Q \approx \int_0^{t_{imp}} i_b(t)dt = \frac{\left(U_{C_e}^{max} - U_b\right)C_{cap}C_8}{C_{cap} + C_8}, \qquad (2.19)$$

where $i_b(t)$, t_{im} is the beam current and its duration; U_{C6}^{max} is the maximum voltage on the capacitance C_6; U_t is the voltage of the autoemission threshold from the edges of the profiled emitter screen; C_{cap} is the capacitance formed by the emitting surface and the plate *11*.

Upon completion of the transfer of energy from C_{n1} to C_5 and C_6 under the action of the voltage on C_5, applied to the secondary

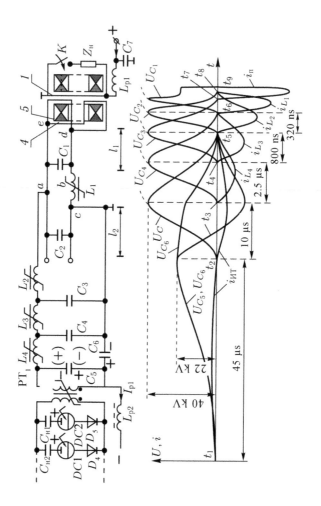

Fig. 2.17. The equivalent scheme for the injector module of the accelerator, the diagrams of currents and voltages in its elements: l_1, l_2 – the geometric length of the FL C_1 and C_2; U_{C_1}, U_{C_2} – change of voltage on the SFL C_1, C_2; U_{C_3}, U_{C_4} – voltage change on the C_3, C_4 capacitors at MPG input; U_{C_5}, U_{C_6} – change of voltage on the capacitances C_5, C_6 of the input link; U_c – change in the voltage at the input of the MPG; i_b – change of current in the load $Z_н$ (beam current); i_{L_1} – change of current in the turn of the magnetic generator L_1; $i_{L_2} - i_{L_4}$ – change of current in saturation chokes of compression links; i_{PT} – change of current in the secondary winding of PT$_1$; t_1 – the beginning of the discharge of the capacitive storage C_{n1} to the PT$_1$ winding; t_2 – the moment of saturation of the core of PT$_1$ and the beginning of inverting of the voltage on the capacitor C_5; t_3, t_4, t_5 – moments of saturation of the cores of the chokes L_4, L_3, L_2; t_6 – moment of saturation of the core of the MC L_1; t_7 – the moment of connection of switching-on of a switch K (the moment of injection of the beam i_b); t_8, t_9 – the moments of the drop to zero of the current in the MC L_1 and the beam current i_b.

winding of the PT_1, its core begins to be remagnetized in the opposite direction. At time t_2, the PT_1 core is saturated and the capacitance C_5 begins to be recharged through its secondary winding. The voltages on the capacitances C_5, C_6 cease to balance each other, and under the action of their difference, $U_C(t) = U_{C_6}(t) - U_{C_5}(t)$, the core of the choke L_4 leaves the state of negative saturation and begins to magnetize. At time t_3 it saturates, but already in the positive region, and the discharge of the series-connected capacitors C_5 and C_6 in the capacitor C_4 begins. Under the conditions $C_5 > C_6$ and $1/C_5 + 1/C_6 = 1/C_4$ and at coincidence of the time instants of the completion of recharging the capacitor C_5 via the secondary winding PT_1 and end of the power transmission to the capacitor C_4 the energy from the capacitors C_5, C_6 is virtually completely (except ohmic losses) transmitted to the capacitor C_4. Ideally, the voltage at C_4 should be doubled with respect to the charge voltage of the capacitors C_5, C_6.

Then the energy is transferred from the *LC*-circuit to the next circuit with a consecutive time compression of the pulses by 3–4 times at each subsequent stage (Fig. 2.17). At the same time, a fast (for ~320 ns) charging of the forming line with capacitor C_1 is achieved, which ensures the operation of a single-turn magnetic commutator L_1 with a relatively small cross-section of the core. The capacity of each subsequent link is 10–15% less than the capacity of the previous one. As a result, the voltage reduction in the links is compensated due to ohmic losses. Under the influence of increasing voltage on C_1, the core of the MC L_1 is magnetized. When it is saturated in the time interval $t_8 - t_6$ the line with capacitor C_2 is discharged to a line with C_1. The forming line C_1 is directly connected to the magnetizing coils 5 of the induction system. The voltage thus induced through the case *1* and the load circuit Z_H is applied to the terminals of the open switch *K* (Fig. 2.17). In the accelerator, the latter means the localization of the voltage in the cathode–anode gap and at the ends of the vacuum paths *16* of the accelerating modules. At the indicated time, the induction systems of the modules operate in the 'idle' mode, since there is no beam in the accelerator.

At the time t_7, close to the end of the charging of the C_1 line, the thyratron *T* is activated. A dielectric emitter is triggered and the beam is injected. The work of the emitter will be described in detail in Chap. 3. Further, the accelerated beam, entering the accelerating modules, sequentially switches them from the 'idle' mode to the matched load mode. In the equivalent circuit (Fig. 2.17), this means that all the voltage induced by the module is applied to the load Z_H.

A circuit is used to feed the emitter circuits and synchronize the moment of its activation. This circuit is 'rigidly' attached to the MPG both by the voltage of charging its circuits and by controlling the moment of switching on its commutator. In this scheme, the emitter charge is carried out from a pulse of microsecond duration fed to the MPG input, which, according to (2.19), automatically maintains a predetermined proportion between the injection voltage and the charge applied to the emitter surface. Turning on switch of the emitter – hydrogen thyratron – is carried out by applying a voltage pulse from the MPG output to its grid via the trigger system (SC). At the same time, the instantaneous switching on of the emitter and the increase in the injection voltage to a desired value are rigidly synchronized, since the moments of supply to the modules the voltage from the output of the MPG and the appearance of the injection voltage divide a strictly defined time interval associated with passing the signal through the passive delay circuits – the forming lines of the modules.

The energy input from the last stage of the MPG to the C_2 line occurs at point a (Fig. 2.17), in which it is connected to the C_1 line and to the magnetic commutator L_1. This is done to ensure the joint parallel discharge of the capacities of the last stage of the MPG and the C_2 line to the C_1 line with the minimum possible parasitic inductances and to reduce the flux linkage of the switch L_1. The processes occurring in this process are indicated in Fig. 2.18. In contrast to the conventional operating mode of the compression links, when each subsequent link transfers energy from only one previous one, the described accelerator uses a mode of energy transfer to each subsequent link from the two previous ones. For this, the magnetic commutator L_1 is switched on (saturated) not at the instant t_7^*, corresponding to the switch angle

$$\omega_3 T_s = \sqrt{\frac{C_2 + C_3}{C_2 C_3 L_2 T_s}} = \pi,$$

and with the advance – at the time t_6 at $\omega_3 T_s = 2.4–2.5$ rad. This allows, while maintaining the energy transfer efficiency, to reduce the flux linkage of the magnetic commutator L_1 by 1.8 times (by the amount of the shaded area in Fig. 2.18). The dimensions and inductance decrease proportionally. It should be noted that in the interval $t_8 - t_7 = t_{imp}/2$ in addition to the C_1 line, the C_2 line, which provides partial compensation of the non-linear change in the magnetization current of the induction system, takes part in the

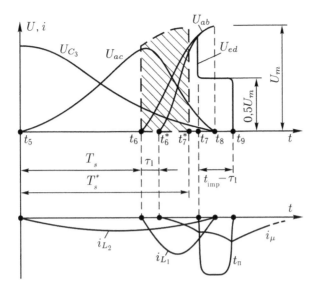

Fig. 2.18. Diagrams of currents and voltages when the line C_2 is discharged to the line C_1; i_μ – change of magnetizing current of the cores.

formation of the beam current. In order to transmit the maximum possible fraction of the energy stored in the lines to the beam, it is necessary that the end of the energy transfer from the capacitors of the last MPG link and the line C_2 to the line C_1 coincide with the time of the electromagnetic wave travelling along the C_1 line from the points of its connection to the turns of the induction system (points *e*, *d* in Fig. 2.17) to the points of connection of the magnetic commutator (points *a*, *b*) (time t_8).

The regime with advanced activation of the chokes was used in all parts of the MPG. Reduction of their flux linkages by an average of 1.8 times allowed to reduce the mass of the choke cores ($L_1 - L_4$) by 340 kg.

All the described processes take 60 µs. Then the cycle of formation of accelerating voltages and the beam is repeated again, but already from the discharge of the capacitive storage C_{st2}, etc., by the number of storages.

Figure 2.19 shows the oscillograms of the beam current pulses in LIA 4/2, obtained with the charge voltage of the line C_2 $U_{C_2} = 42$ kV. The beam was formed in an increasing magnetic field at an injection voltage of 1.2 MV, followed by acceleration up to 2.4 MeV. The induction of the magnetic field in the plane of the emitting surface of the cathode was close to zero, and in the paths of the anode and

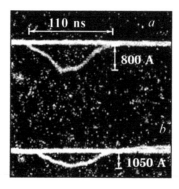

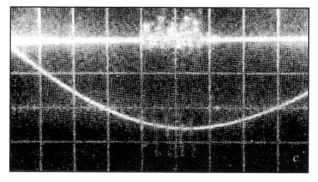

Fig. 2.19. Oscillograms of beam current at the outputs of the accelerator (*a*) and its injector part (*b*), as well as the oscillogram of the packet of 5 beam current pulses against the background of the focusing magnetic field pulse (*c*). Vertical scale 300 A/division, horizontally – 1 ms/division.

accelerating modules it was ~0.22 T. Characteristic is the presence on the peaks of pulses of a flat part with a duration corresponding to the double electrical length of the forming line C_1. The significant duration of the pulse front (~35 ns) is a consequence of the high inductance of the thyratron–dielectric emitter circuit. Measurements of the current transmission coefficient of the beam, defined as the ratio of the currents at the outputs of the accelerator and its injector part, yielded an average value of 0.75.

The carried out researches have shown that the main elements determining the maximum possible repetition rate of the beam current pulses in the packet and the stability of their amplitude–time parameters are magnetic pulse generators charging the lines of the modules. The following basic requirements must be met for them: 1) in the pause between the pulses of the packet, all the cores of the magnetic elements must be magnetically reversed to the original state; 2) all the transient processes associated with reverse magnetization reversal must be completed by the time the subsequent operating pulse begins to form.

Theoretically, the minimum possible value of the time interval between the pulses is $T \geq (t_1 - t_0) + (t_3 - t_2) + \Delta t_3$, wherein $t_1 - t_0$ is the charging time of the capacitances C_5, C_6 from the storage units $C_{st1} - C_{st5}$; $t_3 - t_2$ is the time of voltage inverting on C_5 through the secondary winding PT_1; Δt_3 is the time of reverse magnetization of cores PT_1 and chokes $L_1 - L_4$. Analysis shows that the value T can be significantly reduced only by reducing the value of Δt_3 with increasing demagnetization current I_{d2}. This current during the interval Δt_3 is closed through the capacitance C_6, and under the influence of the increasing voltage of this capacitor the magnetization of the cores is reversed:

$$\int_0^{t_3} U_{C_6}(t)\,dt = \frac{I_{d2}\Delta t_3}{C_6}. \qquad (2.20)$$

On the oscillograms obtained at a voltage $U_{C_6}^{max} = 42$ kV and a demagnetization current $I_{d2} = 10$ A, the time intervals indicated are $t_1 - t_0 = 45$ µs, $t_3 - t_2 = 10$ µs, $\Delta t_3 = 70$ µs, whence $T = 125$ µs. However, in practice, it is not possible to obtain pulses in a packet with an interval of 125 µs due to the presence of oscillatory processes that decay in a time much longer than T. These processes are associated with the scattering of energy accumulated in the capacitor C_6 during the time Δt_3. In reality the attenuation corresponds to a time of 250–300 µs. Without special measures, attempts to generate packets with a shorter time interval between pulses lead to a sharp deterioration in the stability of the amplitude–time parameters of the pulses being formed, since in the presence of such oscillations the initial value of the magnetic induction in the cores is not determined. One of the ways to combat oscillations is the output of the energy spent for reverse magnetization reversal, using valve circuits with the absorption of excess energy on the resistive load.

During the operation of the accelerator in the packet mode, a N beam current pulses with frequency F is symmetrically positioned with respect to the apex of the magnetic field pulse of an acceleration path quasi-homogeneous along the length (~3 m). The circular frequency ω_f of this field is determined from the condition that the permissible deviation B/B_τ from the pulse to the pulse in the packet does not exceed

$$\omega_f < F\frac{\pi - 2\arcsin\left(1 - B/B_t\right)}{N} , \qquad (2.21)$$

where B_t is the magnetic field. Figure 2.19 *c* shows the waveform package of five beam current pulses following a sinusoidal pulse on the background of the focusing magnetic field in a packet with a frequency of 3.3 kHz. This frequency provides a value of $B/B_t <$ 0.03, sufficient for transporting the beam current while maintaining the amplitude–time parameters from pulse to pulse. A beam current of 0.8–1 kA was obtained with a duration of 110 ns at an electron energy of 2.4 MeV and a repetition rate of five pulses in a 3.3 kHz packet.

In the opinion of the authors of Ref. [3], based on the results of the projection and studies of LIA 4/2 with magnetic elements, they have practically realized a new method of accelerating the beam in the LIA, in which the influence of the inductance of the switching circuits is excluded due to the use of controlled injection. The time characteristics of the beam current pulse and its energy spectrum are determined by the parameters of the strip lines, the induction systems of the modules, and the characteristics of the cathode. In this case, the last switching step is carried out within the cathode–anode gap by the section with the highest voltage and the lowest current (beam current). This makes it possible to generate beam current pulses with the minimum possible fronts. The injection of electrons at the initial moment is carried out in the idle mode of induction systems of modules, when the accelerating voltage in 1.8–2 times higher than the operating voltage. As a result, low-energy electrons are absent on the front of the beam current pulse and the fluctuations of the beam current along the length of the accelerator are eliminated.

2.7. Linear induction accelerators on magnetic elements for relativistic microwave generators

At the end of the past and the beginning of the present century, several LIAs on magnetic elements were designed and manufactured at the Tomsk Polytechnic University intended for feeding relativistic microwave devices. Their main feature was that they were injector modules designed to form electron beams of relatively small energy (300−500 keV) at a current of 3−6 kA. Since they have also been used to supply relativistic magnetron microwave generators, we will consider them in more detail, describe a technique for their engineering calculations and present the results of computer simulation.

The layout and principal electrical diagram of such an accelerator (LIA 04/4000) are shown in Fig. 2.20 [17]. In a cylindrical case with a diameter of 700 mm and a length of 1600 mm there is an induction system of fifteen cores and a magnetic commutator L_1, which is a single-turn saturation choke. Over the cores there are electrodes of a single strip forming line with capacitance C_1 and a capacitor of the last MPG compression unit with capacitance C_2 (made using the single-forming line technology). Common high-voltage electrodes C_1 and C_2 are connected to the last stage of MPG.

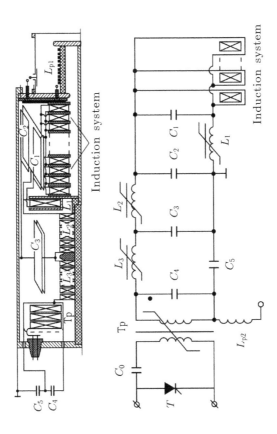

Fig. 2.20. Layout and circuit diagram of LIA 04/4000.

The other two electrodes are interconnected by the winding of the magnetic commutator L_1. A high-voltage electrode and a cylindrical insulator are installed along the axis of the induction system of the module. Thus, in contrast to LIA 4/2, in the general case of LIA 04/4000 there are also elements of a magnetic pulse generator, which makes it possible to reduce the inductance of the connection of the elements, to increase reliability, and to reduce the weight and size parameters of the installation.

The magnetic pulse generator has three stages of compression: on saturation chokes L_3, L_2, pulse transformer Tr and capacitors C_5, C_4, C_3. To reduce the inductance, the capacitor C_3 is also made using the SFL technology. The high-voltage capacitors C_4, C_5 of the type K75-74 0.1 µF, 40 kV are installed outside the case. The pulse transformer Tp simultaneously performs two functions: increases the voltage of the capacitor and at saturation it ensures the charge of the capacitor C_4 through the secondary winding. The cores of the saturation chokes of the MPG were made of a 0.02 mm rolled strip of permalloy 50 NP, and the switch cores and the induction system were made of permalloy 50 NP with a thickness of 0.01 mm.

The induction system is demagnetized through the single-layer inductance L_{p1}, connected to the primary winding of the pulse transformer, and, in addition, by the charge current of the capacitors of the first stage of MPG compression. The core of the pulse transformer is demagnetized from the external source through the inductance L_{p2}. The inductance terminal is shunted by the capacitors.

The principle of LIA operation on magnetic elements is as follows. Initially, the required current determining the magnetic state of the core is established in the circuit Tr, and rectifiers are added to charge the capacitor C_0. With the arrival of a control pulse (at time t_0) to the thyristor block T, the capacitor C_0 is connected to the primary winding of the pulse transformer. The voltage variation on the circuit elements is shown in Fig. 2.21. The charging of C_5, C_4 begins and the demagnetization current of the induction system is formed. This process continues for approximately 28–30 µs depending on the value of the residual voltage of the storage unit C_0. The capacitor C_4 is charged directly from the secondary winding Tp, and the windings of the saturation chokes (L_3, L_2, L_1) and the magnetization coils of the induction system, whose cores are demagnetized, are included in the C_5 charge circuit.

The discharge time interval C_0 must correspond to the duration of magnetization reversal of the core of the pulse transformer:

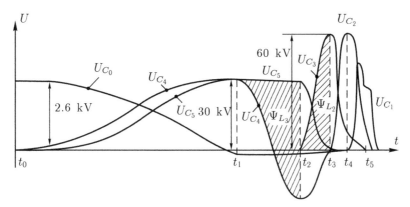

Fig. 2.21. Variation of voltage on the elements of the LIA.

$$\pi\sqrt{L_0 C_0 / 2} \approx \frac{\Psi_{Tp}}{\langle U_{C_0} \rangle}, \tag{2.22}$$

where L_0 is the discharge circuit inductance; $\Psi_{Tp} = W_{Tp} S_{Tp} \Delta B$; W_{Tp} and S_{Tp} are the number of turns of the primary winding and the cross section of the steel of the transformer; ΔB is the induction increment in the core; $\langle U_{C_0} \rangle \approx U_{C_0}/2$ is the average voltage acting on the windings of the transformer; U_{C_0} is the amplitude of the charging voltage of the capacitor C_0.

The action of the capacitor voltage C_0 results in saturation of the core of Tp, and the capacitor C_4 is recharged through the inductance of the secondary winding of the transformer. In this case, the sum of the voltages of the capacitors C_5, C_4 is applied to the turns of the saturation choke L_3. The magnitude of the flux linkage of the saturation choke L_3 is chosen such that, at the time t_2 to which the capacitor C_4 is completely recharged, saturation of the core occurs, i.e., the following condition is satisfied:

$$\pi\sqrt{L_{Tp} C_4} \approx \frac{\psi_{L_3}}{\langle U_{C_3} \rangle}, \tag{2.23}$$

where L_{Tp} is the inductance of the secondary winding of the pulse transformer in the saturated state; $\psi_{L_3} = W_3 S_3 \Delta B$; W_3 and S_3 are the number of turns and the cross section of the steel of the saturation choke L_3; $\langle U_{C_3} \rangle \approx (U_{C_4} + U_{C_5})/2$ is the average voltage acting on the choke turns L_3; U_{C_4}, U_{C_5} are the amplitudes of charging voltage of the capacitors C_4, C_5.

The capacitors C_4 and C_5 connected in series are discharged into C_3. In the time interval $t_2 - t_3$, the core of the choke L_2 is magnetically reversed. When it is saturated (time t_3), C_3 begins to discharge to the capacitor C_2. In the interval t_3-t_4, the core of the magnetic commutator L_1 is reversed, and C_2 is discharged into the capacitance of the forming line C_1. The latter, discharged through the magnetizing turns of the induction system, forms a high-voltage pulse. When operating on an ohmic load, the voltage and current pulses are bell-shaped. In this case, the SFL C_1 plays the role of a matching line. When working on the electron diode by delaying the appearance of current in the explosive electron emission SFL C_1 is discharged in a state close to the 'idle' running forming voltage and current pulses with a flat top. The appearance of the LIA 04/4000 is shown in Fig. 2.22.

2.7.1. Power supply for LIA on magnetic elements

The basic electrical circuit for supplying power to the LIA 04/4000 is shown in Fig. 2.23. It is divided into two functional parts: the input of energy into the LIA and the charging of the capacitive storage.

The energy enters the accelerator by discharging primary storage capacitors C_{01}, C_{02} and C_{03}, consisting of low-inductance high-frequency capacitors 1 μF, 5 kV, and through a saturated choke when

Fig. 2.22. Appearance of LIA 04/4000.

three parallel channels are connected from three serially connected fast thyristors (VS_9–VS_{11}). The principle of operation is as follows. A pre-demagnetized choke L_7 (L_8, L_9) delays the discharge current by 3–4 µs. The current amplitude in the output bus is 13.4 kA with the charging voltage of the storage unit 2.6 kV. The duration of the current pulse is ~30 µs. After the start of operation of the accelerator, some of the energy is returned back and after 20–30 µs it emits an emf of opposite polarity at the input terminals of the LIA. This is due to the use of the effect of overlapping phases in the MPG compression links and with incomplete matching of the forming line and the load.

To increase the efficiency of the power system, the duration of the control pulses on the thyristors VS_9–VS_{11} is increased so that at the time of energy return they remain conductive. The bulk of the energy is returned to C_{01}–C_{03}, creating a second pulse of the charging current of the storage unit. The energy remaining after switching off the thyristor is dissipated in the resistors R_1–R_3 and diodes VD_4–VD_6.

Charging of the capacitive storage devices C_{01}, C_{02}, C_{03} to a maximum voltage of 2.6 kV with a cycle frequency of up to 320 Hz is performed from a device supplied by a three-phase network with a power of 90 kW. The device circuit realizes the principle of oscillatory charging of the capacitor from a constant voltage source. The controllable rectifier VS_1–VS_6 is fully open in the operating condition. The rectified voltage is $U_{rec} \approx 1.5$ kV. The main control function is to switch off the control pulses from the thyristors in case of a short circuit or overload. Phase control is used to smoothly charge the capacitor of the filter C_1, as well as for manual adjustment of the rectified voltage.

The capacitors C_{01}, C_{02}, C_{03} are charged via the choke L_2 when VS_7 is switched on. The duration of the sinusoidal half-wave current is ≈1.5 ms, and the amplitude of the charging current is ~300–500 A, depending on the presence of reverse voltage on C_{01}–C_{03}. The charging voltage is controlled by interrupting the charging process with a control depth of ~50% of the maximum. The thyristor VS_8 turns on when the charging voltage of the primary drives reaches a predetermined level. The current through L_2 switches to C_2 and VS_7 turns off. The fast-recovery diode VD_3 reduces the amplitude of the reverse voltage to VS_8. The voltage increase on C_2 is limited by the circuit VD_1–the secondary winding L_2 at the level $2U_{red} \approx 3$ kV, since the winding transformation ratio is one. The energy remaining in L_2 returns through this circuit to the capacitor filter C_1, and the

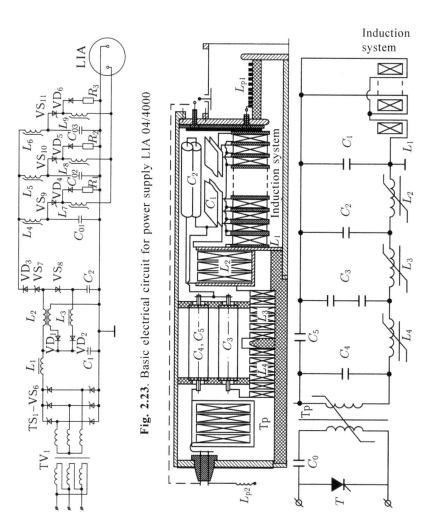

Fig. 2.23. Basic electrical circuit for power supply LIA 04/4000

Fig. 2.24. Layout and circuit diagram of LIA 04/6

thyristor VS_8 is de-energized and switched off. The energy stored in C_2 also returns to C_1 oscillatorily through L_3 and VD_2. Switching on VS_7 is delayed by 1 ms from the beginning of the discharge. By this moment, the discharge circuit is already completely de-energized, and the $VS_9 - VS_{11}$ thyristors restore the locking capability.

The thyristors $VS_9 - VS_{11}$ with a large area of the structure have a leakage current of up to 10 mA. This current, and also the current through the padding resistors, discharge the storage capacitors, creating an error that increases sharply with decreasing frequency of the charge cycles. To compensate for the leakage of the charge, an additional 150 W power source with a maximum voltage of 3 kV is used. The source is made according to the scheme of a single-cycle voltage converter and is included in the general stabilization circuit. It starts working only after the main power supply has charged the storage unit to the specified voltage.

The operational experience of the LIA 04/4000 showed the need to limit the voltage of the $C_{01} - C_{03}$ storage units to increase the reliability of the thyristor unit. It was decided to reduce the charging voltage to 1000 V, that is, to the level of the operating voltages of one thyristor. In this case, the network transformer is excluded from the circuit and the voltage of the three-phase network is fed to the rectifier input. The energy stored in the primary storage device is increased due to the use of a capacitance of 1000 µF (capacitors 20 µF, 1.6 kV). This extends in time the discharge process and leads to an improvement in the operating conditions of the thyristors in terms of the rate of current rise. The amplitude of the current commutated by one thyristor is reduced by dividing the discharge circuit into 6 channels. In order to charge the MPG capacitors to their operating voltage, the transformer ratio of the pulse transformer is made equal to 30. A compression link is added to maintain the value of the compression ratio of one link within 3−4 and maintain a high efficiency of energy transfer to the MPG. The layout and principal electrical diagram of such an accelerator (LIA 04/6) are shown in Fig. 2.24.

The increase in the time intervals of the charge transfer processes in the first links of the MPG compression enabled the use of industrial low-inductance high-frequency capacitors 0.1 µF, 40 kV, installed in parallel for 6 pieces for C_4 and C_5 and in series-parallel for 12 pieces for C_3 and C_2. As a result, the capacitor of the first MPG compression link was also placed in the common case, greatly simplifying the process of assembling the accelerator. Changes in the design of the accelerator related to the use of solid turns of magnetization of the ferromagnetic cores of the induction system (previously, turns in the form of copper strips were used). This made it possible to reduce the inductance of the discharge circuit, while simultaneously increasing the 'parasitic' capacitance of the induction system. Such a decision was made on the basis of the analysis of the results of [21], in which the influence of switching and explosive emission processes, as well as the inductance and capacitance of the discharge circuit on the parameters of the generated pulses of the high-current electron accelerators (HCEA)was investigated. In particular, a decrease in the inductance of the discharge circuit while simultaneously increasing the 'parasitic' load capacitance makes it possible to reduce the voltage surge caused by the delay in the explosive emission of electrons on the cathode surface, which facilitates the operation of the high-voltage insulator of the LIA.

2.7.2. Engineering calculations of LIAs on magnetic elements

Let's consider the technique of engineering calculation of elements using the example of LIA 04/6 and justify the choice of their parameters. First of all, it is necessary to determine the total energy compression coefficient, given by the product of the compression coefficients of individual MPG links:

$$n_{comp}^{k} = \frac{\Delta t_N}{\Delta t_1}, \tag{2.24}$$

where n_{comp} is the compression ratio of one link; k is the number of links; Δt_N is the time of energy transfer from the primary storage device to the capacitors of the first MPG compression link; Δt_1 is the energy transfer time from the capacitors of the last link of the MPG to SFL. These data make it possible to calculate the number of MPG links and, correspondingly, the weight and size parameters of the LIA. To estimate the first time interval, it is necessary to know the capacitance of the primary storage and the inductance of the discharge circuit, which includes the inductance of the turns of the transformer and the supply busbars.

As already noted, the primary storage voltage is selected not exceeding the value of $U_B \sim 1000$ V. The maximum repetition rate ($F = 200$ Hz) is limited by the amount of power consumed from the network. Suppose that it can not exceed $P_1 = 100$ kW. Consequently,

$P_1 \le C_0 U_B^2 \, F/2$ and the primary storage capacitance is $C_1 = 10^{-3}$ F. Let the transformation ratio of the pulse transformer be $K_{tr} \approx 30{-}32$. Then the capacitors C_4, C_5 can be charged up to 30 kV (their nominal voltage in the pulse–periodic operation mode). Thus, the capacitor capacity will be $C_4 = C_5 = C_0/(2K_{tr}^2) \sim (0.55{-}0.6) \cdot 10^{-6}$ F. It can be obtained using 6 parallel type capacitors 40 kV, 0.1 μF.

The inductance of the discharge circuit L_0 is formed by the inductances of scattering of the pulse transformer, capacitors, chokes and input busbars. According to the calculations it is $L_0 \sim 0.8 \cdot 10^{-6}$ H. Therefore, the discharge time interval C_0 is equal to $\Delta t_0 = \pi\sqrt{L_0 C_0 / 2} \sim 68{\cdot}10^{-6}$ μs.

Capacitance C_1 in the last link of compression of the MPG should be charged for the time not longer than the time which can provide flux linkage of the magnetic commutator ψ_k. To reduce the size of the magnetic commutator and the inductance of the magnetizing coil, it is advisable to perform it with a single-turn using one core with the same dimensions as that of the cores of the induction system. So, we have

$$\Delta t_1 \leqslant \frac{\psi_k}{\left\langle U_{C_2} \right\rangle}$$

where $\left\langle U_{C_2} \right\rangle \approx U_{C_2}/2$ is the average voltage acting on the coils of the magnetic commutator; U_{C_2} is the amplitude of the charging voltage of the last MPG capacitor.

To obtain high energy transmission efficiency, the value of n_{comp} should be selected in the range 3–4 [14, 22]. On the other hand, we have

$$n_{comp} = \frac{\psi_{n+1}}{\psi_n}, \tag{2.25}$$

where ψ_{n+1}, ψ_n are the flux linkages of the saturation chokes of the $(n + 1)$-th and n-th MPG links.

The weight and size parameters of the LIA on magnetic elements are reduced by the effect of overlapping the discharge phases of the capacitor of the previous MPG compression link and the charge of the next link capacitor (Fig. 2.25).

If the value of the flux linkage of the choke ψ'_n is less than necessary: $\psi_n = (1.1{-}1.3) \, \psi'_n$, then the efficiency of energy transfer is reduced. However, in this case the cross section of the steel core of the saturation choke is noticeably reduced, which means that its

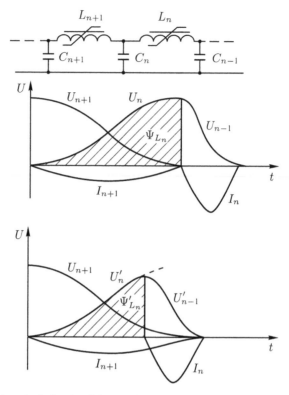

Fig. 2.25. The electrical circuit of the MPG compression links. Diagrams of voltage and current in MPG: *a*) without overlapping phases; *b*) with overlapping of the discharge phases of the capacitor C_{n+1} and the charge of the capacitor C_n.

weight and energy losses for magnetization reversal decrease. At the same time, the inductance of the saturation choke turn decreases and the time of charge–discharge processes of the capacitors decreases. When this effect is used, a non-zero difference appears between the time of energy transfer in the $(n+1)$-th link and the time of magnetization reversal of the saturation choke in the n-th MPG link:

$$\Delta t_{n+1} = \pi \sqrt{\frac{L_{n+1}C_{n+1}C_{n+2}}{C_{n+1}+C_{n+2}}} \leq \frac{\psi_n}{\left\langle U_{C_{n+1}} \right\rangle} = \Delta t'_n. \tag{2.26}$$

We choose $\Delta t_{n+1}/\Delta t'_n \sim 1.1$–$1.3$, realizing overlap of the charge phase C_n and discharge phases C_{n+1}. In this case, the transfer coefficient of the voltage amplitude from the capacitance C_{n+1} to the capacitance C_n at $C_n = C_{n+1}$ will be

$$K_t = \frac{U_n}{U_{n+1}} = \frac{1}{2}\left|1 - \cos\frac{\pi}{1.1 \div 1.3}\right| = 0.98 \div 0.87, \tag{2.27}$$

where $U_n < U_{n+1}$ – are the voltage amplitudes on the capacitors. At higher values of $\Delta t_{n+1}/\Delta t'_n$ the voltage and energy losses become unacceptable.

With a decrease in the charging voltage of the primary storage (the operation of the LIA is carried out with reduced output parameters), there is no overlap of the phases and the transfer of energy from one compression link to the other occurs without loss.

So, there are two conditions for choosing the choke parameters of the n-th link:

$$\psi_{n+1} = (3-4)\psi_n; \tag{2.28}$$

$$\pi\sqrt{L_{n+1}\frac{C_{n+1}C_{n+2}}{C_{n+1}+C_{n+2}}} = (1-1.3)\frac{W_n S_n \Delta B}{\langle U_{C_{n+1}}\rangle}. \tag{2.29}$$

With their help, it is possible to conduct estimated calculations for the preliminary selection of the elements of the LIA.

In order to eliminate voltage losses in the transfer of energy from the capacitor of the last stage of compression of MPG to SFL, the condition (2.29) must have the form

$$\pi\sqrt{L_2\frac{C_2 C_1}{C_2+C_1}} = \frac{W_k S_k \Delta B}{\langle U_{C_2}\rangle}, \tag{2.30}$$

where L_2 is the sum of the inductances of the saturation choke turns of the last stage of compression and the current leads to it, as well as the intrinsic inductance of capacitance C_2; W_K, S_K – the number of turns and the cross section of the steel of the magnetic commutator.

To increase the power released on the load, it is proposed in [23] to use the condition $C_1 < C_2$. In this case, the forming line is charged to a higher voltage, and since its capacitance becomes smaller, it is discharged in a shorter time (Fig. 2.26). If we use the simplest equivalent circuit which is a series connection of a capacitor discharged to inductance and load resistance, then, depending on the ratio of the characteristics of the listed elements, the power growth can reach 40%. Computer modelling (see below), taking into account 'parasitic' inductances, capacitances, losses in the core steel,

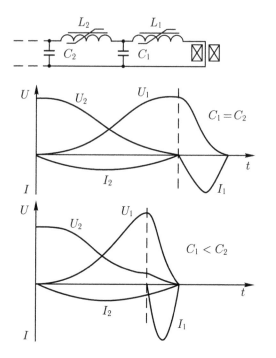

Fig. 2.26. The electrical scheme for connecting the MPG to the SFL. Voltage and current diagrams for discharging the capacitor of the last MPG compression link to the SFL and SFL to the load: at $C_1 = C_2$; at $C_1 < C_2$

etc., shows an increase in power by approximately 30% when the accelerator operates at an ohmic load of 100 ohms. The dependence of the output power of the LIA on the ratio C_2/C_1, obtained by calculations on a computer model, is shown in Fig. 2.27. For the LIA 04/6, the ratio $C_2/C_1 = 1.5$ was chosen, since a decrease in C_1 to 0.15 µF leads to an increase in the charging voltage of the forming line, which necessitates the use of additional insulation. In addition, the value of the residual voltage C_2 increases.

An important point in the implementation of this technical proposal ($C_2/C_1 = 1.5$) is the residual voltage of the capacitor C_2 after the transfer of energy to the SFL. Therefore, it is useful to introduce the energy transfer coefficient ζ, which is defined as the ratio of the energy stored in C_2 to the moment of saturation of the choke L_2, to the energy stored in the forming line at the time of saturation of the magnetic commutator. In general, the transmission coefficient depends on the energy loss in the winding L_2, in the core L_2, the capacitances C_2 and C_1, and also on the ratio of the capacitances

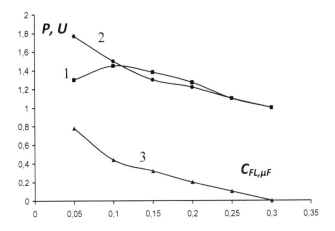

Fig. 2.27. Dependences of the output power of the LIA (*1*), the charging voltage of the SFL (*2*) and the residual voltage of the capacitor of the last MPG compression link (*3*) on the SFL capacitance (C_2 = 0.3 µF).

$\xi = C_2/C_1 = 1.5$. Assuming there are no losses in the winding and capacitors, we can write

$$\zeta = \frac{4\xi}{(1+\xi)^2} \approx 96\%.\qquad(2.31)$$

Nevertheless, an increase of 30% in this important parameter of LIA, like pulsed power, with insignificant losses of energy (~4%) makes it expedient to use the unbalance of capacitances. The process of increasing the pulse power is accompanied by a significant increase in the charging voltage of the forming line, which can exceed the breakdown values. The superposition on the magnitude of the flux-linkage of the magnetic commutator of the condition [24]

$$W_k S_k \Delta B = \langle U_s \rangle \sqrt{\frac{(C_1+C_2)}{L_2 C_2 C_1}} \arccos \frac{C_1}{C_2}\qquad(2.32)$$

leads to the following. Once the capacitor of the last link of the magnetic pulse generator is discharged, the value of the charging voltage of the forming line reaches the charging voltage of the capacitor (~$2\langle U_s \rangle$), the core of the magnetic commutator becomes saturated and the capacitor and the forming line begin to discharge together to the load. Thus, in this case, the overlapping phases of the discharge of the capacitor and the charge of the forming line are simultaneously used and the unbalance of the capacitances is applied. The pulse power of the accelerator under condition (2.32) was calculated using the computer model of LIA. Changes in C_1 relative to C_2 were 2–4 times. The inductance of the discharge circuit comprising the inductance of the saturation choke of the last link of compression of the MPG was assumed equal to $L_2 = 23.5 \cdot 10^{-9}$ H, which corresponds to the actual accelerator parameters of the LIA 04/6. The highest calculated power was obtained with the equality $3C_1 = C_2$ and the condition (2.32). In comparison with the case without phase overlap at the same capacitance ratio, the power increase was 11%.

Note that since the value of the capacitance of the forming line has significantly decreased, the time of discharge–charge processes in the capacitor C_2–forming line circuit also decreased. This makes it possible to reduce the magnitude of the flux linkage of the magnetic capacitor, which automatically leads to a reduction in its size, and hence to a decrease in the inductance of the coil of the magnetic commutator. In the calculations, the parameter varied from $18.7 \cdot 10^{-9}$ H to $16 \cdot 10^{-9}$ H, which caused the increase in power to the load by another 3%.

Thus, in comparison with the traditional LIA circuits with magnetic elements having in all links of MPG compression a set of capacitors of the same capacitance equal to the capacitance of the forming line, execution of the accelerator in accordance with the recommendations of [24] causes an increase in the power released on the load by ~40%. This option was implemented when creating LIA 04/6.

In the process of computer calculation of a linear induction accelerator, the parameters of its elements and their design can be changed to adjust the LIA for maxima in terms of the pulse power, efficiency, the quality of the electron beam, etc. (see section 2.8).

2.7.3. The pulse repetition rate of the LIA on magnetic elements

When creating LIA on magnetic elements, the question of the limiting pulse repetition frequency is important. In general, the limitations here relate to the reverse magnetization reversal of the ferromagnetic cores of the MPG saturation chokes and the pulse transformer, as well as to the decay of the interpulse oscillations [25]. The process of reverse magnetization reversal must be completed by the time the switch of the primary energy storage is switched on. Otherwise, the stability of the amplitude–time parameters of the output pulses is violated. Therefore, the task of analyzing the process of reverse magnetization is to estimate its duration. The maximum flux linkage is possessed by the saturable pulse transformer of the first compression link and the saturation choke L_4 of the second compression link. It is the magnetization reversal of these two elements that uses most of the energy and time. In addition, these elements have the largest number of turns in comparison with others and are magnetized by smaller currents. With a coercive force of $H_c \sim 25-30$ A/m of the 50 NP with a thickness of 0.02 mm, currents of 1 and 1.5 A are sufficient to convert the transformer cores and the saturation choke L_4 to a saturated state. Thus, the minimum value of the demagnetization current must be at least 1.5 A. We recall that the saturation chokes $L_4 - L_2$ are demagnetized by the charge current of the capacitor C_5. After completion of the transfer of energy from the capacitors C_4 and C_5 to C_3 at the time t_3 under the effect of voltage on the capacitor C_3, the core L_4 begins to magnetize in the reverse direction and the core L_3 in the forward direction. As a result, the demagnetization current of the pulse transformer is closed through the capacitance C_5, forming on it a negative magnetization reversal pulse.

By analogy with the investigations carried out in [25], one can write the expression for the limiting pulse repetition rate in the form

$$F \leqslant \frac{\sqrt{k}\sqrt{F_1}n_{comp}^2}{\sqrt{2}t_p'(t_1/2+t_p)}, \qquad (2.33)$$

where $F_1 = \pi^2 \mu_0 \mu_{sat} H_0/(16\Delta B)$; k is the number of compression links; μ_{sat} is the relative magnetic permeability of the core in the saturated state; $t_p' = 1-1.5$ is the parameter of filling the period with intermittent oscillations, t_1 is the time of charge of capacitances C_5 and C_4 (see Fig. 2.21); t_p is the duration of control pulses on thyristors VS_9-VS_{11} (see Fig. 3.8) after the end of the charge of the capacitors C_4 and C_5 to restore the valve properties of the primary storage switch. After substituting the numerical values of the parameters for LIA 04/6, we obtain the limiting pulse repetition rate: $F \leq 3230$ Hz. Naturally, such modes of operation of the accelerator due to the huge power consumption are possible only with the use of several pre-charged storages (each with its own switch) discharged in series to a pulse transformer (as is realized in the case of LIA 4/2).

A possible factor limiting the pulse repetition rate and the number of pulses in a continuous series can be heating of the elements of a linear induction accelerator. The thermal conditions of the elements of the LIA determine the following types of losses: 1) ohmic losses when current flows through conductors; 2) losses in the dielectric material of capacitors; 3) losses due to eddy currents and magnetizing current in the cores of the MPG saturation chokes, the magnetic commutator and the induction system. Their evaluation will be presented below. The existing computer program makes it possible to calculate the losses for both the individual element of the accelerator and for the entire installation as a whole. The increase in the temperature of the cores of the magnetic elements for a certain time interval at different pulse repetition rates is determined using the technique of calculating conventional transformers. In general, depending on the matching of the accelerator with the load and the magnitude of the charging voltage, the efficiency of energy transfer from the primary storage is ~40–50%. Naturally, all the remaining energy is released in the form of heat in the bulk of the accelerator (about 1800–2000 kg). The operating time of the LIA before heating such a mass to a temperature of 60°C, representing the threshold for the considered industrial insulators, at a pulse repetition rate of 80 Hz without taking the heat sink into account exceeds $4 \cdot 10^3$ s (i.e. more than 1 hour).

2.8. Modelling the work of LIA

Quite often, the LIAs are used to form voltage pulses or electron beams for devices of relativistic high-frequency electronics (for more details, see Chapters 3 and 4). One of the main tasks of relativistic microwave electronics is to increase the power and energy of microwave radiation pulses from high-current accelerators, when the efficiency of the application and the efficiency of the devices depend substantially on the physical characteristics of the voltage pulse. So, for example, to reduce beam losses during its transportation in a slow wave structure of the relativistic tube of the backward wave, a rectangular pulse of accelerating voltage is required to ensure a powerful monoenergetic electron beam. In a number of cases it is necessary that the shape of the accelerating voltage pulse be varied according to a predetermined law. The peculiarity of the voltage pulse in relativistic magnetrons is due to the fact that in these regions the formation of the electron beam and the interaction of the beam with the slow waves of the anode block coincide spatially. For such devices, the magnitude of the operating current is determined not so much by the geometry of the electrodes and by the magnitude of the electric field strength as by the intensity of the high-frequency fields, which depends on the properties of the resonator system. In addition, for the effective operation of a relativistic magnetron, the ratio of the voltage by the source $U(t)$ in the cathode–anode gap of the magnetron and a constant magnetic field H must satisfy the synchronism condition. At the same time, the value of the anode current essentially determines the value of $U(t)$, since the power source has limited power, i.e., the modes of operation of the magnetron and the power supply are interdependent. Therefore, such a device should be considered as a common system with strong feedback [26]. Such a regime imposes stringent requirements on power supplies and microwave generators on the energy conversion efficiency. Since the power source of the relativistic magnetron is laborious in manufacturing and an expensive installation, the stage of its design proves to be decisive.

The task of modelling is the study of physical processes in the power supply–non-linear load system the resistance of which depends both on the intrinsic characteristics and on the parameters of the supply pulse. In this case, a unified model of a power source suitable for modelling any relativistic microwave generator is considered and allows an optimal choice of the parameters of the elements entering into it. Optimization is carried out by means of numerical modelling, since the system as a whole has a strong feedback and the calculation of transient processes on the elements is a rather difficult problem related to solving non-linear differential equations. We also note that the model is constructed with the involvement of the physics of the processes of pulsed magnetization reversal of steel of ferromagnetic cores of the LIA (saturation chokes, magnetic commutator, induction system).

The practical goal of modelling is to determine the optimal parameters and the choice of the design of the installation elements. In this connection, it is necessary to create a joint model for calculating the processes taking place in the LIA and in a load with non-linear characteristics which is the relativistic magnetron. In this section, we demonstrate the feasibility of solving this problem by modelling based on the representation of LIA units by equivalent circuits. The non-linear elements of the equivalent scheme of a relativistic magnetron are calculated in accordance with the theory of averaged motion [27].

When solving the modelling problem, a two-step approach is used. The first stage of the development of the outline design of the installation using a simplified equivalent scheme, includes the first-order calculations. In this case, the transient processes, energy characteristics and parameters of the installation are analyzed in general form without unnecessary detail. The operating mode of the power supply and the microwave generator is optimized by a sequential search of the parameters. After clarifying the main regularities of processes and clarifying the parameters of the elements of the installation, the second stage of modelling is carried out. To do this, more accurate non-linear and parametric representations of individual units are used, numerical experiment planning methods are used to find optimal conditions, and a final calculation is made.

It should be emphasized that various parameters can serve as criteria for estimating the optimal mode for the power source and the microwave generator: full or electronic efficiency, output power, electron beam spectrum on the load, stability of operation.

2.8.1. Construction of the model and selection of the parameters of the equivalent circuit

To solve the problem, the real electrical circuit of the LIA and load is represented by an equivalent circuit (Fig. 2.28) and the parameters of the elements are determined. Differential equations for voltages and currents are recorded [26]. The processes in the computer model are considered from the moment when the capacitances of the magnetic pulse generator C_{m4} and C_{m5} are charged, the switch K_1 is closed and C_{m4} begins to be recharged through the inductance of the secondary winding of the saturated pulse transformer L_{m5}. Switching on the keys K_1-K_5 in the equivalent scheme of the accelerator simulates the transition of the cores of the saturation chokes from the unsaturated state to the saturated one.

The upper part of Fig. 2.28 shows the equivalent circuits of the forming line, the induction system and the load, the lower part the equivalent MPG scheme, formed by four compression links. The derivation of the formulas used to calculate the parameters of the elements will be given in section 5.6.1 of the book.

1. *The first MPG compression link* consists of the capacitors C_{m4}, C_{m5} and the secondary winding of the pulse transformer (the

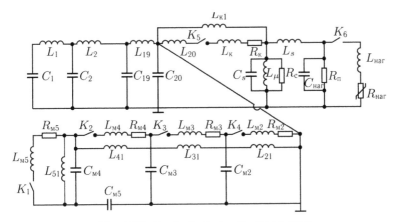

Fig. 2.28. Equivalent circuit of LIA 04/6.

transformer performs two functions: increases the voltage to 30 kV when the primary storage is discharged, and at saturation (switch K_1 turns on), the capacitor C_{m4} is recharged. In this scheme:

 – L_{m5} is the inductance of the secondary winding of a saturated pulse transformer, defined as [28]

$$L_{m5} = \frac{1}{2\pi} a_s W_s^2 \ln \frac{D_{ext5}}{D_{int5}}, \qquad (2.34)$$

where a_s is the linear size of the secondary winding; D_{ext5}, D_{int5} are external and internal diameters of the winding; W_s is the number of turns;

 – R_{m5} is an ohmic loss equivalent which includes the resistances of the winding, the connecting current leads and contact connections, as well as capacitor losses;

 – L_{51} is the equivalent inductance of the magnetization of the transformer. The circuit with inductance L_{51} describes the process of the magnetizing current flowing in the transformer core when charging C_{m4}, C_{m5}. The presence of the magnetizing current leads to a partial discharge of the capacitors and incomplete energy transfer from the previous MPG compression link to the subsequent one. Since this computer model does not cover the capacitor charging process, and the voltage inverting occurs with the saturated core of the transformer, the loss in the steel is not taken into account and it is possible to accept $L_{51} \to \infty$.

2. *The second–fourth compression links of MPG* (saturation chokes L_{m4}–L_{m2}, capacitors C_{m5}–C_{m2}) have almost identical equivalent circuits and consist of:

– inductances L_{m4} – L_{m2} of the windings of the chokes in the saturated state. They are determined by formulas analogous to (2.34);

– ohmic equivalents R_{m4}–R_{m2} of the total losses (in the steel of the chokes in case of their magnetization reversal, in the copper of the windings, supply circuits and contact connections, in the discharged and recharged capacitors);

– the inductances L_{41}–L_{21} of the magnetization of the cores of the saturation chokes. The core of the saturation choke of the second compression link is magnetized during the recharge time of C_{m4} under the action of the total voltage at the capacitors ($U_{C_{m4}}+U_{C_{m5}}$). The cores of the chokes of the third and fourth compression links are remagnetized when charging C_{m3} and C_{m2}, respectively. The magnitudes of the magnetization inductances of the cores are determined by the following formula:

$$L_{i1} = \frac{2W_i^2 B_s l_c K t_i \left(D_{i\,ext} - D_{i\,int} \right)}{\pi \left(D_{i\,ext} + D_{i\,int} \right)\left[H_0 t_i + 2S_{oe} + S_{oo} \right]}, \qquad (2.35)$$

where W_i is the number of magnetizing turns; l_c is the width of the core steel; B_s is the saturation induction of the ferromagnetic material; K is the coefficient of filling the steel volume of the core; t_i is the duration of the magnetization reversal process; $D_{i\,ext}$, $D_{i\,int}$ are the external and internal diameters of the cores; H_0 is the start field; S_{oe} is the switching coefficient due to the action of eddy currents; S_{oo} is the switching coefficient due to magnetic viscosity. Knowing the characteristics of the ferromagnetic material used (permalloy 50 NP) [29], we can determine L_{i1}.

Turning of the switches K_4 – K_2 occurs after acquiring the necessary voltseconds by the i-th saturation choke:

$$\psi_i = \langle U_i \rangle t_i = 2W_i S_i B_s K , \qquad (2.36)$$

where $\langle U_i \rangle \approx U_{max\,i}/2$ is the voltage applied to the turns of the chokes; $U_{max\,i}$ is the amplitude of the applied voltage; S_i is the cross section of the steel of the saturation chokes.

3. *Magnetic commutator.* The equivalent circuit of the magnetic commutator is similar to that discussed above. It consists of the following elements:

– equivalent inductance L_{s1} of the magnetization of the magnetic commutator:

$$L_{s1} = \frac{2B_s l_s K(D_{ext.s} - D_{int.s})t_2}{\pi(D_{ext.s} - D_{int.s})[H_{0s}t_2 + 2S_{ae} + S_{ao}]}, \tag{2.37}$$

where $D_{ext.s}$, $D_{int.s}$ are the external and internal diameters of the turn of the magnetic commutator; l_s is the width of the magnetizing turn of the switch; t_2 is the charge time of the forming line. In contrast to (2.35), here $W = 1$ and is not present in the formula;

– inductance L_s, which is the sum of the inductances of the magnetizing coil of the magnetic commutator, the current leads to it, the inductance of the magnetizing turns of the induction system and its leads. The inductance of the magnetizing turn of the magnetic commutator is defined similarly to (2.34). The rest of the components of L_s depend on the concrete structure and are calculated using [28];

– ohmic equivalents R_s of the losses in the magnetic commutator. The activation of the switch K_s simulates the discharge of the forming line through the induction system to the load under a condition similar to (2.36).

4. *The induction system* consists of the following elements:

– the dynamic capacitance C_s of the induction system, which depends on the type of the magnetizing coil of the ferromagnetic inductor, i.e., whether the turn consists of individual strips ('inductive' inductor) or is made in the form of a disk ('capacitive' inductor); an intermediate version of the inductor is also possible.

In general, the capacitance of the inductor consists of four components: the turn–steel of the 'own' inductor; coil–steel of the neighbouring inductor; turn–turn of the neighbouring inductor; inductor–inductor. All of them are calculated using [30];

– leakage inductance of the inductor L_s, determined by the type of inductor ('capacitive' or 'inductive').

For a 'capacitive' inductor

$$L_s^{(1)} = \frac{\mu_0 l_c}{2\pi} \ln \frac{D_{ext.c}}{D_{int.c}} \tag{2.38}$$

where l_c is the linear dimension of the inductor core; $D_{ext.c}$, $D_{int.c}$ are the external and internal diameters of the magnetizing coil of the inductor.

In the case of an 'inductive' inductor, setting $\mu \to \infty$, it can be assumed that the leakage field is lumped in the volume between the coil and the core of the neighbouring inductor. Outside this volume, the magnetic fields of adjacent turns partially compensate each other and the leakage inductance is

$$L_s^{(2)} = \frac{\mu_0 \left(D_{ext.c} - D_{int.c}\right)}{n} \cdot \frac{\left(l_1 + 2l_2\right)}{h},$$

(2.39)

where l_1 is the thickness of the insulation between the core steel and the magnetizing coil; l_2 is the distance between the magnetizing coils of the neighbouring cores; n and h are the number and width of the strips of the magnetizing coil.

To the inductance $L_s^{(1)}$ (or $L_s^{(2)}$) it is necessary to add the inductance of the leads. Assuming that the leakage field is enclosed between the ferromagnetic material and the copper of the forming line closest to the cores, a formula is used to calculate the inductance of the rectangular turns [28]:

$$L_{lead} = \frac{\mu_0}{nl_{lead}} \left[\frac{\pi \left(D_{ext.lead}^2 - D_{int.lead}^2\right)}{4n} - \frac{h\left(D_{ext.lead} - D_{int.lead}\right)}{2} \right]$$

(2.40)

where $D_{ext.lead}$, $D_{int.lead}$ are the outer and inner diameters of the leads; l_{lead} is the lead terminals length.

In general, the leakage inductance of the inductor is $L_s = L_s^{(1)} + L_{lead}$ ('capacitive' inductor) or $L_s = L_s^{(2)} + L_{lead}$ ('inductive' inductor);

– the inductance of the magnetization of the inductor L_μ, defined similarly to $L_{41} - L_{21}$, taking into account the equality $W = 1$;

– the ohmic losses in the cores of the induction system R_c, which are calculated using the equation for magnetization reversal of steel under the action of a rectangular voltage pulse [29]:

$$R_c = \frac{2\left(D_{ext.c} - D_{int.c}\right)l_c KB_s}{\pi \left(D_{ext.c} + D_{int.c}\right)\left(S_{\omega e} + S_{\omega 0}\right)}$$

(2.41)

5. *The load with a cathode holder*, is represented by an equivalent circuit including a load capacitance C_{load} (the sum of capacitances of the cathode holder within the induction system, the cathode holder capacitance in the high voltage insulator and the load capacitance), a load inductance L_{load} including the inductance of the cathode holder and the load resistance R_{load} (linear or non-linear). This includes:

– the capacitance of the cathode holder within the induction system consisting of N_c ferromagnetic cores is determined by the formula

$$C_{cat} = \frac{2\pi\varepsilon_m\varepsilon_0 l_{is}}{3\ln\left(D_{int.lead}/D_c\right)},\qquad (2.42)$$

where D_c is the diameter of the cathode holder; l_{is} the length of the induction system; ε_m is the dielectric permittivity of the insulation (transformer oil);

– the capacity of the cathode holder in the region of the high-voltage insulator consists of two components:

$$C_{cat}^{(1)} = C_{cat}^{(2)} + C_{cat}^{(3)},\qquad (2.43)$$

where $C_{cat}^{(2)}$ is the capacitance of the cathode holder between its outer surface and the spiral of the insulator; $C_{cat}^{(3)}$ is the capacitance between the outer surface of the spiral and the case of the vacuum chamber. For calculating $C_{cat}^{(2)}$ and $C_{cat}^{(3)}$, formulas analogous to (2.42) are used;

– The capacity of the load is calculated by the formula for a coaxial conductor divided into i sections of length l_i with different external $(D_{ext\,i})$ and internal $(D_{int\,i})$ diameters:

$$C_{load} = \frac{2\pi\varepsilon_0 l_i}{\ln\left(D_{ext\,i}/D_{int\,i}\right)};\qquad (2.44)$$

– the inductance of the load, including the cathode holder, is equal to the sum of the inductances in individual sections. It is calculated by the formulas for a coaxial conductor:

$$L_{load} = \sum_i L_{ni} + L_{ch} = \sum_i \frac{\mu_0}{2\pi} l_i \ln\frac{D_{ext\,i}}{D_{int\,i}} + L_{ch},\qquad (2.45)$$

where L_{ch} is the inductance of the cathode holder within the induction system due to currents flowing in the opposite direction in the cathode holder and along the inner surfaces of the magnetizing coils. Neglecting the intervals between the magnetizing coils, one can write

$$L_{ch} = \frac{\mu_0 l_{is}}{2\pi} \ln\frac{D_{int.lead}}{D_c}.\qquad (2.46)$$

6. *Strip forming line.* The LIA on magnetic elements uses strip single forming lines – uniform two-wire lines with distributed parameters.

The equations of the transient processes for the switching of the SFL can be obtained on the basis of the scheme for its equivalent, which is a chain of similar *RLC*-links [31].

When $\Delta t \rightarrow 0$ these equations have the form

$$L_{\text{li}} \frac{\partial i(x,t)}{\partial t} + R_{\text{li}} i(x,t) = -\frac{\partial U(x,t)}{\partial x}; \qquad (2.47)$$

$$C_{\text{li}} \frac{\partial U(x,t)}{\partial t} + G_{\text{li}}(x,t) = -\frac{\partial i(x,t)}{\partial x}; \qquad (2.48)$$

where C_{li}, L_{li}, R_{li}, G_{li} are the linear capacitance, inductance, resistance and conductivity of the forming line. They are known as telegraph equations. The number of elementary cells of the line is assumed equal to 20, starting from the condition that the discharge time of the elementary capacitance of one cell per neighbouring elementary capacitance must be much less than the duration of the front of the propagating wave.

To account for parasitic leakage currents from the cathode holder, breakdowns along the insulator, etc., resistance loss R_{lo} shunting the load is introduced into the equivalent circuit. In addition, the switch K_6 is introduced into the equivalent circuit, simulating the operation of the accelerator on a controlled cathode [32], a peaker or a shock line [33].

7. *The load resistance* R_{load} in the computer model can be represented as: 1) linear resistance $R_{\text{load}}(t) = \text{const}$; 2) electron diode, i.e. non-linear resistance which varies according to the three-halves power law: $R_{\text{load}}(t) = P^{2/3}/I^{1/3}$, where P – perveance; I – diode current; 3) non-linear parametric resistance $R_{\text{load}}(t)$, which at each integration step is determined as the result of calculating the model of the relativistic magnetron. For the second and third circuits, the turn-on time of the switch K_6 corresponds to the achievement of a voltage at the cathode of a value equal to the threshold voltage of the explosive electron emission.

Consider the process of modelling the design and operation of LIA in the case of the most complex load – the relativistic magnetron. These devices are described in detail in Ch. 4. Here we present the elements of the theoretical model used for modelling. Two currents flow in the magnetron. It is the end current [34], which is defined as the limiting current from the end of a magnetized cathode of radius r_c:

$$I_{cur} = \frac{mc^3}{e} \frac{(\gamma^{3/2}-1)^{3/2}}{1+2\ln(r_{tube}/r_c)}, \quad \gamma = (1-\beta_\phi^2)^{-1/2}, \tag{2.49}$$

where r_{tube} is the inner radius of the drift tube, and the anode current of the magnetron I_a which, in accordance with the parameter $\alpha = 1 - \beta_\phi/\beta_e$ (the difference of the electron rotational velocity around the cathode β_e and electromagnetic wave β_ϕ) formed from the electrons, falling into the region of correct phases. In this case, the electrons can get to the anode only in the case when the quantity α satisfies the condition

$$A\ln(A+\sqrt{A^2-1}) - \sqrt{A^2-1} - ApR_L - \text{sh}(pR_L) \leqslant 0, \tag{2.50}$$

where $p = 2\pi/(\lambda\beta_\phi\gamma)$ is the transverse wave number; λ is the wavelength of the radiation; $A = a\gamma\text{sh}(pd)E_0/E_f; E_0$ is the static electric field strength; E_f is the effective amplitude of the basic spatial harmonic of the microwave field on the surface of the anode block; d is the cathode–anode distance; $R_L = E_N/(H_N^2 - E_N^2)$ is the Larmor radius of rotation of the electron in crossed electric ($E_N = eE_0/(mc^2)$) and magnetic ($H_N = eH_0/(mc)$) fields. Calculating the drift velocity of electrons from the region of incorrect phases to the region of correct ones in the same way as was done in [35], the following estimate for the anode current of the magnetron can be obtained in the cathode current limitation regime with a space charge:

$$I_a = \frac{mc^3}{e} NL \frac{E_f}{E_0} E_N \frac{\beta_\phi[A_1 + \text{ch}(pR_L)]}{8\pi\gamma\,\text{sh}(pd)(1-\alpha\gamma^2)}, \tag{2.51}$$

where L is the length of the device; N is the number of resonators of the anode block of the magnetron; A_1 is

$$A_1 = \begin{cases} A & \text{at } A > \text{ch}(pR_L), \\ \text{ch}(pR_L) & \text{at } A \leq \text{ch}(pR_L). \end{cases} \tag{2.52}$$

For the LIA to work in the optimal mode, it is necessary to ensure that the magnetron is a matched load that on the generated voltage pulse. This is achieved by selecting the appropriate values of the

Hartree threshold voltage (U_{th}) and the buildup time (rise time) of the oscillations (τ_{set}), which in turn depend on the geometric dimensions of the magnetron and the magnitude of the constant magnetic field.

The rise time of the oscillations can be estimated by considering the perturbations of the near-cathode flux of electrons as a result of their interaction with the synchronous harmonics and using the excitation theory of resonators [35]:

$$\tau_{set} \approx \frac{N/2+7}{30} \frac{sh(pd)}{sh(pR_L)} \sqrt{\frac{2\gamma V}{E_N S_c}} \qquad (2.53)$$

where S_c is the area of the cathode; V is the volume of the resonator.

From the time t_0 when the generated voltage reaches the value U_{th}, the radial (anode operating) current begins to flow through the magnetron diode, and the resistance of the magnetron changes exponentially:

$$R_{load}(t) = R_{r0} \exp[-\frac{(t-t_0)}{\tau_{set}\delta}], \qquad (2.54)$$

where R_{r0} is the resistance of the magnetron, determined by the end current I_{cur} at time t_0; t is the current time; δ is a constant determined from *a priori* data. Thus, until the resistance of the magnetron R_{load} becomes equal to R_c (self-consistent resistance of the LIA), the condition (2.50) is not satisfied and there is no radiation. When $t_m = t_0 + \tau_{set}\delta \ln(R_{r0}/R_c)$ the excitation conditions of the magnetron begin to be satisfied, and subsequently its dynamic resistance is determined by the sum of the anode and end currents.

The general equivalent circuit of a linear induction accelerator and a relativistic magnetron for which the differential current and voltage equations are compiled, was shown in Fig. 2.28.

2.8.2. Modelling of LIAs on magnetic elements with a relativistic magnetron

The model problem was solved consistently: an ohmic resistance, an electron diode, and a relativistic magnetron were connected to the output of the LIA. With the use of the ohmic resistance and the electron diode, the results of numerical simulation (amplitude and time parameters of output pulses) corresponded with practically 100% accuracy to the results of experimental studies of the simulated accelerator. This circumstance made it possible to draw a conclusion about the adequacy of the description of the computer model of the LIA of the processes that take place, and to proceed to the investigation of the LIA–relativistic magnetron system.

Preliminary calculations have shown that due the strong feedback in the power source–magnetron load system generates an oscillatory regime, characterized in oscillations of the microwave power, voltage and current of the accelerator. This result is a consequence of the application of the analytical formulas of the stationary theory of the relativistic magnetron for the calculation of non-stationary processes. Disruptions of microwave generation are associated with a fast exit of the device from the synchronism regime, which is experimentally observed only with non-optimum settings. This behaviour of the model has the following explanation: as soon as the voltage on the magnetron exceeds the threshold value, the oscillations begin to rapidly increase. This causes the appearance of a high anode current, which reduces the dynamic resistance of the relativistic magnetron. Since the LIA is a power source with limited power, reducing the load resistance leads to a reduction in the induced voltage below the threshold level and the exit of the device from the synchronism mode. Introduction of the *a priori* constant δ for the generating device allowed to smooth out these oscillations in the computer model. Physically, the value of δ determines the inertia of the processes of accumulation and leakage of the space charge in the interelectrode gap of the magnetron. Thus, the use of the model allows us to impose limitations on the Q-factor of the resonator system.

Practical interest in the 'LIA–RM' computer model is due to the possibilities: 1) optimizing the efficiency and output power of the accelerator and magnetron parameters; 2) calculation of the amplitude and envelope of microwave pulses of a relativistic magnetron; 3) estimating the effect of a dynamic load of the RM type on discharge processes in the forming line. Modelling allows determining the optimal geometric characteristics of a relativistic magnetron (the diameters of the anode block and the cathode, the length and type of the resonator system), the design parameters of the linear induction accelerator (the type and number of cores of the induction system, the characteristics of the forming line, the energy reserve of the primary storage, etc.) the value of the induction of the magnetic field. Besides, a model experiment made it possible to rethink the concept of creating LIA from the point of view of the balance of capacitive (C_s) and inductive (L_s) characteristics of the induction system.

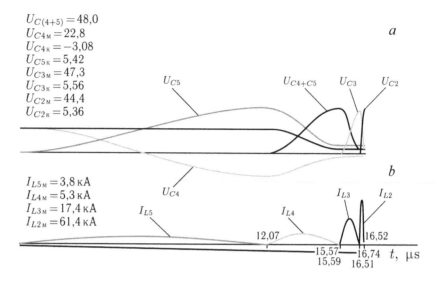

Fig. 2.29. The results of the calculations of processes in the magnetic pulse generator of LIA 04/6: the voltage diagrams on the capacitors (*a*) and currents through the windings of the saturation chokes (*b*); amplitude value of voltage ($U_{C\max}$) and residual voltage ($U_{C_{\kappa}}$) on capacitors of compression stages; current amplitudes ($I_{L\max}$) in the windings of the saturation chokes; characteristic time intervals.

Below are the results of the application of the model and their comparison with the experimental data for an installation based on a 6-resonator relativistic magnetron powered by a LIA on magnetic elements. The results of calculations are presented in tabular and graphical form. Figure 2.29 shows the processes in the magnetic pulse generator of the LIA (diagrams of currents and voltages, amplitude values of currents and voltages, characteristic time intervals for the 'ideal' configuration of the elements of the circuit). From the figure, in particular, it can be seen that the value of the voltage on the capacitors of the magnetic pulse generator circuits remains practically unchanged and the energy is compressed by increasing the current whose amplitude increases. The computer model determines the coefficients of energy transfer from one compression link to another, which makes it possible to calculate the thermal processes of the individual elements of the accelerator. Figure 2.30 shows the results of the calculation of the output pulses of the voltage and current of the LIA and the microwave power of the 6-resonator relativistic magnetron.

The results of calculations are tested by comparing them with the experimental data. Exceptions were also made to such extreme cases

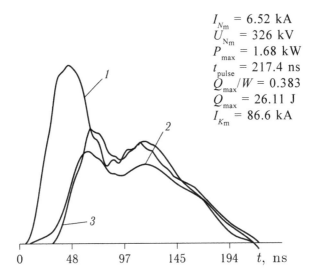

$$I_{N_m} = 6.52 \text{ kA}$$
$$U_{N_m} = 326 \text{ kV}$$
$$P_{max} = 1.68 \text{ kW}$$
$$t_{pulse} = 217.4 \text{ ns}$$
$$Q_{max}/W = 0.383$$
$$Q_{max} = 26.11 \text{ J}$$
$$I_{K_m} = 86.6 \text{ kA}$$

Fig. 2.30. The results of computer simulation of LIA and 6-resonator relativistic magnetron: diagrams of voltage (*1*), current (*2*), microwave power (*3*); I_{N_m} and U_{N_m} are the amplitudes of the current and voltage of a relativistic magnetron at the instant of maximum microwave power; Q_{max} and P_{max} – the energy and maximum power of the electron beam on the load; W is the energy stored in the primary storage; t_{pulse} is the duration of the current pulse on the load; Q_{max}/W is the efficiency of transfer of energy from the primary storage to the load; I_{K_m} – is the amplitude of the current through the winding of the magnetic commutator.

as the operation of the accelerator in regimes close to short circuit and idle running (magnetic field is absent or much higher than the synchronous value), and also when the electron beam is formed by a magnetically insulated diode (only the end current of the magnetron is available). In these cases, the discrepancy between the measured and calculated values of voltage and current (amplitude and duration of pulses) did not exceed 10%. The results of calculations of the level of generated power and the electronic efficiency of a relativistic magnetron in the region of synchronous magnetic fields differed from the measured values by no more than 20 and 10%, respectively. The latter can be related to the fact that the output device of the microwave radiation from the anode block is not taken into account.

Thus, computer simulation was carried out to investigate the processes in the power supply–microwave oscillator system with strong feedback. The modelling of such a complex system made it possible to consider in greater detail the physical processes taking place in its various elements and analyze the effects of the mutual influence of the load and the power source. The results are obtained using a model based on equivalent power source circuits, as well as formulas of the analytical theory of the averaged motion of a relativistic magnetron. The load (RM) is represented in the form of a non-linear parametric resistance, which at each step of integration of differential equations is determined as the result of calculating the relativistic magnetron model. Note that an analytical description of processes in such a system is hardly possible. Therefore, the implemented method of computer simulation is very fruitful.

This method allows one to quickly adjust the microwave source to the extremes of output power, electronic and full efficiency. The application of the model made it possible to justify the use of 'capacitive' type inductors in linear induction accelerators intended for feeding relativistic magnetrons, and also to use the unbalance of capacitances of the last link of MPG compression and the forming line [23]. Note that the power supply in the model can also be represented by a more complete equivalent circuit, but its complexity should be justified, since in the actual design of the plant there are difficult to take into account parameters (the leakage current from high-voltage electrodes mentioned above, the wide variation in the operating characteristics of ferromagnetic inductors, etc.).

2.8.3. Modelling of LIAs with multichannel dischargers

Let's consider an LIA the elemental base of which includes a multichannel spark discharger switching the forming line with a low (units or fractions of Ohms) wave resistance. Usually, these accelerators use several parallel spark dischargers, which are not structurally connected, but are united by a common start circuit. In this case, the time constant of the switch–line circuit considerably exceeds the switching time of the discharger. Thus, we can assume that at the initial time (units of nanoseconds) discharge processes are determined by the properties of the discharger itself, and its spark channel is formed due to the discharge of interelectrode capacitances. In the future, the law of the change in the current of the discharge circuit is determined by the values of its resistance R_s and inductance L_s, as well as the wave resistance of the line ρ_{li}. Therefore, to calculate the transient processes a model is chosen from transferring the switching from a non-conducting state in a finite time $t_{fr} = 2.2\tau_{sw}$ where τ_{sw} is the switching time during which the voltage across the discharger falls from the level of 0.9 to 0.1 level. The change in voltage is approximated by formula

$$U_s(t) = U_0 \exp\left(-\frac{t}{\tau_{sw}}\right), \tag{2.55}$$

where U_0 is the charge voltage of the forming line.

The current I_s switched by the discharger is described by the parametric equation

$$L_s \frac{dI_s}{dt} = U_c(t) - U_s(t) - R_s I_s, \tag{2.56}$$

where $U_c(t)$ is the voltage change at the end of the electrode of the forming line to which the switch is connected.

Simulation of the magnetron is carried out as described above. Figure 2.31 shows the results of calculations using the model considered for the case of a 6-resonator RM generator powered by a LIA with a multichannel discharger.

In conclusion, we note that the linear induction accelerators on magnetic elements presented in this chapter provide a pulse repetition rate of hundreds of hertz with a high repeatability of the amplitude and shape of the output pulses. The use of the original layout scheme, application of the effects of phase overlap and unbalance of capacitance of the last link of the compression of the magnetic pulse generator and the capacity of a single forming line make it possible to significantly reduce the weight parameters of accelerators in comparison with the existing analogues. The pulse-periodic relativistic magnetron generators created on their basis can reliably and efficiently operate with a high average microwave power, have a long service life, and have a high stability of generated oscillations. With the use of relativistic magnetron powered by linear induction accelerators it is possible to create compact radiating installations, including on a mobile platform.

Figure 2.32 compares the mass and dimensions parameters of the two types of LIA. As follows from the figure, LIA on magnetic elements can generate electron beams of much higher average power due to high pulse repetition rate, although they are inferior to LIAs with multichannel surge dischargers. We emphasize that LIAs on magnetic elements have significant weight and high cost due to the need to use magnetic pulse generators to charge the forming lines.

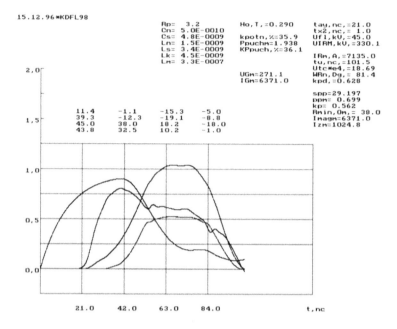

Fig. 2.31. Results of simulation of a relativistic magnetron with a LIA with a multichannel discharger.

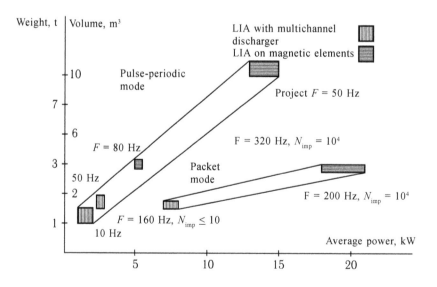

Fig. 2.32. Comparison of the mass-size parameters of various types of LIA (including the power system) indicating the average power of the electron beam generated by them (N_{im} is the number of pulses in the series).

References

1. Furman E.G., Prib. Tekh. Eksper., 1987, No. 5. Pp. 26–31.
2. Vasil'ev V.V., et al., *ibid*, 1985, No. 6. Pp. 19–23.
3. Furman E.G., et al., *ibid*, 1993, No. 6. Pp. 45–55.
4. Vintizenko I.I., Furman E.G., Izv. VUZ, Fizika, 1998, No. 4. Application. P. 111–119.
5. Lebedev N.N., Zh. Teor. Fiz., 1958. V. 28, No. 6. Pp. 1330–1339.
6. Schwartz L., Mathematical Methods for Physical Sciences, Moscow: Mir, 1965.
7. Kiselev Yu.V., Cherepanov V.P., Spark dischargers, Moscow: Sov. Radio, 1976.
8. Furman E.G., Vasil'ev V.V., Prib. Tekh. Eksper., 1988, No. 1. P. 111–116.
9. Vasil'ev V.V., Furman E.G., *ibid*, 1992, No. 6. Pp. 158–164.
10. Vintizenko I.I., Lukonin E.I., Furman E.G., in: Proc. doc. 8th All-Russian Sympos. on High-current electronics. Sverdlovsk. 1990. Part 3. P. 133–135.
11. Forman E.G., Vintizenko I.I., in: Proc. 12 Int. Conf. on High Power Particle Beams. 1998. Israel. Tel-Aviv. P. 107.
12. Vintizenko I.I., Shlapakovski A.S., in: Proc. on Int. Power Modulator Conference. Hollywood. USA. 2002. P. 510–512.
13. Meerovich A.A., et al., Magnetic pulse generators, Moscow: Sov. Radio, 1968.
14. Dolbilov G.V., et al., Prib. Tekh. Eksper., 1984, No. 4. Pp. 26-31.
15. Birx D.I., et al., IEEE Transactions on Nuclear Science. 1985. V. NS-32. P. 2743–2747.
16. Vintizenko I.I., et al. in: Proc. 12 Int. Symposium on High Current Electronics. Tomsk, 2000. V. 2. P. 255–258.
17. Butakov L.D., et al., Prib. Tekh. Eksper., 2000, No. 3. 159, 160.
18. Butakov L.D., et al., *ibid*, 2001, No. 5. 104–110.
19. Harjes H.C., et al., in: Proc. 9 Int. Conf. on High-Power Particle Beams. Washington. USA. 1992. V. 1. P. 333-340.
20. Ashby S., et al., in: Proc. 8 Int Conf. on High-Power Particle Beams. Washington. USA. 1992. V. 2. P. 1855–1860.
21. Mesyats G.A., Movshevich B.Z., Zh. Teor. Fiz., 1989. Vol. 59, No. 5. P. 39–50.
22. Vintizenko I.I., Izv. VUZ, Fizika, 2007, No.10/2. Pp. 136–141.
23. Vintizenko I.I., Linear induction accelerator, Patent of the Russian Federation for invention No. 2178244. BI. 2002, No. 1.
24. Vintizenko I.I., Linear induction accelerator, Patent of the Russian Federation for invention No. 2185041. BI. 2002, No. 19.
25. Gordeev V.G., et al., Prib. Tekh. Eksper., 1980, No. 5. Pp. 117–119.
26. Vasil'ev V.V., et al., Izv. VUZ, Fizika, 2003, No.10. Pp. 14–23.
27. Nechaev V.E., et al., Pis'ma v ZhTF, 1977. Vol. 3. Vyp. 15. P. 763-767.
28. Kalantarov P.L., Tseitlin L.A., Calculation of inductances, Leningrad, Energoatomizdat, 1986.
29. Vakhrushin Yu.P., Anatsky A.I., Linear induction accelerators, Moscow: Atomizdat, 1978.
30. Iossel' Yu.A., et al., Calculation of the electric capacity, Leningrad, Energoizdat, 1981.
31. Ginzburg S.G., Methods for solving problems on transient processes in electrical circuits, Moscow, Vysshaya shkola, 1967.
32. Tomskikh O.N., Furman E.G., Prib. Tekh. Eksper., 1991, No. 5. Pp. 136–138.
33. Dubiev A.I., Kataev I.G., *ibid*, 1979, No.4. 172, 173.
34. Sulakshin A.S., Zh. Teor. Fiz., 1983. V. 53, No. 11. P. 2266–2268.
35. Nechaev V.E., et al., in: Relativistic high-frequency electronics, Gorky: IPF Academy of Sciences of the USSR. 1979. P. 114–130.

3

Relativistic microwave devices with the electron beam formed by the linear induction accelerator

Introduction

The successes of recent years in the development of methods for generating high-power coherent radiation indicate the possibility of a number of new applications of microwave sources. The variety of generators and amplifiers produce quasi-monochromatic radiation (for heating the plasma at electron cyclotron resonance) and broadband radiation (for visualizing objects and generating radio interference). The designs of relativistic microwave devices are largely similar to classical sources of coherent radiation. In O-type instruments, the magnetic field focusing the electron beam is directed along the axis of the device, i.e., it coincides with the direction of the electric field generated by the accelerating voltage. In such fields a rectilinear electron beam is formed interacting with various microwave structures and transmitting its energy to them. In M-type instruments, the magnetic field is chosen to be normal to the electric field and is used to distort the beam trajectories. Relativistic M-type devices (or devices with crossed fields), using high-current linear induction accelerators as power sources, will be considered in Ch. 4.

This chapter presents information on the principle of operation and design of relativistic microwave devices with rectilinear electron beams, and the designs of LIAs forming electron beams for powering such devices are considered. The action of O-type instruments is based on the dynamic method of current control. As a rule, only a constant accelerating field acts on the space charge at the cathode, and the occurrence of an alternating field occurs in the region remote from the cathode, because of the interaction of the beam electrons with the microwave field. As a result of the interaction, the electrons acquire a high-speed modulation. With further movement, they shift relative to each other and form electronic bunches. Thus, the kinetic energy of the electron beam goes over into the potential energy of the electron compaction. The slowing-down of electron bunches by the microwave field converts the energy of the electron beam into the energy of microwave oscillations, selected by the resonator creating the slowing-down field.

The interaction of electrons with a microwave field can have a discrete or continuous character. The first variant is realized in klystrons, which are a set of resonators separated by drift regions. Continuous interaction is possible when synchronizing the moving of the microwave field with the motion of electrons. It occurs in such devices as a running (RWT) or backward (BWT) wave tube using slowing-down systems.

Some scientific centres are developing sources based on a new technology that allows to extend the wavelength range and increase the power level while maintaining their high efficiency. In the future, this will be beneficial for such applications as sources of spectroscopy, accelerators of advanced type, shortwave radar, plasma heating in fusion reactors.

3.1. Undulators and free-electron lasers

Classical sources of coherent radiation (magnetron, klystron, wave tubes), even with super-power sources, have limited output power and efficiency at high frequencies. To overcome these limitations, both new mechanisms and modifications of traditional approaches were proposed. Two types of sources are now of great interest, first seen around 1960: a free-electron laser (FEL) [1] and a cyclotron resonance maser (see section 3.3.1). The common feature for both devices is the presence of a beam of relativistic electrons.

The term 'free-electron laser' is used because of the mechanism of action of the device, associated with the emission of electrons during their accelerated motion. If a beam of moving electrons is subjected to an electromagnetic field with a certain polarization and phase,

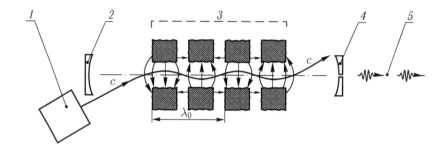

Fig. 3.1. The FEL scheme with an alternating magnetic field. The trajectory of the particle lies in a plane perpendicular to the figure; the arrows indicate the directions of the magnetic lines of force; λ_0 is the length of the particle trajectory period in the undulator.

the electrons will be accelerated in such a way that they begin to radiate coherently. The condition of coherence is that the emission of electrons must enhance the superimposed electromagnetic field. The name 'FEL' comes from the opposition to traditional lasers, in which electrons are in a bound state in an atom, and the radiation occurs during their transitions between energy levels. In FEL interaction of the electron beam with microwave waves of electrodynamic systems or with fields of magnetic devices occurs in vacuum, as in traditional microwave devices, and the electron can be considered free. A coherent source is called the amplifier if the field is applied from the outside, and the generator, if the field of spontaneous emission of individual electrons acts as an applied field.

The scheme of the free-electron laser device includes an electron beam, an external pump field, and an applied radiation field (Figure 3.1). As a pumping field, any field that leads to electron oscillations in the transverse direction (usually a static periodic magnetic field) can be used. Although relativistic effects do not underlie the radiation mechanism, electrons must be known to be relativistic for radiation in the short-wavelength region. The emission wavelength of an FEL, unlike most classical sources, is not rigidly related to the size of the system. Consequently, large structures in principle can emit short waves at high powers.

Undulators. Let us consider an ordinary pump field – a periodic magnetic field in space created by a wiggler magnet, or an undulator (a device for building an electron beam). The components of such a field are located perpendicular to the direction of radiation and the motion of electrons (Fig. 3.1). The undulator is a device in which periodic fields are created that act on the charged particles passing through it with a periodic force with zero mean over the period value. A moving charged particle, hitting an undulator, performs a periodic oscillatory–translational motion, that is, it is an oscillator moving uniformly and rectilinearly. Such a particle emits undulator radiation. The most common particle trajectories in the undulator are sinusoids and spirals.

On the basis of the type of fields created, the undulators are divided into two types. In undulators of the first type, the fields periodically change in space or in time (alternating magnetic field – Fig. 3.1, helical magnetic field, high-frequency electric field, electromagnetic wave field, etc.). In undulators of the second type, static focusing magnetic and electric fields act (a homogeneous magnetic field, crossed homogeneous electric and magnetic fields, a quadrupole electric field, etc.). The length of the trajectory period of a particle in the undulator of the first type is given by the period of the undulator field, depends on the angle and the coordinate of the particle's entry into the undulator and in the relativistic case does not depend on its energy. In undulators of the second type, the length of the particle trajectory period is determined by the focusing properties of the fields (gradient, magnitude), the amplitude of the particle oscillation (given by the angle and the coordinate of its entry into the undulator), the particle energy.

Different types of undulator radiation sources, consisting of an accelerator or an electron storage and undulator accumulator, can emit spontaneous incoherent, spontaneous coherent and induced undulator radiation. The particle velocity in the undulator can be represented as a sum of a constant (v) and periodic variable ($\Delta v(t+T) = \Delta v(t)$) velocity, where T is the period of oscillations of the particle in the undulator; t is time). A single accelerated particle passing through the undulator emits a train of electromagnetic waves whose duration Δt depends on the angle Θ between v and the direction of observation. At distances $R \gg q\lambda_0$ (where λ_0 is the length of the particle trajectory period in the undulator, q is the number of periods), we have

$$\Delta t = \frac{q\lambda_0}{c\beta_\phi}(1 - \beta_\phi \cos\Theta) \qquad (3.1)$$

where $\beta_\phi = v/c$. The particle emitted by the particle train contains K periods. The circular frequency of the fundamental harmonic of undulator radiation is

$$\omega_n = \frac{n\Omega}{1 - \beta_\phi \cos\Theta} \qquad (3.2)$$

where $\Omega = 2\pi\beta_\phi c/\lambda_0$ is the particle oscillation frequency in the undulator. For $\Theta = 0$, the frequency of undulator radiation is maximal. Due to the finite duration of the trains the undulator radiation emitted by a particle in a certain direction, is distributed over the frequency interval $\Delta\omega_n$, which determines the natural spectral line width:

$$\frac{\Delta\omega_n}{\omega_n} \approx \frac{1}{nK}$$

In general, the wave trains of the undulator radiation in the interval Δt are not harmonic. Radiation occurs on several harmonics, multiple of the fundamental harmonic. However, for $K \gg 1$, the undulator radiation observed at a given angle Θ is monochromatic and has a frequency (3.2). The major part of the energy emitted by the relativistic particle is concentrated near the direction of its instantaneous velocity v in a narrow range of angles:

$$\Delta\Psi \approx \frac{mc^2}{E} = \sqrt{1 - \beta_\phi^2} \qquad (3.3)$$

where E is the particle energy.

The sources of undulator radiation have an important advantage over synchrotron radiation sources, gas lasers and other sources of infrared and optical ranges – the possibility of smooth regulation of the radiation frequency by changing the magnitude of the undulator magnetic field and the energy of the beam particles. Spontaneous undulator radiation can be used in the same fields of research as synchrotron radiation: in X-ray microscopy, X-ray structural analysis, atomic and molecular spectroscopy, crystal spectroscopy, radiography, lithography, medicine, etc. Compared with synchrotron radiation, it has higher intensity, directionality, degree of monochromaticity and polarization.

In the case of using high-current electron beams, when the Langmuir frequency of the beam is comparable to the pump wave frequency, there is a three-wave parametric process in which amplification of the backscattered wave occurs as a result of interaction with a pumping wave or a space-charge wave of a beam or a cyclotron wave. Dispersion relations for these two cases are written as follows [2]:

$$k\upsilon_z + k_0\upsilon_z - \omega_p\gamma^{-1/2} = \left(\omega_{\text{crit}}^2 + k^2c^2\right)^{1/2}; \qquad (3.4)$$

$$k\upsilon_z + k_0\upsilon_z - \Omega_c\gamma^{-1} = \left(\omega_{\text{crit}}^2 + k^2c^2\right)^{1/2}, \qquad (3.5)$$

where k is the wave number; $k_0 = 2\pi/\lambda_0$; ω_{crit} is the critical frequency of the waveguide; $\omega_p = (4\pi n e^2/m_0)^{1/2}$ is the plasma frequency; $\Omega_c = eB_z/(m_0 c)$ is the cyclotron frequency; $\gamma = (1 - \beta_\phi^2)$ is the relativistic factor.

It follows from (3.4) and (3.5) that the guiding magnetic field affects the radiation frequency only when interacting with the cyclotron mode.

At the same time, the results obtained during the first experiments with high-current electron beams were significantly different from the theoretical estimates of the radiation frequency and the efficiency of conversion of the electron beam energy to radiation energy. This is primarily due to the fact that an exact theoretical consideration of the processes of stimulated emission in the undulator with an allowance for the real parameters of the electron beam and the installation is extremely difficult, and the accepted simplifications often do not correspond to the experimental conditions. It was shown in [2] that the efficiency of the interaction of the pump wave with the electron beam increases with the resonance of the pump wave and the cyclotron frequency of the electrons in the guiding magnetic field. In paper [3], the effect on the generated power of the amplitude of the pump field and the length of the undulator was studied. A comparison of the obtained data with the results of [2] makes it possible to determine the effect of the plasma frequency of the oscillations of the electrons ω_b on the generation frequency in the resonance region. Thus, an experimental study of the influence of the main installation parameters on the output power and the generation frequency is of great interest. One of the first instruments installed on the LIA was an undulator developed at the Tomsk Polytechnic University in 1981. An accelerator was used for the experiments, constructed a year earlier on the basis of the original layout considered in Ch. 2 [4]. The scheme of the experiment is shown in Fig. 3.2. The electron energy could vary from 0.55 to 0.8 MeV with the addition of the accelerating section of the LIA. An uniform beam of electrons 32 mm in diameter with a current of ~0.5 kA entered a circular waveguide

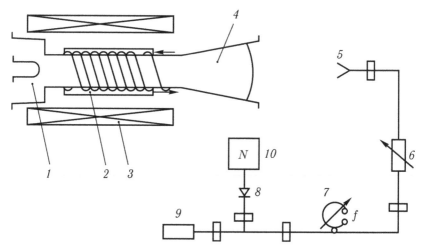

Fig. 3.2. Schematic of the experiment: *1* – electron gun; *2* – undulator; *3* – the solenoid; *4* – transmitting antenna; *5* – receiving antenna; *6* – attenuator; *7* – the filter; *8* – detector; *9* – the load; *10* – oscilloscope.

having an internal diameter of 35 mm. Pumping was carried out by the magnetic field of a spiral undulator made in the form of a conductor with a current wound around a waveguide with a period $\lambda_0 = 4$ cm. The magnitude of the undulator field $B_\perp$ varied from 30 to 500 G. A waveguide with the undulator was placed in a solenoid, which created a guiding magnetic field B_z with a value of up to 0.9 T.

Figure 3.3 shows the dependence of the radiation power P on the magnetic field for various parameters of the system. Curve *1* was obtained for electrons with an energy of 0.55 MeV, and curves *2–4* – 0.8 MeV. The maximum radiation power, as in [2], was found to be due to the resonance of the pump wave and the cyclotron frequency. With increasing electron energy, the cyclotron frequency of their rotation in a magnetic field decreases. Therefore, in order to preserve the resonance, it is necessary to increase the amplitude of the guiding magnetic field, which was confirmed in the experiment. With increasing energy, the wavelength of the generated radiation varies from 9 to 8 mm. Investigation of the influence of the amplitude of the pump field $B_\perp$ on the radiation power has shown that there exists a range of values $B_\perp$ (from 80 to 250 G), at which the maximum output power of radiation within the limits of measurement accuracy remains practically constant (curve *2*). With increasing $B_\perp$ up to 450 G, the radiation power decreases sharply (curve *3*). Curves *1–3* were obtained with the undulator length 50 cm. Curve *4* was recorded for $B_\perp = 450$ G and the length of the undulator

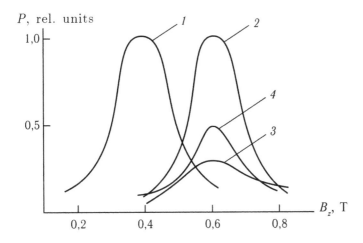

Fig. 3.3. Dependence of the radiation power on the guiding magnetic field.

was 40 cm. It is seen that with increasing $B_\perp$ the maximum radiation power is achieved with a shorter undulator length (curves *3* and *4*). The latter is due to the fact that increasing $B_\perp$ decreases the optimal length of the device.

In the experiment there was no change in the radiation frequency of electrons of the same energy for different $B_\perp$ and B_z. At the same time, the electron density in the beam has a significant effect on the generation frequency. The latter is clearly seen when compared with the results of [3]. In both cases, the uniform beam was used but in [2] $\omega'_b =$ $4.67 \cdot 10^{10}$ s^{-1} and in [3] $\omega''_b = 0.73 \cdot 10^{10}$ s^{-1}. In this case, the measured wavelengths of the microwave radiation at the same energy of the beam electrons (~0.8 MeV) were 13 and 8 mm, respectively. The numerical values of the frequencies were obtained from the formula used in [5]:

$$\omega_S = \frac{2\pi V / l_0 - \omega_n / \gamma^{3/2}}{1 - \beta},$$

where ω_S is the generated frequency; V is the velocity of the electrons; $\gamma = = (1 - \beta^2)^{-1/2}$; $\beta = V/c$. They do not coincide with the experimental frequencies, since this formula is derived for free space and does not take into account the influence of the waveguide. At the same time, it can be used to calculate the change in the generation frequency due to the above change in ω_b:

$$\Delta\omega_S = \omega_S'' - \omega_S' = \frac{\omega_b' - \omega_b''}{\gamma^{3/2}(1-\beta)} = 11.64 \cdot 10^{10} \text{s}^{-1},$$

which corresponds to a value $\Delta\lambda \approx 6.5$ mm, which differs little from the difference in the generation wavelengths obtained experimentally.

The maximum power recorded in this experiment was 150 kW. Such a low efficiency of converting the energy of an electron beam into radiation energy can be explained by the large energy spread of the beam electrons and by the low beam current density.

Free-electron lasers. In the development of free-electron lasers (FELs), it was assumed that they could be used to create sources of high-power microwave radiation ranging from short centimeter waves to soft X-rays. To generate the millimeter and submillimeter range, high-current relativistic electron beams of relatively low energy (2–5 MeV) can be used as an active medium, and for pumping a powerful electromagnetic wave or a spatially periodic magnetic field, for example, the magnetic field of an undulator. Immediately after the creation of the first operating model of the FEL in the United States, it was proposed to use it for military purposes to combat ballistic missiles during the acceleration phase and on the flying part of the trajectory. Since FELs are heavy and large, in the near future it is not possible to place them in space. It is assumed that the facilities will be based on ground in the form of several territorially scattered 'farms'.

The advances in the short-wave range for the FEL are mainly due to the energy capabilities of the accelerators. At present, most of the experiments are performed on accelerators designed for the needs of nuclear physics, which have high energies and small beam currents. However, for the FEL it is necessary to create special accelerators, which allow changing the energy of the beam electrons and thereby to change the frequency of the output microwave radiation. In addition, significant currents of the electron beam are needed to obtain coherent radiation. These requirements are met by linear induction accelerators.

At the Lawrence Berkeley Laboratory, the ETA linear induction accelerator (Experimental Technology Accelerator) operates with up to 5 MeV and a current of up to 10 kA, and experiments are conducted to generate powerful millimeter radiation using a free-electron laser. The experiments [6, 7] with the FEL were carried out using an electron beam with an energy of 3.5 MeV to correspond to the 34.6-gigahertz resonance of the undulator magnet. The ETA injector contained an explosive emission cathode which provided an electron beam current of 4 kA. An emitter filter was used to reduce the current and emittance of the beam introduced into the three-meter undulator (the beam brightness was maintained at $2 \cdot 10^4$ A/(cm $\cdot$ rad)^2 at a current of 850 A). The beam pulse duration was 30 ns. When the electron beam was transported to the undulator, the fluctuations in the voltage amplitude during the pulse resulted in the loss of the rising and falling edges of the beam pulse. As a result, the beam duration in the FEL was 10–20 ns.

The effective length of the undulator could be changed by adjusting the currents in the magnetic coils to create a resonant magnetic field $B_\perp$ at the required length. The currents in other coils at this time decreased to 40% of the maximum current value, providing a field of approximately 0.4 $B_\perp$ for the rest of the undulator. A reduced field was necessary to transport the electron beam from the interaction region to the current control device at the end of the undulator without changing the effective length of the FEL. The output microwave signals from the FEL amplifier were output to a vacuum chamber in which directional couplers, variable attenuators, rejection filters, and crystal detectors were installed.

To study the operating characteristics, a homogeneous undulator 1 m long was used in the first stage. The magnitude of the output signal of the amplifier was determined as a function of the undulator magnetic field. At the undulator length of 1 m the amplifier was not saturated. As a result, a tuning dependence of the output power on the magnitude of the magnetic field was found. For a beam of electrons with an energy of 3.5 MeV, the gain was maximum for an undulator magnetic field of 3.72 kG. Using this value, the dependence of the FEL power on the length of the undulator, which has a resonance character, was investigated. With an input signal of 50 kW, the output microwave signal grows exponentially at the first 1.3 m at a rate of 34 dB/m. With an undulator length of 1.3 m, the amplifier is saturated. The output power in this case reaches 180 MW with an efficiency of 6% (the power of the electron beam is 3.0 GW).

In [8, 9], the results of the experiment with FEL were reported, which had two features. Firstly, there was no guiding magnetic field in the interaction region, although the electron current was relatively high (~200 A). As a consequence, cyclotron radiation, which was added in other experiments, was absent. As a result, the operation modes of the FEL could be uniquely identified and analyzed. Secondly, the pulse duration of the beam was 2 μs, that is, approximately 40 times longer than the pulse duration in previous

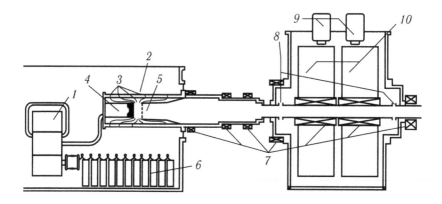

Fig. 3.4. Scheme of the LIA of the National Bureau of Standards: *1* – pulse transformer; *2* – electron gun; *3* – electrodes; *4* – cathode; *5* – anode; *6* – modulator; *7* – focusing coils; *8* – acceleration gap; *9* – modulators; *10* – cores.

experiments with FELs with high-current electron beams. A long pulse duration made it possible to obtain valuable information about the FEL saturation mechanism and other non-linear effects.

The electron beam in these experiments was created by the LIA, which was designed and built at the National Bureau of Standards (NBS) of the United States (Fig. 3.4). Its unique feature is the pulse duration of ~2 µs (see Table 1.1 in section 1.2.2). The injector generates a voltage on the cathode-holder of 350–450 kV and includes an electron gun from the Astron II accelerator. The cathode is located in a practically zero magnetic field ($\leq$5 Gs). A number of solenoids are used to focus the beam and carry it through the rest of the accelerator. Two accelerating gaps, excited by individual induction modules, increase the voltage by 100–150 kV, and the beam energy to 550–750 keV.

The accelerator uses a specially designed electron gun. In terms of obtaining a homogeneous high-current beam, two types of cathodes were tested: thermionic and multi-point cathodes. The multi-point cold cathode formed both short ($\leq$100 ns) and long (up to 2 ms) current pulses. It was made in the form of an aluminium plate 1 cm thick and 17 cm in diameter, through which, at a distance of 1 mm, graphite filament 10 µm thick protruding 1 cm above the surface of the plate (like a brush) were located at a distance of 1 mm from each other. At an electric field strength of ~30 kV/cm, the current density from the cathode was equal to ~3.7 A/cm². The pulsation of the amplitude of the voltage and current during a pulse with a duration of 1.5 µs was from 5 to 10%, which is certainly worse than a 2% pulsing when using a thermionic cathode, but much better than a 30% pulsing for explosive emission cathodes. In the case of a thermionic gun, about 95% of the electrons reach the collector, while for the brush-shaped cathode the current transmission was only 30%. However, with an additional magnetic field, it was possible to optimize transportation and bring the current transmission to 70%.

It should be noted that the main problems of obtaining a small emittance and a high current density of the electron beam, in addition to the need to optimize the electron beam control system, are incorporated in the beam formation system at the cathode (explosive emission cathodes do not allow obtaining a small emittance value). Usually, to reduce it, part of the electron beam is cut out by the diaphragm, but in this case the emittance of high-current accelerators is more important than the emittance of linear waveguide accelerators. Even the thermionic cathodes give emittance values which are more than necessary for the FEL. Therefore, photocathodes have now become widespread; their surface is irradiated with laser radiation, as a result of which a uniform electron beam with a high current density is formed. Such cathodes have already been tested at the National Laboratory in Los Alamos (LANL). They will also be used on the ETA-2 installation at the National Laboratory in Livermore (LLNL).

The graphite-brush cathode proved to be very reliable and durable, withstood thousands of pulses without any obvious deterioration in the emission properties. The voltage and current of the electron gun were constant with an accuracy of 5% during a pulse duration of ~1.5 µs. The normalized beam emittance was ~0.25π cm · rad which is only approximately twice the value of the measured emittance using the thermionic cathode.

The undulator was a bifilar (two-wire) spiral having a period of 4 cm and a diameter of 4 cm. The spiral was made of two layers of thin copper wire 1 cm wide and 0.12 cm thick. The total length of the undulator was 128 cm, including 6-period adiabatic transitions at each of its end. A large perturbation of the magnetic field at the input end of the undulator was reduced to <3% of the maximum value of the field by means of an external compensating winding. A stainless steel pipe with an internal diameter of 3 cm and a wall thickness of 0.75 mm passing through the undulator enabled a pulsed magnetic field of the undulator (rise time ~100 µs) to penetrate to the axis almost without distortion.

One of the main advantages of a spiral undulator is that it provides radial beam focusing. It was necessary in the experiment, since the electron beam entering the undulator expanded and its current dropped until it reached the undulator section with a constant amplitude of the magnetic field. The reduction of current at the transition depended on the magnetic field, but in the field of ≥300 G there was no noticeable loss of current through the undulator. By the end of the undulator the beam was deflected to the wall of the electron beam output channel. The measurements showed that at a distance of ≈30 cm approximately 2/3 the beam was lost. If the undulator field was absent, the beam current dropped almost to zero. The undulator field stopped this process. With its magnitude of ~250 Gs the current ceased to decrease, remaining constant.

Measurements of the position of the beam and its size in the undulator were performed using a smoothly moving molybdenum target with an attached scintillator, which was photographed from the opposite end of the undulator. The beam radius was usually ~0.5 cm, but the deviation of the position of its centre of gravity from the axis of the instrument reached 0.5 cm. Consequently, the beam electrons, passing through the undulator, had a spiral component of motion. The radiation was extracted from the FEL through a conical horn and a teflon vacuum window 15 cm in diameter. The high-frequency output power was measured by a gas-breakdown spectrometer and a calorimeter. Within the accuracy of diagnosis the results were consistent. The maximum radiation power was 4 MW. It was obtained at an electron energy of 700 keV, beam current of 200 A, and a magnetic field of 625 G. In the experiments, a strong interrelation between the radiation power, the beam current and the magnetic field of the undulator was noted. A distinct boundary of the onset of microwave generation was observed at $B \approx 400$ G. The beam current and microwave power reached a maximum at $B \sim 625-750$ G. The ratio of the microwave power to the electron beam current remained practically constant at magnetic fields from 500 G to 1 kG.

The experiment unequivocally testified to the operation of the FEL in the generation mode. Its results can be summarized as follows. After passing through an accelerator of length ~3 m, an electron beam with an energy of 700 keV and a duration of 2 μs entered the interaction region. Since there was no cyclotron radiation, the observed microwave signal was definitely the radiation of the FEL. The frequency was well tuned by the energy of the beam electrons. The duration of the microwave pulse was 2 μs. The shape of the microwave pulse indicates that saturation has not occurred for at least half the pulse duration of the beam current. Measurement of the frequency of the output radiation was carried out by beyond-cutoff filters, a spectrometer and a Fabry-Perot interferometer. For electron energies of 650 keV, the results of measurements on the Fabry-Perot interferometer showed radiation with a frequency of 31 GHz, and for 575 keV with a frequency of 23 GHz. The band width of the radiation was 10%. Measurements with the help of beyond-cutoff filters showed the presence of higher-frequency components with frequencies from 40 to 60 GHz. The mode structure of the radiation was determined by a waveguide of the appropriate wavelength range, which moved along the exit window. The distributions of the radial and azimuthal components of the electric field were removed. Both components of the field were independent of the azimuthal coordinate, which corresponds to the mode TE_{11}, which has a circular polarization, as expected for a spiral undulator. At a beam energy of 650 keV and an undulator field of 1 kG, radiation was obtained at frequencies in the range from 30 to 33 GHz, smaller than theoretical predictions. These frequencies correspond to values of the transverse velocity from $0.25c$ to $0.36c$ (or the value of the undulator field from 1.5 to 2.2 kG). At the same time, a transverse velocity value of about $0.16c$ was obtained in the experiment.

The measured microwave power (~4 MW) corresponded to an interaction efficiency of ~3% (with a theoretical estimate of ~10%). It was impossible to achieve saturation in the single-pass mode of operation of the FEL. Therefore, with this scatter of velocities and propagation of the beam near the undulator axis, this result can be considered satisfactory. The need to carry out work to improve the quality of the beam and centre it in the undulator was noted. This should increase the excitation of the TE_{11} mode and increase the efficiency of the free-electron laser.

The US Naval Laboratory conducted a two-section FEL study. In an ordinary FEL an electron beam passes through a magnetic undulator, where conditions are created for coherent amplification of the radiation at wavelength $\lambda = d/(2\gamma^2)$. However, if the electromagnetic wave has a power density of about 10 MW/cm², it can easily replace the magnetic undulator. If E_0 is a transverse electric field with frequency ω_0, and v_{ph} is the phase velocity of the electromagnetic wave, then it gives the required magnetic field of the undulator: $B_{it} = E_0(1 - v/v_{ph})/v$. In this case the undulator period is $d_{un} = 2\pi c/\omega_0(1 - v/v_{ph})$. Figure 3.5 shows the scheme of the corresponding experiment, as well as the area of interaction of the

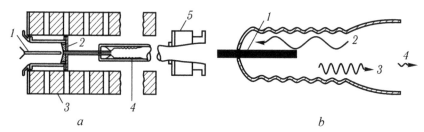

Fig. 3.5. General view of the installation for a two-section FEL: *1* – cathode; *2* – anode; *3* – magnetic field coils; *4* – corrugated waveguide; *5* – horn (*a*). Area of interaction of FEL: *1* – relativistic beam; *2* – backward wave; *3* – scattered wave; *4* – output radiation (*b*).

FEL. The electron beam passed through a relativistic backward wave tube (BWT), which generated radiation at a frequency of 12.5 GHz (λ = 2.4 cm) with a power of about 500 MW. This radiation was then used as an electromagnetic undulator for the second FEL section operating at a frequency of 20 GHz (λ = 1.5 cm). The slow-wave structure for the BWT was a 25 cm long corrugated surface with a period of 1.6 cm. Its outer diameter was 5 cm, the inner diameter was 4 cm. The beam was formed by a diode system with a current of 1 kA and an energy of 900 keV. The diameter of the beam was about 6 mm for a longitudinal velocity dispersion of less than 1%.

The electromagnetic wave (TM_{02} mode) had a phase velocity equal to $7 \cdot 10^8$ m/s, which corresponds to an effective undulator period of 1.7 cm. The dependences of the low- and high-frequency radiation of the FEL on the magnitude of the undulator magnetic field were experimentally obtained. The low-frequency radiation of the TM_{01} mode (at 8 GHz) appears due to Doppler stimulated scattering, and the high-frequency (140–260 GHz) mode due to inverse stimulated Raman scattering.

A scheme was proposed to optimize the two-section FEL in which the regions of the FEL and BWT are spaced and regulated by different values of the magnetic fields B_1 and B_2 (Fig. 3.6). This allows one to get rid of the above disadvantages.

Thus, the conducted experiments have shown that the FEL with an electron beam by the LIA can generate radiation in different wavelength ranges, providing a tuning of the radiation frequency in a wide range, and use for this purpose different types of undulator pumping. These devices cover the range from 10 GHz to 1 THz. The maximum power is now attained at LLNL (Livermore, USA): P = 2 GW at 140 GHz with an efficiency of 13% and a duration of

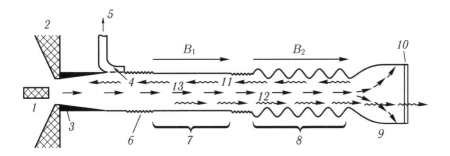

Fig. 3.6. Optimized design of the two-section FEL: *1* – cathode; *2* – anode; *3* – microwave absorber; *4* – directional coupler; *5* – output of microwave to the detector of the X-range; *6* – directional reflector; *7* – interaction region on the backward wave; *8* – area of BWT; *9* – collector of electrons; *10* – the output window; *11* – backward wave (BWT, $f = 12$ GHz); *12* – direct wave (FEL, $f = 200$ GHz); *13* – electron beam.

20 ns. The electron beam in these experiments had an energy of 6 MeV and a current of 2.5 kA.

Potentially, the FELs have high radiation powers, and their efficiency can reach 20%.

In conclusion, we describe one of the possible applications of the technology of free-electron lasers and induction accelerators – the creation of high-energy accelerators (~1 TeV) [10]. It is based on the idea of a *two-beam accelerator* (TBA), proposed by A.M. Sessler in 1982. The TBA has two channels for transporting the electron beam. This is the central channel of a linear accelerator with an accelerated high-energy beam and a channel passing parallel to it, serving to generate high-power microwave radiation. The central channel of the accelerator is similar to the channel used in the Stanford Linear Accelerator Center (SLAC). It is a waveguide fed by microwave energy. The peculiarity of TBA is that in the second channel elements of the FEL, which is a source of powerful microwave radiation, are installed. The FEL power is periodically discharged along the undulator and feeds the resonators of the accelerator central channel. A free-electron laser is designed so that the increase in its microwave power per unit length is equal to the average power taken from a unit length.

The advantage of the FEL, in addition to its relative simplicity, is the unique ability to efficiently generate high power at very high frequencies. This eliminates the need for thousands of individual microwave generators, existing with the traditional execution of accelerators. Although the advantage of operating at higher frequencies has long been recognized, until the recent appearance of gyrotrons and FELs, there were no suitable powerful sources in the wavelength region ~1 cm. Recently, gyrotrons have been developed intensively and can in time turn out to be real powerful sources for accelerators. However, their maximum output power in the 1 cm range of wavelengths is likely to remain below 2 MW. Consequently, an accelerator of 1 TeV would require thousands of such sources synchronized properly. The idea of using FEL is precisely based on their ability to provide more than 100 MW of average power in a 1 cm range of wavelengths.

A schematic diagram of a two-beam accelerator is shown in Fig. 3.7. It consists of a high-gradient accelerator structure (HGAS), periodically connected by waveguides to the output elements of the FEL. Curiously, there is an inverse relationship between the basic functional concepts of FEL and HGAS. Indeed, the electromagnetic field of the FEL receives the input power from the electron beam, while the electron beam of the HGAS receives the input power from the microwave field. These relationships are similar to the principle

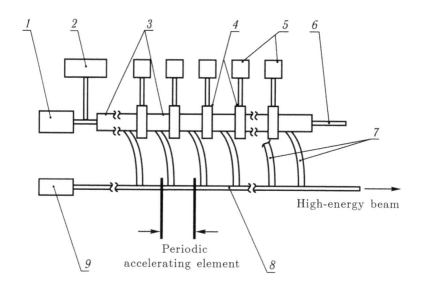

Fig. 3.7. Schematic diagram of TBA: *1* – low-energy electron beam accelerator; *2* – microwave source of low energy; *3* – modules of the undulator of an FEL; *4* – modules of the induction accelerator; *5* – energy sources for induction modules; *6* – low-energy beam collector; *7* – waveguides; *8* – high-gradient accelerator structure (HGAS); *9* – high-energy linear accelerator.

of the transformer: the microwave field is analogous to the magnetic field in it.

In the process of generating powerful electromagnetic radiation, an electron beam in the FEL loses its energy. However, the energy can be replenished by induction accelerators located periodically along the length of the FEL undulator.

Further research in the field of FEL applications in accelerators will be conducted in three directions. First, it is necessary to optimize the design of the tract for accelerating the FEL beam. Its total length in the beam transport path should probably be a few centimeters to avoid a serious reduction in the high rate of acceleration of the TBA. Secondly, it is required to improve the brightness of the beam. When accelerating individual electron bunches, it is difficult to achieve the desired brightness of the high-energy beam. The situation is simplified in the case of the acceleration of packets from many bunches. This mode of operation must be fully analyzed and optimized. Thirdly, the problems of phase stability and control must be solved, since an automatic phase-stabilization scheme with an almost instantaneous response is required. Perhaps this is the most difficult task in the design of TBA.

3.2. Relativistic klystrons

3.2.1. Klystron amplifiers and generators

The klystron is an electrovacuum microwave device, in which the conversion of a constant electron flux into a variable density flux is accomplished by modulating the electron velocities by a high-frequency electric field [11, 12]. Modulation occurs during the passage of electrons through the buncher gap of the cavity resonator and subsequent grouping into bunches due to the difference in their velocities in the drift space free from the microwave field. Two classes of klystrons are available: floating-drift and reflex. The reflex klystrons have not found application in relativistic high-frequency electronics because of the presence in their design of elements such as a resonator and a reflector, the possibility of operating at high power levels seems unlikely.

In the floating-drift klystron, the electrons consistently fly through the gaps of the tuned resonators (TRs). In the gap of the input TR, the electron velocities are modulated: the periodic electric field in it accelerates half-way, and the next half-cycle slows down the motion of the electrons. In the drift space, the accelerated electrons catch up with the delayed ones, as a result of which bunches form. Passing through the gap of the output TR, the electron bunches interact with its electric microwave field. Most of the electrons are inhibited and part of their kinetic energy is converted into the energy of microwave oscillations.

The idea of converting a constant electron flux into a flux of variable density due to the fact that accelerated electrons catch up with the slowed down ones was put forward by the Soviet physicist D.A. Rozhansky in 1932. The method of obtaining powerful microwave oscillations based on this idea was proposed by the Soviet physicist A.N. Arseneva and the German physicist O. Haile in 1935. The first designs of floating-drift klystrons were realized in 1938 by American physicists V. Khan and G. Metcalf, and, independently of them, R. Varian and Z. Varian.

Most of the transit klystrons are multi-cavity amplifying klystrons (Figure 3.8 *a*). Intermediate TRs, located between the input and output TRs, make it possible to expand the bandwidth, improve efficiency and gain. Amplifying klystrons are available for operation in narrow frequency sections of the decimeter and centimeter wavelength bands with an output power from several hundred W to 40 MW in pulsed and from several W to 1 MW in continuous modes. The gain of klystrons is usually 35−60 dB, the efficiency is from 40 to 60%, the bandwidth is less than 1% in continuous and up to 10% in the pulsed mode. The main applications of the klystrons are Doppler radiolocation, communication with artificial satellites of the Earth, radio astronomy, television (continuous mode klystrons), linear accelerators of elementary particles, final power amplifiers of long-range radar stations and high resolution (klystrons of pulse mode of operation).

A small portion of the commercially produces klystrons comprise generating klystrons with continuous operation (Fig. 3.8 *b*). They usually have two cavity resonators. In these devices, a small fraction of the microwave power produced in the second TR is transmitted through the coupling slot to the first TR for modulating the electron velocities. Their output power is about 1 to 10 W, efficiency – less than 10%. The generator klystrons are used mainly in parametric amplifiers, as well as radio beacons of centimeter and millimeter wave bands.

The theoretical dependence of the efficiency on the microperveance of the electron flux η, as well as experimental data on the efficiency

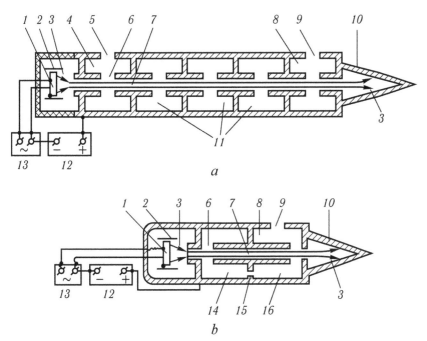

Fig. 3.8. Constructions of amplifying (*a*) and generator (*b*) klystrons: *1* – cathode; *2* – focusing cylinder; *3* – the electron beam; *4* – input of tuned resonators (TR); *5* – hole for the input of microwave energy; *6* – buncher gap; *7* – the drift space; *8* – output of the TR; *9* – a hole for the output of microwave energy; *10* – collector of electrons; *11* – intermediate TR; *12* – source of constant anode voltage; *13* – source of cathode heating voltage; *14* – the first TR; *15* – the coupling slot, through which part of the microwave energy passes from the second resonator to the first; *16* – second TR.

of single-beam and multi-beam klystron amplifiers are shown in Fig. 3.9. The above dependence was proposed by A.N. Sandalov of the Physics Department of the Moscow State University in 1975 (published in 1983 [14]) on the basis of calculations using the Klystron-MSU software complex. This complex makes it possible to estimate the possibility of obtaining the maximum value of the efficiency at different values of the microperveance. The experimental data shown in the graph were obtained on klystron amplifiers (single-beam and multi-beam) developed in various laboratories of the world. For many years, up to the publication by the researchers of the Istok Research Institute in 1993 of the work on multi-beam klystrons, Sandalov's dependence was regarded as not corresponding to the physics of processes in klystron amplifiers. However, at the present time this is the most popular dependence of the efficiency on microperveance, which reflects real physical processes occurring in

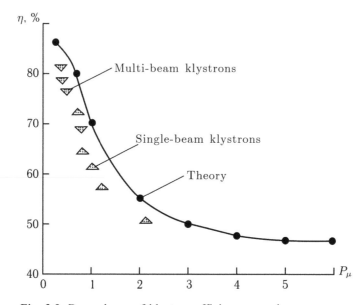

Fig. 3.9. Dependence of klystron efficiency on microperveance.

single-beam and multi-beam klystron amplifiers. As can be seen from the graph, the maximum efficiency and the amplification band of the klystron amplifiers are achieved at values of the microperveance less than 0.5 and the optimal choice of the drift tube lengths, the electrodynamic characteristics of the resonators and the parameters of the space charge forces of the electron beam ω_p/ω and $k = \alpha v/(\omega r_b)$, where ω_p is the plasma frequency of an infinitely wide electron flux; α is the solution of the transcendental equation for an electron beam of radius r_b in a drift tube of radius r_{tube}; v is the speed of the electron beam). The parameter α varies from 1.2 to 2.0. In one-dimensional models of the electron beam, it is usually assumed to be equal to 2. The value of this parameter plays an important, if not decisive, role in the comparative analysis of the experimental and theoretical characteristics of powerful devices. These considerations served as the basis for calculating the relativistic klystron amplifier. For the first time, studies of klystrons with relativistic electron energies of the beam were carried out in the early 90s of the last century. A relativistic klystron (RK) of a 3 cm wavelength range was tested with a design output microwave power of about 40 MW on a linear induction accelerator [13]. The klystron was calculated and designed at the Physics Department of the Moscow State University, and experiments were performed on the LIA of the Tomsk Polytechnic

University. The experimental setup is described below and the results of the first experiments with this device are presented.

3.2.2. KMT-3 relativistic klystron amplifier

When developing the klystron, it was taken into account that relativistic electron fluxes are mainly circular. The increase in the voltages and beam currents is automatically accompanied by an increase in the transverse dimensions of both the electron beam and the electrodynamic system of the devices. On the other hand, an increase in the accelerating voltage is accompanied by an increase in the total length of the devices (proportional to γ^2), which can complicate the system of electron beam formation.

The relativistic klystron amplifier KMT-3 (Fig. 3.10) was designed for experiments in the 3 cm wavelength range. In the early 90s of the last century, this range was rarely used by devices with explosive-emission cathodes, but was well mastered by klystrons with thermionic cathodes. The klystron was designed to work at an accelerating voltage of 400 kV and a beam current of 200 A, formed by a dielectric cathode to obtain a better quality of the electron beam than in the case of an explosive-emission cathode.

To study a high-efficiency klystron amplifier, it is necessary to analyze physical processes in all its parts from the cathode to the collector: in an electron gun, a linear amplifier, a non-linear buncher, an output section and a collector. For these purposes, the Klystron-MSU and Arsenal-MSU software packages were developed at the Physics Department of Moscow State University [15−17]. The Klystron-MSU software is 1.5-dimensional. The interaction of the electron beam and the electromagnetic field is realized in it by means of the disk–circular model of the electron beam. The Klystron-MSU was created by A.N. Sandalov for studying physical processes and designing powerful narrowband and broadband klystron amplifiers

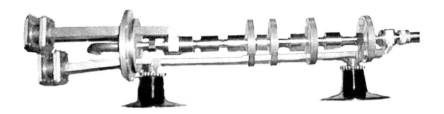

Fig. 3.10. The KMT-3 relativistic klystron.

with high efficiency. On its basis, the Moscow State University created the 2.5-dimensional software complex Arsenal-MSU, which allows analyzing klystron amplifiers, containing electron guns (both thermionic and explosive emission), linear amplifiers and non-linear klystron bunchers, distributed output structures, convection collectors and collectors with recuperation.

These software products were used to calculate the relativistic klystron KMT-3 (Fig. 3.11). It contains one input single-gap resonator. The linear amplifier forms a band of amplified frequencies and provides the required gain level. It consists of single-gap resonators tuned to the fundamental frequency and its second harmonic. The number of resonators and the length of the device determine the gain level and the output power. Klystron KMT-3 contains three buncher resonators (including one at the second harmonic of the fundamental frequency) and a two-gap output resonator. Waveguides of standard cross-section are introduced to the entrance and exit ceramic windows, which are led to the area of the klystron collector. The collector of electrons in the klystron KMT-3 is of the convective type of an axially symmetric construction. The latter should ensure a uniform distribution of the exhausted electrons over the surface of the collector. The focusing system of the klystron amplifier performs the formation and transportation of the electron beam under the action of a solenoidal magnetic field. The drift tubes must ensure the transport of the electron beam without current to the walls of the electrodynamic structure. Therefore, the coefficient of filling of the drift tube by an electron beam $k = r_b/r_{tube} = 0.7-0.8$. It depends on the focusing magnetic field: $B = B_z/B_B = 2.0-3.0$, where B_B is the value of the Brillouin magnetic field (the minimum magnetic field necessary to focus the electron beam).

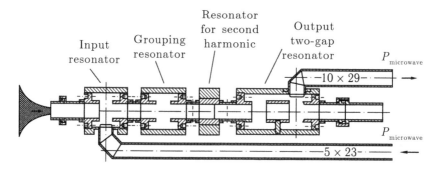

Fig. 3.11. The construction of the KMT-3 relativistic klystron.

The results of modelling KMT-3 are shown in Fig. 3.12. The maximum design efficiency of the relativistic klystron amplifier KMT-3 was 40%.

Initial experiments have shown that the most difficult moment for the relativistic klystron is the extraction of high-frequency power from the output resonator. To avoid shortening the pulse duration and very high gradients of the high-frequency field of the electromagnetic wave, the interaction space between the wave and the electron beam must be extended (larger than for conventional one- or two-gap resonators).

The extended output circuit can be a multi-gap resonator or various types of resonators with diaphragm coupling loaded onto the waveguides. The output structure of KMT-3 was designed as a two-gap resonator. Preliminary estimates were made of the electric field strength in a single-gap and two-gap resonators. Figure 3.13 shows the results of this calculation using the Arsenal-MSU software (the output system is represented here as four- and six-pole network). In the presence of a single-gap resonator, the efficiency of the klystron was 46%. The electric field strength at the tip of the gap was 600 kV/cm. With a two-gap resonator, the efficiency of the klystron increased to 55%, and the electric field strength dropped 1.5 times. The electric field distribution in the output resonators is shown in Fig. 3.13 on the right. In accordance with the calculations, to improve the reliability of the output mode, a two-gap design of the output resonator is adopted.

After the manufacture of the klystron, cold tests were carried out. The resonator frequencies were set close to the theoretical values.

3.2.3. Dielectric emitter of the injector of the LIA

Experiments with the klystron were performed on the LIA, designed to form and further use high-current relativistic electron beams formed by a cathode with full or partial magnetic insulation. The corresponding experimental setup is shown in Fig. 3.14. Its accelerator contains the injector and accelerating sections, made on the basis of the layout scheme of the Tomsk Polytechnic University (see Chapter 2).

In the injector section, a high-voltage electrode is located along the axis, while in the accelerating section there is an anode drift tube with an impulse solenoid placed on the outer surface. A klystron amplifier is installed inside the anode drift tube.

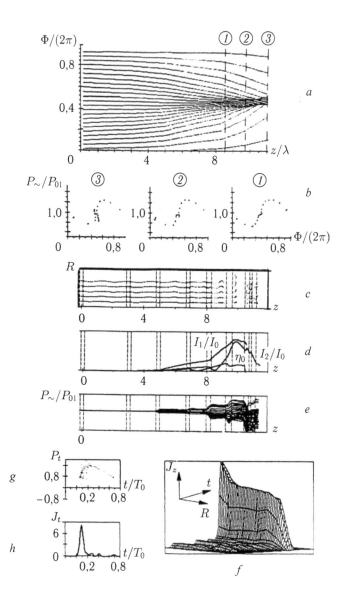

Fig. 3.12. Results of klystron KMT-3 calculations: *a*) phase diagram; *b*) formation of a bunch of electrons in the last drift tube; *c*) a schematic drawing of the interaction space and instantaneous photograph of the position of the particles in it; *d*, *e*) the process of bunching, increasing the flow of microwave power and energy exchange along the drift tube; *e–h*) the resulting parameters of the grouped electron beam at the entrance of the output resonator (axial distributions of particles (*f*), current density (*g*) and the shape of a bunch of electrons in three-dimensional space in the form of the dependence of the density on the radial position and time during the period of the microwave field (*h*).

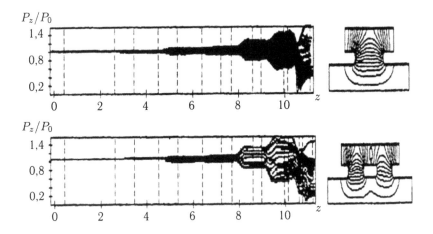

Fig. 3.13. Results of calculations of KMT-3 with single-gap and two-gap resonators.

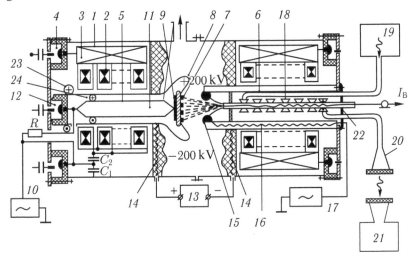

Fig. 3.14. Scheme of installation for Klystron KMT-3 studies.

The induction system of each section of the LIA contains seven ferromagnetic inductors *2* with the dimensions of 250 × 110 × 50 mm made from a tape of the permalloy 50 NP with a thickness of 0.01 mm. Before the operating cycle, the cores are demagnetized by pulsed currents through the spiral inductance *14* from the source *13*. The high voltage is formed when the strip DFL *3*, with a wave resistance of about 0.3 ohms. Forming lines are commutated by multichannel (18 channels) dischargers *4* with forced division of the current between the channels, which are located on the ends of the sections. The electrodes of the dischargers are in a compressed nitrogen medium. An induced voltage of 400 kV at the discharge of the DFL on the cathode holder *5* and on the drift tube *6* is applied between the cathode and the anode. The geometry of the cathode–anode gap is similar to the system with Pierce optics with a fully shielded cathode. A fundamentally new cathode, a dielectric emitter (DE), was used as the electron source [18]. When a high voltage appears on the cathode–anode gap, electrons form in the electron-optical system. Then, through the aperture of the anode *15*, they enter the region of the increasing magnetic field produced by the solenoid *16*, powered by a pulsed power supply *17*. The input microwave power for the relativistic klystron amplifier *18* comes from the magnetron generator *19* (Tesla 58SP52). The power from the output resonator is emitted by the antenna *20* and received by the detector *21*. The Faraday cylinder *22* serves to register the current of the electron beam, and Rogowski's coils *23* and *24* are used to measure the total current of the accelerator and the discharge current of the dielectric emitter.

Dielectric emitter in the injector of a linear induction accelerator. The name 'dielectric emitter' was given to the cathode by E.G. Furman, who proposed a description of the principle of its work based on the preliminary accumulation of a charge of electrons on the dielectric surface and their subsequent injection [18]. This cathode is used to form electron beams, including for the relativistic klystron amplifier. It allows to obtain a maximum current of 1.5 kA with an accelerating voltage of 400 kV.

The basis for the development of a new type of cathode was the fact that the effective application of accelerator technology, including for relativistic electronic devices, is largely determined by the capabilities of electron sources. The thermionic cathodes have limitations in current density, low efficiency, are difficult to operate, and traditional explosive-emission cathodes do not allow the formation of high-quality electron beams. Advantages of these types of cathodes are united by a dielectric emitter (DE), which makes it possible to obtain electron beams of good quality at low energy and operating costs. Required for generating an electron beam pulse of duration τ with the current amplitude I_b the electron charge $Q = \int\limits_0^\tau I_b dt$ is collected on the surface of the DE and is held there by the Coulomb forces of the positive charge of the opposite metallized DE plate. When an accelerating voltage pulse is applied to the cathode–anode gap, the accumulated electrons are removed from the DE surface and an electron beam is formed. Structurally, the dielectric emitter is a ferroelectric disk (barium titanate $BaTiO_3$) with a metal plate 9 (Fig. 3.14). The emitter is made on the basis of the capacitor (40 kV, 10 nF, $\varepsilon \sim 1300$), at which one cover is removed. The working surface diameter 7 of the DE is 80 mm. At its edge, a ring is left from the metallized lining, which acts as a field electron emitter for depositing a charge on the surface. The ring is connected to the focusing electrode 8. To prevent breakdown along the side surface, the edges of the DE are filled with an epoxy compound.

Figure 3.15 shows oscillograms of current and voltage in the dielectric emitter circuit. When the impulse source is switched on, the charge of the capacitors of the DFL begins (curve U_0). In parallel, the potential U_{em} is applied to the 9 DE plate through R. Starting from time t_1, the electric field strength at the edge of the electrode 7 reaches a critical value ($U_{em} \approx 200$ V) and a charge is deposited due to the field emission current I_{em} from the tip on the surface of the DE 7. Electrons are held at the surface 7 by a positive charge of the electrode 9. The charge accumulated in the device under consideration on the DE surface is

$$Q = \varepsilon\varepsilon_0 E_{em} S = C_{em}\left(U_{em} - U_{th}\right), \qquad (3.6)$$

where ε, E_{em} is the permittivity of the DE material and the electric field strength in it; $S = 50$ cm^2 is the surface area of the DE; C_{em} is the capacity of the DE; U_{em} is the voltage on the DE plate; $U_{th} -$ is the autoemission threshold.

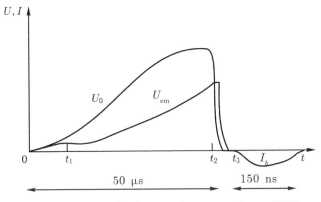

Fig. 3.15. Oscillograms of voltage and current pulses of DFL and DE.

To form a beam current pulse with an amplitude I_b and a duration τ on the DE surface, it is necessary to accumulate a charge Q, the value of which for a given U_0 is regulated by a current-limiting resistor R. When charging the forming lines to the required voltage and accumulating sufficient charge on the DE surface (time t_2) multichannel dischargers *3* of the cathode and anode sections are triggered. After a time, an electromagnetic wave run on the arm of the DFL ($\tau = 50$ ns) in the cathode–anode gap induced by the accelerating electric field E_{acc}, the plate of the DE *9* is connected to the accelerator case and, consequently, to the cathode holder *5* and the focusing electrode *8*. At the same time, the following electric fields act on the electrons accumulated on the working surface of the DE: the accelerating electric field E_{acc} on the side of the anode; the field of positive charges E_{em} of the plate *9* from the dielectric. If condition $E_{acc} > E_{em}\sqrt{\varepsilon}$ is satisfied, the accelerating field compensates the field of the positive charges of plate *9*, the electrons are removed from the DE surface and are trapped in acceleration.

Experiments have shown that the duration of the beam current pulse was approximately equal to the time of double travel of the voltage wave along the DFL. The duration of the pulse front was determined by the inductance of the switching circuit of the DE. With a charging voltage $U_0 = 29$ kV and an accelerating voltage $U_{acc} = 400$ kV, the beam current reached a value of 1.5 kA. The magnetic field was 1200 G in the drift tube, 150 G at the anode cutoff, 20 G at the cathode. A beam injected from a cathode 80 mm in diameter into a drift tube with a 68 mm aperture was compressed to a diameter of 15 mm in an increasing magnetic field and transported practically without loss at a distance of 1 m. The maximum current density obtained from the cathode was 30 A/cm².

The authors carried out a number of experiments, confirming, in their opinion, the described principle of the DE operation. At $U_{em} = 0$ (the circuit of the plate *9* is open or open and shorted to the cathode holder *5*), the DE surface is not charged; the beam current reaching the collector is 10–50 A, in spite of the fact that the voltage in the cathode–anode gap reaches up to 700 kV. In this mode, the beam current is due to spurious emission from the metal surfaces of the cathode assembly.

On the basis of experimental studies, the following conclusions were drawn. The dielectric emitters allow the construction of guns that form electron beams with a current of one kiloamperes in a nanosecond pulse duration range. To obtain electron beams, one can use Pierce's electron-optical systems with a suitable choice of electric and magnetic fields. At the same time, at minimal energy costs, it is quite simply that emission surfaces with an area of hundreds of square centimeters of practically any configuration are realized.

In conclusion, we note that in the late sixties of the last century an original method was proposed for creating a dense plasma with an excitation independent of the accelerating field [19]. It consists in creating an incomplete discharge on the dielectric surface in a vacuum. The main element of the emitter is disk *2* of barium titanate (Fig. 3.16). On the one hand, the metal needle *3* is pressed against the disk, and the contact metal layer is applied to the other. The discharge occurs when a pulsed voltage U_p, exceeding a certain threshold value, is applied. Owing to the high dielectric permittivity of the dielectric, the discharge voltage is several hundred volts. It was shown in [20] that when a negative potential is applied to the point, the discharge arises when the dielectric evaporates under the action of bombardment by electrons emitted by the tip due to field emission. Plasma discharge is created in the vapour of the destroyed surface layer of the dielectric. After ignition of the discharge, the plasma propagates along the dielectric surface at a velocity $v = AU_p$, where $A = 21$ m/(V · s) with a negative polarity of the needle relative to the contact layer. Simultaneously with the motion of the plasma along the dielectric, its propagation is perpendicular to the dielectric surface at a velocity of 2×10^6 cm/s. The expanding plasma is an effective emitter of the electrons extracted and accelerated by means of the electrode positive relative to the needle *4*. The start of emission coincides with the accuracy to 10^{-9} s with the occurrence of the plasma at the tip. The current extracted from the plasma rises

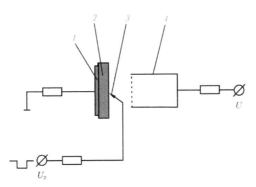

Fig. 3.16. Scheme of the plasma source: *1* – contact layer; *2* – barium titanate disk; *3* – the needle; *4* – accelerating electrode.

with increasing voltage U_p, which is explained by an expansion in the plasma surface emitting the electrons.

The use of a small-structure grid instead of the needle pressed against the dielectric surface allows one to simultaneously create a large number of emitting centers due to discharges over the surface of the dielectric at the points of contact with its grid. As a result, the total area of the plasma increases. An electron accelerator with a cathode coated with a grid 40 mm in diameter was described in [21]. With an accelerating voltage of 500 kV and a pulse duration of 25 ns, it provided a current of 10 kA.

The described mechanism of operation of the dielectric emitter seems more plausible than that presented in [18]. In addition, in the future it was experimentally discovered that there is no direct relation between the discharge current of the DE and the beam current. First, if the discharge current was of a constant shape, then the beam current varied from shot to shot. Secondly, the charge carried away by the electron beam was 8−10 times less than the charge that occurs when the DE is discharged. Thirdly, the pulses of the beam current appeared much later than the discharge current of the DE. The current removed from the DE was maximal when its origin coincided with the end of the discharge current. In this case, the deviation of the beginning of the discharge current in either direction led both to a decrease in the maximum value of the electron beam current and to a deterioration in the shape of the pulse. All these data indicate that the part of the charge, accumulated on the surface of the DE, is not involved in acceleration and leaves it in a different way. The only way in which the charge can leave the emitter, avoiding acceleration, is the DE charge circuit. At the same time, a surface discharge appears on the surface of the DE, resulting in the formation of a plasma. It is electrons from this plasma that are involved in the acceleration process. This explains why the beam current is maximal when its origin coincides with the end of the discharge of the DE when its beginning coincides with the end of the discharge of the DE.

Regardless of what mechanism actually takes place, the DE showed its efficiency. The dielectric emitter was used in experiments with a relativistic klystron amplifier.

3.2.4. Measurement of the characteristics of the electron beam

The geometry of the electron gun of the klystron and the distribution

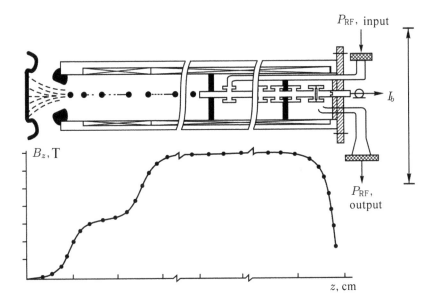

Fig. 3.17. The geometry of the electron gun and the axial distribution of the magnetic field.

of the magnetic field on the axis of the accelerator are shown in Fig. 3.17. The results of the calculations are shown in Fig. 3.18. As can be seen from the graphs, the radial amplitude of the electron oscillations in the beam is very high. The transverse components of the electron pulse in it are also high. These facts indicate a high longitudinal dispersion of energy in the electron beam, as well as a poor quality of the electron beam. Thus, it can be noted that the design of the electron gun was not optimal, it required computer optimization and modernization. Measurements of the distribution of the magnetic field were performed without the klystron. The magnetic field of the solenoid could increase to 0.7 T with similar characteristics of the axial distribution. It was possible to adjust the length of the anode–cathode gap from 40 to 70 mm (see Fig. 3.14).

Figure 3.19 shows electronic prints on targets made of plastic at various points of the anode channel. All prints were made for ten current pulses with the same mode of operation of the accelerator. The letters A, B, C, D, E, F, G correspond to the points shown in Fig. 3.17, and characterize the position of the target in the channel of the anode section of the LIA. The magnitude of the axial magnetic field in these measurements was 0.7 T (the cathode–anode gap was 70 mm). The measurements showed that a decrease in the magnetic field to 0.35 T does not lead to changes in the prints (the beam current does not change). If the magnetic field is less than 0.35 T, significant changes in the structure of the electron beam are observed.

The electron beam is characterized by a density distribution in the cross section of the aperture of the anode drift tube. This distribution was measured experimentally (Fig. 3.20). At the point $G = 58$ cm, the peripheral part of the electron beam was cut by successively

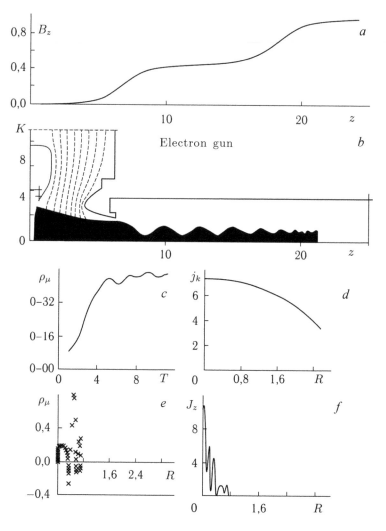

Fig. 3.18. The results of the calculation of the electron gun of the LIA: *a*) the profile of the axial magnetic field; *b*) position of electrodes, electron trajectories and equipotential lines; *c*) the dependence of microverveance on the iteration number; *d*) actual density of current at the cathode; *e*) the radial component of the kinetic energy of the electrons; *f*) radial distribution of current density in the outlet section.

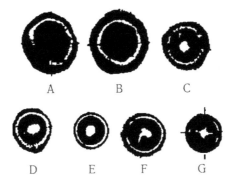

Fig. 3.19. Beam prints along the anode channel.

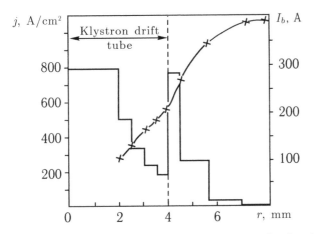

Fig. 3.20. The distribution of the electron beam current density along the radius.

setting diaphragms of different diameters. The part of the electron beam that passed the aperture of the anode and the internal aperture of the diaphragm was measured by a Faraday cylinder installed after the diaphragm. All presented results are averaged over ten pulses. The pulsed electron current deviated in amplitude from pulse to pulse by no more than 15% of the mean value. The total current was 400 A, which corresponds to the calculated data. Two regions of the electron beam were observed: central and annular with a larger current density at diameters of 8–9 mm.

Figure 3.21 shows the energy spectrum of the beam electrons, measured by the absorption of electrons in a foil. A set of titanium foils 20 μm in thickness was used for spectral measurements. These measurements do not claim to be highly accurate, since the

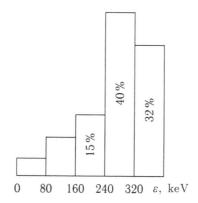

Fig. 3.21. Energy spectrum of the beam.

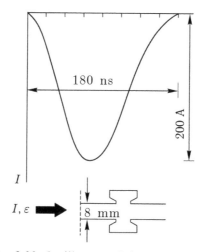

Fig. 3.22. Oscillogram of the current pulse.

beam electrons had transverse velocities. Figure 3.22 shows the characteristic oscillograms of the current pulses at the input of the relativistic klystron (point *G*, diaphragm 7.9 mm in diameter). The current of the electron beam passed through the relativistic klystron was measured with a Faraday cylinder with a diameter of 7 mm installed at a distance of 2 cm behind the output resonator. If the measured current at the input of the relativistic klystron was 200 A, then at the output of the relativistic klystron it was equal to 80–100 A. When evaluating the electron beam transport factor through the relativistic klystron, it is important to take into account the fact that the output current measurements were made by a Faraday cylinder placed at a point where the focusing magnetic field was smaller than in the output cavity.

In the experiments, the electron beam was transported through the relativistic klystron in the absence of an input microwave signal. At the same time, 'parasitic' microwave oscillations with a power of 500 W were detected in the output resonator of the relativistic klystron. These oscillations existed when the relativistic klystron current exceeded 200 A. When the input signal from the magnetron had a power of 50 kW with a frequency of 9.28 GHz and a duration of 1.2 μs, the peak power of the relativistic klystron output signal reached 500 kW with an electron beam current of 70 A. The output frequency of the relativistic klystron was unstable, since the calculated characteristics of the resonators (9.4 GHz) and the operating frequency of the magnetron (9.28 GHz) did not coincide.

Experiments to transport the electron beam have revealed the need to improve the electron-optical system of its formation. The current-carrying coefficient of the electron beam was ~50%, which is clearly insufficient for the optimal operation of the relativistic klystron amplifier.

The characteristics and design features of the KMT-3 klystron are given in Table 3.1.

3.2.5. Relativistic klystron amplifier of the Physics International Company

A high-current relativistic klystron amplifier was tested at the Compact Linear Induction Accelerator (CLIA) at the Physics International Company, USA [22]. The researchers were faced with the task of obtaining a peak power of 1 GW in a packet of 100 pulses at a repetition rate of 100 Hz. As in the experiments of the Tomsk Polytechnic University and the Moscow State University, the uniform beam, usually used in klystrons, was replaced by a thin coaxial electron beam with a radius almost equal to the radius of the drift tube. In addition, the microperveance of the electron beam was chosen much higher. In this case, the beam has a significant fraction of the total energy stored in the electrostatic potential of the space charge. The interaction of the beam with an intense high-frequency field may well have a pronounced nonlinear character, since the beam current is close to the current, limited by the space charge. A tube electron beam with a current of 5 kA, an average radius of 1.8 cm, and a thickness of 0.2 cm was formed by a magnetically insulated diode with a voltage of 500 kV from the CLIA to the graphite cathode. The beam passed through two chambers of the

Table 3.1. Characteristics and design features of KMT-3 klystron

Voltage (project / experiment), kV	400/400
Beam current (project / experiment), A	200/170
Type of cathode	Dielectric
Microperveance	0.7
Rated frequency, GHz	9.4
Frequency (input / output), GHz	9.4 / 9.28
The pulse duration of the beam / microwave radiation, ns	120/120
Number of resonators of frequency ω / frequency $2\,\omega$	6/1
Output section	Two-gap
Focusing system	Solenoid
Magnetic field, kG	5
Type of collector	Conic
Efficiency (theory / experiment),%	50/5
Gain, dB	30

output resonator of the relativistic klystron [23], provided by M. Friedman and V. Serlin from the US Naval Research Laboratory (Fig. 3.23). The radius of the drift tube was 2.38 cm. A 30 cm magnetron with a power of 500 kW provided an input signal with a frequency of 1.32 GHz. The input cavity was $3\lambda/4$. Transportation of the beam by a magnetic system was provided by a longitudinal magnetic field with an induction of 0.6–1 T. The length of the solenoid was 1 m, the internal radius was 0.11 m. The current modulation of the beam current was measured by a point induction sensor installed on the drift pipe section.

The initial efforts of the experimenters were aimed at optimizing the position of the resonator to obtain the maximum modulation of the beam and elucidating possible constructive limitations of the work of the relativistic klystron in the frequency regime. Since the frequency mode of operation requires a low-pressure vacuum system throughout the whole packet of pulses, one of the main features of the relativistic klystron is a good vacuum technique. Copper seals were used in almost all compounds. Hard-to-reach places were pumped through specially made holes. In the joints, a good electrical

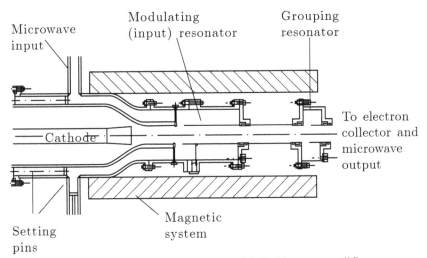

Fig. 3.23. The construction of a relativistic klystron amplifier.

contact was provided to prevent sparking. As a result, the vacuum in the system was maintained at 10^{-8} Torr.

The electron beam collector used in the operation of the relativistic klystron in a single-pulse mode was a flat graphite disk installed in the drift tube directly at the resonator outlet. In the pulsed-periodic mode, instead of the disk, the beam was deposited on the inner surface of a stainless steel tube 15 cm in diameter, which was cooled by liquid nitrogen. In order to transfer the electrons from the cathode to the input cavity, a conical transition was developed taking into account the profile of the magnetic field and the parameters of the electron beam. The opening angle was chosen to be large enough so that the beam propagated along the magnetic field lines without falling onto the walls, but not too large, so that the beam was confined to its own space charge creating the virtual cathode.

Another element designed for a single-pulse mode – a set of radial wires in the input cavity to select the operating mode – proved to be unsuitable for operation in a packet mode [24]. At the front and in the edge of the current pulse through the wires, the reverse current was passed due to the inductor of the resonator ~20 nH (usually in the experiment at the location of the wires with $dI/dt = 5 \cdot 10^{11}$ A/s a voltage ~10 kV was generated). The resulting heating caused degassing, embrittlement and rupture of wires. Broken wires were sources of arc or electrical breakdown. As a result, the incoming microwave signal measured in the first resonator gradually became shorter during the burst of pulses. As experiments have shown, the wires can be removed without appreciable deterioration of the beam bunching.

In the course of the experiments, the position of the resonators was optimized to obtain microwave radiation with reduced electric parameters of the power source: 5 kA and 500 kV (CLIA can operate at a current of 10 kA and a voltage of 750 kV [25]). The optimum position of the second resonator with respect to the input cavity (33 cm distance) was determined as the point of maximum bunching. This distance has a significant effect on the work of the relativistic klystron, which is well modelled by the linear theory and is confirmed both by the previous [22] and by the described studies. Another important parameter affecting the process of bunching of electrons is the radius of the beam. It was 1.8 cm. Taking into account the magnetic field profile of the solenoid, it was possible to control the radius of the beam by changing the position of the magnet relative to the cathode. Such a method of changing the beam diameter is much simpler than replacing the cathode with a cathode of a different radius.

The authors of [22] measured the modulation of the electron beam at a distance of 10 cm from the second resonator in the direction of propagation of the beam. For this purpose, a high-frequency output resonator was designed, which was installed inside the magnet in front of the conical collector. The modulated beam current was measured by a point B-detector on the drift tube section. The high-frequency signal of the sensor was filtered by band-pass filters and was recorded directly on a high-frequency oscilloscope with a bandwidth of 6 GHz. In another measurement method, the sensor signal was sent to the detector and recorded with a digital oscilloscope with a bandwidth of 350 MHz. The results obtained by these two methods are in good agreement (Fig. 3.24 *a*). The modulated current was 2.5–3.0 kA with a full injected current of the electron beam of 4.5 kA. The authors of [22] note a good optimization of the length of the gap between the resonators.

These results were also repeated well in the pulse-periodic regime of the relativistic klystron. Figure 3.24 *b* shows the modulated current for several pulses when the klystron operates at a frequency of 100 Hz in a burst mode of 100 pulses at a current of 4.5 kA. After some transient process, typical for the first about 20 shots, the output pulses became fully reproducible with a current modulation of 3.0–3.5 kA. During the experiments, the number of pulses was increased tenfold (up to 1000), and the frequency of their repetition was doubled (up to 200 Hz) while maintaining high reproducibility of the results. The modulated current was 2.0–2.5 kA at a total current of 3.5 kA. The decrease of the modulated current in comparison with that shown in Fig. 3.24 *b* was due to a decrease in the voltage at the cathode and, consequently, of the total current at a higher pulse repetition rate. In addition, the decrease in the modulation efficiency was due to the fact that the gap between the resonators was not optimized for these beam parameters. The results obtained,

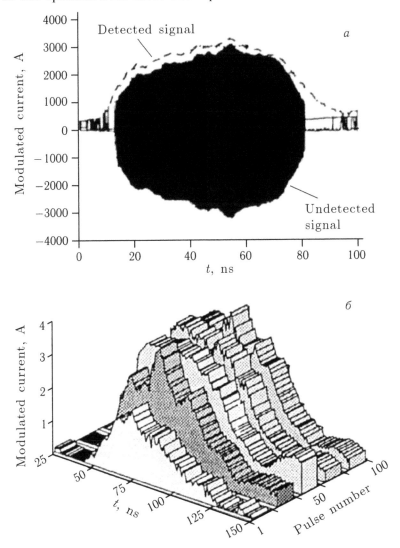

Fig. 3.24. The current measured by the high-frequency oscilloscope and the semiconductor detector (*a*). The amplitude of the modulated current for the selected packets (100 pulses with a frequency of 100 Hz) at a total current of 4.5 kA (*b*).

however, clearly demonstrate the absence of any restrictions on the beam modulation process in the generation of a long packet at a relatively high pulse repetition rate in a relativistic klystron. The pulse power of the relativistic klystron amplifier reached 200–250 MW with a pulse duration of 80 ns at a frequency of 1.3 GHz with a repetition rate of 200 Hz [24].

3.3. Resonant relativistic microwave generators

Section 3.1 describes an experiment on the LIA, in which the stimulated undulator radiation of particles moving in a spatially periodic magnetic field was used. In this section, based on the publication of Ref. [26], test experiments were carried out at relatively low particle energies in one section of the LIA of the collective electron accelerator of the Joint Institute of Nuclear Research [27], in which stimulated Cherenkov and cyclotron radiation at wavelengths of 9–12 mm was obtained with the help of relativistic orotron and gyrotron.

3.3.1. Orotron and gyrotron

An electrovacuum device in which, as a result of the interaction of an electron beam with the periodic structure of an open resonator, electromagnetic waves are excited, is called the orotron. The orotron is used as a generator of millimeter and submillimeter waves, mainly in radio spectroscopy for physical and biological research. The orotron was developed by Soviet physicists F.S. Rusin and G.D. Bogomolov in 1965 [28]. It is an open resonator, one of whose mirrors has a periodic structure (Fig. 3.25). The electron flux is in a focusing magnetic field directed parallel to the motion of the

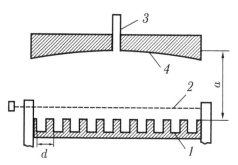

Fig. 3.25. Orotron design: *1* – mirror with periodic structure; *2* – electron flux; *3* – waveguide; *4* – mirror (*a* – distance between mirrors, *d* – period of structure).

electrons. The length of the generated waves λ is determined by the electron velocity v and the period of the structure d ($\lambda = dv/c$), as well as the distance a between the mirrors of the resonator ($\lambda \approx 2$ a/m, where m is the number of half-waves that fit between the mirrors). Thus, the orotron uses the induced diffraction radiation of electrons moving uniformly and rectilinearly along the periodic structure. As in other O-type instruments, in the orotron, the interaction of electrons with the microwave field (their bunching, slowing-down) occurs when the average electron velocity and the phase velocity of the spatial harmonic of the periodic structure coincide. The output of electromagnetic waves from the orotron is carried out either through a waveguide, or directly by radiation into free space. In the latter case, the mirror opposite to a mirror with a periodic structure should be slightly transparent to the generated waves. Frequency tuning is carried out in a band exceeding an octave, or continuously (by simultaneously changing the voltage, accelerating electrons, and moving one of the mirrors, i.e., changing the distance a), or discretely (with m changing). The collector of the orotron collects electrons passing through the resonator.

A high-Q open resonator provides the appearance of a positive feedback between the electromagnetic field and the electronic flow necessary for self-excitation of microwave oscillations. The orotron is self-excited analogously to the O-type backward-wave tube when the beam exceeds the threshold value. The orotron is one of the first electrovacuum devices using a resonant system similar to a laser in the form of a cavity resonator whose dimensions are much larger than the wavelength. Devices of this type are sometimes called generators of diffraction radiation. Advantages of orotrons are high frequency stability and low level of amplitude and frequency noise (an order of magnitude lower than in klystrons or backward wave tubes), disadvantages include a strong unevenness of the output power in frequency with frequency tuning, as well as the complexity of manufacturing reflecting gratings (RGs) in the short-wave part of the millimeter range.

The idea of creating gyrotrons (or masers at cyclotron resonance) was proposed in the early 60s of the last century by Academician A.V. Gaponov-Grekhov, and then realized by a group of Russian scientists (I.I. Antakov, M.I. Petelin, V.A. Flyagin, V.K. Yulpatov and others). A gyrotron is a microwave generator with an open coaxial resonator, in which one of the resonator modes is excited by means of a beam of electrons moving along helical trajectories

(Fig. 3.26). The device is based on the phase bunching under the influence of the electromagnetic field of an ensemble of non-isochronous electron-oscillators, in which the oscillation frequency in a magnetic field depends on the electron energy. Because of this dependence, the beam electrons rotate in a magnetic field with different cyclotron frequencies and, interacting with the wave, move with respect to each other in phases, forming a bunch. The bunch is formed precisely in the decelerating phase of the field, transferring energy to the wave [29]. Among masers with cyclotron resonance (MCR), the most promising are: for weakly or moderately relativistic energies – gyrotrons (in which orbital bunching dominates); at ultrarelativistic energies – the MCR with commensurate efficiencies of longitudinal and orbital bunching. Relativistic effects began to be used in microwave electronics in the late 1950s. The possibility of creating relativistic electron microwave generators of increased power arose in the late 1960s due to the appearance of high-current electron accelerators generating electron beams with energies of 0.5–2 MeV and currents of 1–100 kA.

The gyrotrons consist of an adiabatic gun of a magnetron type, an open resonator with a diffraction output of microwave energy and an output unit that includes an electron collector and, as a rule, a converter of microwave radiation to a linearly polarized wave beam (Fig. 3.26). Strong magnetic fields, necessary for gyrotron operation, are created by superconducting solenoids [30]. The magnetron–injector gun must ensure the formation of a stable helical electron beam with an oscillatory energy share of 60–70% of the total energy at a current of 40–50 A. The speed difference should not exceed 30%. The shape of the gun electrodes is at the first stage of calculations based on the adiabatic theory, and then it is optimized by numerical methods. Transporting an electron beam from a gun through a gyrotron resonator to a collector is a difficult task, since the effects caused by the space charge and its ion compensation play an important role, which are much more pronounced than in the case of rectilinear electron beams. The oversized resonator in the form of a segment of a weakly irregular waveguide (the diameter of the waveguide is much larger than the wavelength) serves as the interaction space of microwave fields with a helical electron beam. As the radius of the resonator increases, the spectrum of the natural waves thickens and difficulties arise in setting the working mode when the gyrotron is turned on. There are also difficulties in removing heat from the walls of the resonator, which arises from

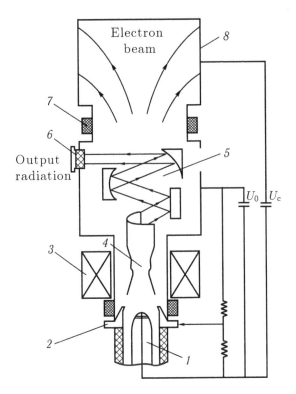

Fig. 3.26. General scheme of the gyrotron: *1* – cathode; *2* – anode; *3* – solenoid; *4* – resonator; *5* – quasi-optical converter; *6* – window; *7* – insulator; *8* – collector.

ohmic losses. The built-in quasi-optical converter transforms the complex spatially developed resonator operating mode to a paraxial wave beam with an optimized structure, separates the microwave radiation from the used electron flux, reduces the harmful effect of the reflected microwave power flows back to the gyrotron. One of the most complex and expensive gyrotron parts is the exit window. During the pulse, it is strongly heated, and its cooling occurs between the pulses. For continuous gyrotrons with an output power of the order of 1 MW, it is necessary to use windows made of artificial diamond, characterized by high thermal conductivity and low losses of microwave radiation.

The gyrotrons are used in installations of controlled thermonuclear fusion for electron cyclotron heating of plasma. In this case, powerful sources of coherent microwave radiation are required in the range from 70 to 170 GHz with a power level of up to 1 MW at a pulse duration from fractions of a second to continuous generation. The highest achievements in the field of creating powerful gyrotrons are as follows: the wavelength range covered by gyrotrons is from 8 to 500 GHz; maximum power at 0.2 MW continuous mode (f = 82.7 GHz, mode $TE_{10,4}$, P = 0.2 MW, efficiency = 52%, FZK, Karlsruhe, Germany, CRPP, Thales ED, GIKOM, Nizhnyi Novgorod, Russia), ~1 MW in modes with a pulse duration of ~1 s (f = 139.8 GHz, mode $TE_{28,8}$, FZK, CRPP, Thales ED, GIKOM) and 1.5−2 MW for pulse durations <0.1 s (f = 170 GHz, mode $TE_{31,12}$, JAEA, Toshiba, Japan) [31].

3.3.2. Relativistic gyrotron and orotron

To carry out experiments with the relativistic gyrotron and orotron [32], the LIA section of the collective electron accelerator of the Joint Institute for Nuclear Research (Dubna, Russia) was refined to obtain a thin coaxial electron beam. The beam was formed in a coaxial explosive-emission diode placed in a strong magnetic field (Fig. 3.27 *a*). The cathode and the anode of the diode were made in the form of thin-walled metal pipes located at a variable distance d_{a-c} from each other. The voltage between the anode and the cathode was obtained by summing the voltage from 24 inductors. The magnitude of the voltage (and, correspondingly, the energy of the particles) could vary within 200−500 kV by changing the voltages at 8 modulators and the distance between the cathode and the anode. The geometry of the diode was chosen in such a way that the electric field strength on the accelerating tube did not exceed 5 kV/cm. The longitudinal magnetic field with a strength up to 20 kG, focused by the particles, was created by solenoids grouped together and fed by three pulsed sources. The non-uniformity of the field in the working space did not exceed 3%. After passing the electrodynamic system of the microwave oscillator in the region of a non-uniform magnetic field, electrons were deposited on the anode tube. The magnitude of the transported current was measured by a reverse current shunt. The maximum value of the current was mainly determined by the value of the internal resistance of the modulators (Fig. 3.27 *b*) and was 2.2 kA at a voltage of 35 kV on the inductors. In experiments with obtaining microwave radiation, the current was regulated in the range 0.4−1.2 kA. The autographs of the electron beam on the viniplast were in the form of a ring with a width of about 1 mm. The duration of the current pulse was close to 200 ns.

The purpose of the test experiments was to obtain information about the quality of the high-current electron beam formed in the

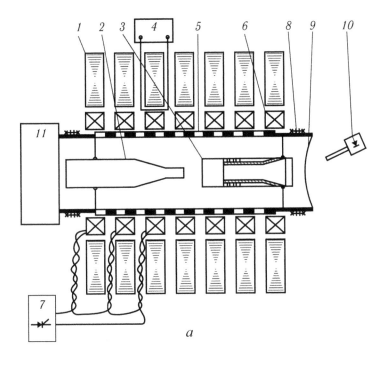

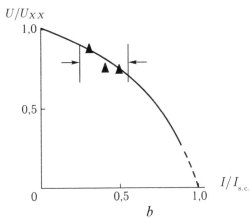

Fig. 3.27. Scheme of the section of the LIA (*a*): *1* – inductors; *2* – cathode; *3* – anode; *4* – modulators; *5* – accelerating tube; *6* – coils for creating a focusing magnetic field; *7* – magnetic field power source; *8* – microwave generator; *9* – vacuum window; *10* – semiconductor microwave detector; *11* – the vacuum pump. Load characteristic of the LIA section (*b*); $U_{xx} = 500$ kV; $I_{s.c.} = 2.2$ kA.

LIA section by comparing the measured and calculated values of the starting currents and the electronic efficiency of the microwave generators. As test generators, the relativistic orotron and the relativistic gyrotron were well theoretically and experimentally studied. As already noted at the beginning of this section, the action of these generators is based on various mechanisms of stimulated emission of electrons: Cherenkov radiation of a rectilinear beam (in an orotron) and cyclotron radiation of a beam of particles moving along helical trajectories in a uniform magnetic field (in a gyrotron).

The main dimensions of the resonators of the orotron and the gyrotron coincided. In this case, a cavity with a smooth lateral surface was used in the gyrotron, and in the orotron an axially symmetric corrugation was applied to the inner wall of the resonator, which served to create a slow spatial harmonic of the resonance oscillation synchronous relative to the electrons. Microwave radiation was extracted from the LIA section through a vacuum window and detected by a semiconductor detector.

Both generators used the same feedback mechanism: the electron beam excites a natural oscillation of the open resonator, formed by waves propagating almost transversely to the direction of the electron beam. Therefore, axially symmetric resonators were used in the form of segments of weakly irregular waveguides (Fig. 3.28). From the cathode end the resonators were limited by restrictions for the working mode, from the output end – by the horns, which carry out the diffraction output of the radiation.

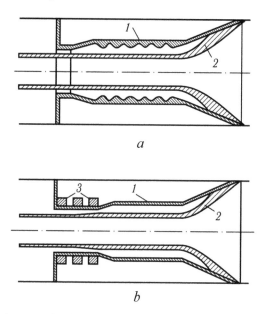

a

b

Fig. 3.28. Schemes of microwave oscillators studied: *a*) orotron; *b*) gyrotron; *1* – resonator; *2* – electron beam; *3* – rings of the electron beam build-up system.

The mode of the orotron was the mode $E_{1,2,1}$ with a wavelength of 8.5 mm. For selection of the 'parasitic' modes, as in the previous experiments [32], longitudinal notches were made in the resonator wall. With a corrugation depth of the order of $\lambda/4$ and Q of the resonator $Q \approx 2000$ for an electron energy of 450 keV, both the calculated and experimental values of the starting current were 200 A. At a current of 800 A, stable generation was obtained with an efficiency of 5–7% (calculated value ~10%). The type of the mode was confirmed by frequency measurements and the radiation pattern investigated in two ways: by moving the receiving horn and by the glow pattern of a neon lamp display. The radiation power was 15–20 MW with a microwave pulse duration of 50 ns. Generation in the orotron almost always occurred near the front of the voltage pulse (Fig. 3.29 a). With a decrease in the electron energy, a loss of microwave power was observed in the $E_{1,2}$ mode and its appearance in the $E_{2,1}$ mode with a wavelength of 11.5 mm. At an energy of 250 keV and a current of 400 A, the power was 10–15 MW, which corresponded to an efficiency of 10–15%

It should be noted that in experiments with an orotron with a much smaller depth of corrugation, with the resonator in addition having no selective cuts, generation was observed at waves of about 4 mm. Apparently, this is also associated with the excitation of

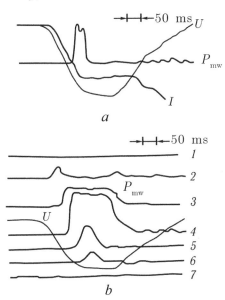

Fig. 3.29. Oscillograms of voltage, current and microwave detector signal for orotron (a) and gyrotron (b) at H [kG]: 9 (1); 10 (2); 11 (3); 13 (4); 14 (5); 15 (6); 16 (7).

orotronic modes, but with their synchronism not with the first but with the second spatial harmonic of the corrugation.

In the gyrotron, the transverse velocity necessary to obtain cyclotron radiation was provided to the particles in a short 'undulator' formed by three copper rings modulating the field of a pulsed solenoid. The distance between the rings was chosen to be resonant (close to the Larmor step of the particles). The working mode was $H_{3.11.1}$ with a wavelength of 12 mm. At the calculated values of the particle energy of 300 keV, the current of 600 A, and the magnetic field of 13 kG, the starting regime corresponded to the transverse particle velocity $\beta_\perp = 0.15$, and for $\beta_\perp = 0.3$, an efficiency of ~5% should be achieved. In the experiment with $\beta_\perp = 0.1-0.15$ radiation was absent (its power was at least below 10 kW). At $\beta_\perp = 0.3-0.4$, stable radiation with a power of 7–10 MW and an efficiency of 3–5% was observed. It should be noted that the duration of the microwave pulse of the gyrotron ($\tau \sim 150$ ns, Fig. 3.29 *b*) was significantly larger than that of the orotron (Fig. 3.29 *a*).

When the magnetic field was varied as the deviation from the resonance value was observed, a change in the shape of the voltage pulse typical for the gyrotron with a displacement of the generation to its fronts was observed (Fig. 3.29 *b*). Simultaneously with gyrotron radiation, Doppler-transformed short-wave radiation with a power of the order of 1 MW in the wave range of 4.0–5.2 mm was measured in this experiment.

The experiments demonstrated a good agreement between the calculated and measured values of the characteristics of the microwave generators, which indicates the satisfactory quality of the electron beam produced by the LIA section. With further acceleration in subsequent sections such an electron beam can be used to obtain shorter-wave microwave radiation.

3.4. Triodes and vircator systems

One of the drawbacks of devices based on the principles of classical microwave electronics (klystrons, direct and backward wave tubes, gyrotrons, free-electron lasers, etc.) is that the operating current in them can not exceed the limiting transport current in the drift tube. Moreover, when the currents approach the vacuum limit current due to the appearance of a beam non-uniformity, the operation mode of these devices can deteriorate substantially. In terms of using all the capabilities of modern high-current power sources, devices with a virtual cathode (VC) are promising [33–36]. Their distinctive feature is the generation of microwave radiation only at currents exceeding the limiting vacuum current, when the condition for the formation of a virtual cathode is fulfilled. On the basis of systems with a virtual cathode, microwave generators can be created in the centimeter and millimeter wavelength ranges. Individual devices with VC have such advantages as the possibility of frequency tuning, design simplicity and compactness, the absence of focusing magnetic systems and, accordingly, the sources of their supply. In a triode with a virtual cathode, the radiation frequency is tuned by changing the cathode–anode gap size, as well as the voltage in it. In a vircator at cyclotron resonance, the frequency of the radiation depends on the magnitude of the external magnetic field. In vircators of the reditron and turbotron type, the frequency of the radiation is varied by changing the density of the electron beam. The design of the devices developed in the Tomsk Polytechnic University, the parameters of power supplies and output characteristics are given in Table. 3.2.

Conventionally, the devices with a virtual cathode can be divided into two types.

1. Devices with VC, in which there are no transit particles, i.e., electrons perform only oscillatory motion between the real cathode and VC. Such devices are called reflex triodes with VC. They differ from the well-known Barkhausen–Kurtz generators, created at the beginning of the last century, by the fact that the electrons in them oscillate between the real and virtual cathodes. Generation of microwave radiation is due to oscillations of the centre of gravity of the beam, accompanied by oscillations of the VC. Such devices dispense with focusing magnetic systems, have an efficiency of up to 10%, power up to 1 GW, and operate in the decimeter and centimeter ranges.

2. Devices with VC, in which, along with the oscillatory motion of electrons between the real and virtual cathodes, there is a stream of electrons passing through the VC into the drift space. Such generators are vircators, reditrons, turbotrons. Vircators have an efficiency of one percent, but they allow one to advance in the millimeter wavelength range.

Investigations of the systems with VC in the Tomsk Polytechnic University began after the creation in 1972 of the Tonus-1 high-current electron accelerator. Almost immediately, high-power microwave radiation was detected in a generating system with a superlimit current [33]. As a result of theoretical and experimental studies it was possible to generate high-power microwave pulses in such devices (10^8–10^9 W) with a duration of 10^{-8}–10^{-6} s in the frequency range ~3 GHz. Relatively small weight-and-size characteristics, constructive simplicity, and the absence of additional energy sources required for creating focusing magnetic fields make promising the use of vircators in creating mobile and autonomous microwave sources with a high level of pulsed power.

Table 3.2. Types of vircators (MC – magnetic coil, t_1 – voltage pulse duration, t_{mw}– microwave power pulse duration)

Type of vircator	The scheme	U, MV	I, kA	t_1, ns	P, GW	f, GHz	t_{mw}, ns
Planar reflex triode with VC		0.65 0.45	14 20	100 1500	1.1 0.45	2.9 2.8	80 1100
Coaxial reflex triode with VC		0.6	62	80	0.2	2.9	70
Vircator on cyclotron resonance		1.0 1.0	20 20	80 80	1.5 0.9	3.1 5.4	30 30
Reditron		0.8	30	80	0.25	16	50
Turbotron		0.4	80	80	0.2	40	40

3.4.1. Reflex triode

Theoretical and experimental studies of reflex triodes have shown that the excitation of electromagnetic oscillations and VC oscillations occurs at a frequency that is a multiple of the frequency of the oscillating electrons, which does not depend on the frequency of the resonant circuit. It is established that the main reason for the instability of the electron beam in systems with VC is the non-linearity of the motion of electrons. The high level of non-linearity of the oscillations of the electrons K_1 and the oscillations of the VC synchronized with the oscillations of the electron beam explains the high level of interaction in the systems with VC (in contrast to systems with a prelimit current). Non-linearity parameter K_1 is determined by the distribution of the potential in the vicinity of the turning point of electrons in the regime of current limitation by the space charge. As the amplitude of the VC oscillations increases, it grows.

The configuration of the system and of the beam affects the level and mode composition of the excited electromagnetic radiation. For example, axially symmetric *E*-waves are predominantly excited in symmetric reflex triodes, and in the asymmetric reflex triodes (systems with a transit current and compression of the electron beam) both *E* and *H* waves can be excited. In addition, due to the oscillations of the VC, oscillations can be excited on the harmonics of the fundamental frequency.

In accordance with the principle of operation of generators with a virtual cathode, it is necessary to use power supplies with high current. The most suitable from this point of view are high-current electron accelerators. However, the Tomsk Polytechnic University also carried out investigations of reflex triodes using a specially designed and manufactured LIA [37]. In addition, in terms of implementing effective work at relatively low currents generated by the LIA, a reflex triode of the original design was investigated.

The aim of the work was to create a compact microwave-radiating complex. The choice of LIA as a power supply for a triode was primarily due to the possibility of a periodic pulse train, as well as its high efficiency and compact dimensions. As noted above, the reflex triode differs from other microwave devices in its simplicity of design and the absence of a magnetic system, which significantly reduces the energy consumption, weight and dimensions of the entire installation. Since the reflex triode is a low-resistance load

for the power source, modifications were made to the design of the
LIA, which made it possible to match the internal resistance of the
accelerator with the impedance of the triode and to realize for the
first time the pulse repetition mode.

The functional scheme of a compact generator of high-power
microwave pulses is shown in Fig. 3.30. The generator consists of
three cylindrical compartments:

1) the power source for charging the primary storage includes a
mains transformer, a high-voltage rectifier, charging chokes and a
switch (ignitron J);

2) the section of the LIA contains an induction system IS, over
whose cores the strip DFLs C_L are wound. Switching of the lines is
carried out by a 24-channel discharger S_0. The discharger is started
using the starting discharger S_1 and the trigger circuit. On the axis of

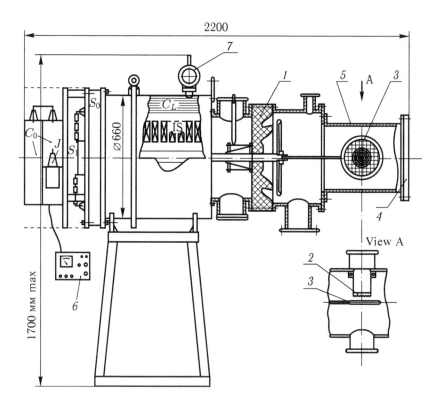

Fig. 3.30. Scheme of installation LIA–reflex triode: *1* – radial high-voltage insulator;
2 – cathode; *3* – grid; *4* – dielectric window; *5* – vacuum chamber of the triode;
6 – the control panel; *7* – oil expansion tank; C_L – forming lines; IS – induction
system; S_0 – control discharger; C_0 – primary storage; J – ignitron.

the section there are pulsed transformer of charge DFL, the Rogowski coil for measuring the total current and capacitive voltage divider;

3) the reflex triode contains a vacuum chamber *5*, a cathode *2*, an anode grid *3*, a dielectric window for microwave output *4*, a horn antenna, a high-voltage radial insulator *1* with a protective screen and an ohmic liquid voltage divider with a correction loop for measuring the voltage on the cathode–grid gap.

The generator is equipped with a portable control panel *6*, designed to turn the unit on and off, control the voltage on the elements of the power scheme, specify a single or frequency pulse repetition.

To reduce the internal resistance of the LIA section, there are 12 parallel DFLs with wave impedance

$$Z = \frac{377d}{mh\sqrt{\varepsilon}} \approx 0.88 \text{ ohm,} \qquad (3.7)$$

where $d = 2 \cdot 10^{-3}$ m is the thickness of the insulation between the line electrodes; $m = 12$ is the number of lines; $h = 0.4$ m is the width of electrodes; $\varepsilon = 3.6$ is the relative permittivity of the insulation between the line electrodes.

The internal resistance of the section in a consistent mode is

$$Z_c = 2ZN^2 = 8.6 \text{ ohm,} \qquad (3.8)$$

where $N = 7$ is the number of cores of the induction system.

The total value of the capacity of twelve DFLs is

$$C_{DFL} = \frac{\varepsilon\varepsilon_0 lhm}{d} = 0.46 \cdot 10^{-6} \, \Phi,$$

where $l = 3$ m is the length of the electrodes.

The magnitude of the switched multichannel charge of the discharger is $C_{DFL} U_{DFL} \approx 27.6 \cdot 10^{-3}$ C where $U_{DFL} = 60$ kV is the maximum charging voltage of the DFL. As noted in Ch. 2, in the frequency mode at low electrode erosion the spark gap can commutate the charge $\sim 10^{-3}$ C. Therefore, a 24-channel discharger with an anode current divider between the channels was used.

The first experiments were carried out by the teams of two laboratories of the TPU with a reflex triode, having the construction shown in Fig. 3.30. The triode had a cylindrical vacuum chamber with a diameter of 350 mm and a length of 500 mm. On the axis of the system there was the anode holder, one end of which was connected to the high-voltage flange of the LIA. At the second end there was a device for installing the triode grid. This device was an assembly of two rings with an outer diameter of 180 mm, between which a stainless steel mesh was clamped with a transparency coefficient of 0.7–0.8. In parallel to the grid at a distance that could change during the experiments there was a stainless steel cathode 100 mm in diameter (flat for single shots or convex for pulse-periodic operation). Ring grooves were made on the surface of the cathode for more uniform emission. On the other side of the anode holder there was a window at the end of the vacuum chamber of the triode to remove the microwave radiation into the free space. Between the LIA and the triode there was an oil-filled section to accommodate the ohmic voltage divider. The corrective loop was located in an additional vacuum chamber, in which a shield was also placed, protecting the radial insulator from the entry of electrons from the cathode–anode gap of the triode.

The generator was tested at a charging voltage of the DFL of 60 kV, which made it possible to form a voltage of up to 350 kV at the cathode–grid gap of the triode at a current of 12 kA with a duration of 160 ns. The inductive voltage drop measured by the correction loop was significant ($\approx$70 kV). The high inductance of the current leads was associated with the presence of an additional chamber for the voltage divider, a vacuum chamber for the shielding screen, and also with the design of the triode generator itself.

Estimates of the level of the microwave power generated by the reflex triode were carried out by integrating the radiation pattern measured by the detector at a distance of 10 m from the window of the vacuum chamber. In the single-mode operation, some microwave pulses reached a level of 400 MW, but the averaged value for 100 pulses did not exceed 150 MW. In the case of a pulse repetition mode with a frequency of 0.5 and 1 Hz, the flat cathode was replaced by a convex one. In this mode of operation, the central part of the grid approached the cathode for a distance smaller than that required for optimal microwave generation of single pulses. When the power supply was turned on, during the first 5–10 pulses, the grid threads were heated and lengthened. The grid acquired a convex shape, and its central part was removed from the cathode. Having experimentally selected the cathode profile corresponding to the profile of the heated grid, at a frequency of 0.5 Hz, it was possible to generate microwave pulses with an average output power of about 150 MW. An increase in the repetition rate of pulses over one hertz reduced the number of pulses of stationary generation due to overheating of the grid and its burnout.

Measurements of the frequency spectrum of the radiation were performed by a bandpass filter. They showed that the central frequency of the radiation was in the range 3300−3400 MHz. Thus, the first experiment demonstrated the possibility of a reflex triode operating when powered by a linear induction accelerator. A large variation in the values of the output power was determined by the non-optimal energy characteristics of the LIA. In addition, the design of the reflex triode, previously used to generate microwave radiation when powered by high-current electron accelerators, was not tuned to the output parameters of the LIA.

The authors of [38, 39] attempted to eliminate the shortcomings of the reflex triode in the axisymmetric construction shown in Fig. 3.31. When developing it, they tried to reduce the inductance of the current leads to the grid of the reflex triode and thereby reduce the inductive voltage drop and reduce the duration of the voltage pulse front, and also ensure the symmetry of grid power. Replacement of the radial insulator with a shielded cylindrical one allowed to reduce the length of the anode holder due to the lack of a screen.

In the traditional design of the reflex triode, the current lead-in to the grid is made along the central anode holder (see Fig. 3.30). The current causes an azimuthal magnetic field that acts on the electrons of the beam and deflects their trajectories from rectilinear. In the axisymmetric design of the reflex triode, the magnetic field

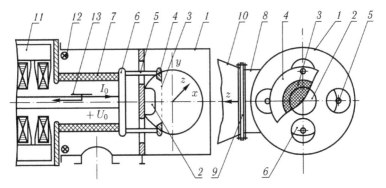

Fig. 3.31. Design of an axisymmetric triode: *1* – vacuum chamber; *2* – cathode; *3* – grid; *4* – grid holder; *5* – current leads to the grid; *6* – high-voltage flange of the LIA; *7* – cylindrical insulator; *8* – outlet for microwave radiation; *9* – dielectric window; *10* – a horn; *11* – LIA; *12* – Rogowski coil; *13* – capacitive voltage divider.

encloses the electron beam, since the current flows through several symmetrically located anode holders. In such a design, the electron beam is not displaced in the cathode–anode gap in a direction perpendicular to the electron oscillations. This makes it possible to reduce beam losses, and the superposition of the magnetic fields of currents flowing along the current leads to the grid forms an azimuthal magnetic field that focuses the electron beam.

The inductance of the current leads to the grid is defined as

$$L_T = \frac{\mu_0}{2\pi} a \ln \frac{D_1}{D_2} \approx 2.3 \cdot 10^{-7} \text{H}, \tag{3.9}$$

where $a = a' + a''$ is the axial length, and D_1, D_2 are the diameters of the external and internal current leads. The inductance for ordinary triode geometry is several times higher.

A circular waveguide with a vacuum dielectric window was used to output the microwave radiation from the triode into free space. A horn antenna could be attached to the end of the waveguide.

When a horn antenna is used, the flux density of the microwave power on the axis is ≈ 200 W/cm^2 at a distance of 10 m from the output end of the antenna using a horn antenna at an optimal value of the cathode–grid gap of 14 mm. The integration of the microwave radiation pattern made it possible to estimate the output power of an axisymmetric reflex triode at a level of 200–300 MW (individual pulses reached a level of up to 400 MW). Measurements with a bandpass filter showed that the radiation frequency was in the range of 3400–3500 MHz and varied with the cathode–grid distance. The bandwidth at the half-power level did not exceed 3%. As in the case of a triode of conventional geometry, a convex cathode was used in the pulse-periodic mode (0.5 Hz), which allowed the generation of microwave pulses with an average output power of ~200 MW.

Comparison of the results of experiments with reflex triodes of two designs makes it possible to give preference to the latter, although it is not free from the drawback associated with the limited grid resource.

The possibility of further increasing the output parameters of a reflex triode with power from the LIA is associated with minimizing the inductance of the current leads, solving the problem of heat removal from the grid and selecting the grid material. The results of calculations presented in [40] testify the fact that in the pulse-periodic mode of operation the anode grid undergoes considerable mechanical loads. In this work the thermal and mechanical modes of operation of the anode grid of tungsten (0.88 transparency coefficient) at a pulse duration of about 20 ns, electron energy 100 keV, current 12 kA, beam diameter 4 cm, is analyzed. Numerical solution of the temperature balance equation for the central wire grid has shown that ohmic heating is extremely small here compared to a thermal shock, and the thermal conductivity for nanosecond pulses can be neglected. Within one second of the installation (i.e., 10 pulses), the temperature of the wire reaches 1000 K. At the same time, thermal stress causes mechanical stresses that can exceed the strength characteristics of the grid material. For the analysis of mechanical stresses, the equation of thermoelasticity for displacements was solved numerically. As a result, it was shown that even at 1000 K these stresses exceed the static strength limit of tungsten. The results of the calculations were used in the experiments (see below).

3.4.2. Vircator based on a 'non-iron' LIA

Experimental studies of the vircator using a 'non-iron' LIA were performed at the All-Russian Scientific Research Institute of Experimental Physics (Sarov), under the direction of V.D. Selemir. In the formulation of the experiments, it was assumed that, like traditional power sources (high-current electron accelerators, inductive storages with plasma current interrupters [41, 42], and explosive-magnetic generators [43]), 'non-iron' LIAs are able to generate high-power pulses. At the same time, among the advantages of LIA, one can note the possibility of operating in the mode of forming high-voltage pulses of both polarities, as well as in the train of pulses. In addition, the feature of the 'non-iron' LIA is the static grounding of both the cathode and the anode, which makes it possible to develop and investigate new types of vircators, for example, a cyclotron resonance vircator [44].

In [45], a vircator realized on the basis of a 'non-iron' LIA is described and the results of its simulation, measurements of microwave characteristics and optimization are presented. One module of the Corvette LIA was used to power the vircator (see section 1.3.4) [46, 47].

This vircator was formed by a coaxial cathode electrode and an anode electrode surrounding it (Fig. 3.32). The cathode electrode consisted of a cathode holder made in the form of a 0.2 mm thick thin-walled stainless steel tube to which a steel substrate was welded from the anode side. Graphite cylindrical elements with a diameter

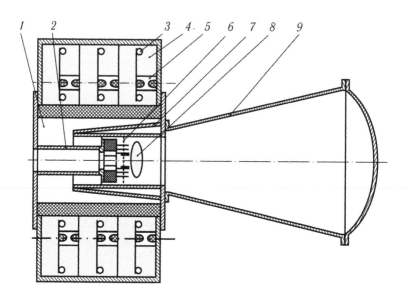

Fig. 3.32. A vircator on one block of LIA inductors: *1* – vacuum volume of the accelerator; *2* – cathode; *3* – high-voltage electrode of the inductor block; *4* – deionized water; *5* – gas-filled controlled discharger; *6* – anode grid; *7* – anode; *8* – VC; *9* – horn of the microwave output.

of 20 mm and a length of 40 mm, which acted as emitters, were attached to the substrate by means of screws. A total of 8, 10 or 12 emitters were placed on the substrate tightly to each other, forming cathodes with an outer diameter of 72, 92 or 112 mm, respectively. The anode electrode was a hollow cylinder with a diameter of 160 mm and a length of 380 mm. A metal ring was placed in the cylinder, onto which an anode grid with square cells 3 mm in size, made of nichrome wire 0.3 mm in diameter, was attached. Moving along the axis of the vircator a ring with an anode grid, it was possible to vary the value of its cathode–anode gap in the range 7–19 mm. The anode electrode ended with a conical horn antenna with an opening angle of the radiating horn of 10°.

The magnitude of the voltage developed on the vircator diode was 900 kV with the maximum charge of the inductor unit. The current in the diode reached 50 kA, which is several times higher than the value of the limiting beam current in the anode cavity behind the grid (15–20 kA). The currents flowing in the vircator were measured with three Rogowski coils installed in it. Measurement of the energy contained in the radiated microwave pulse was carried out using

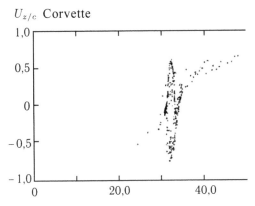

Fig. 3.33. Computer simulation of the vircator: an instantaneous phase portrait of the beam.

a wideband calorimeter. The power of microwave generation, was determined using semiconductor detectors on hot carriers.

Computer simulation of the vircator was carried out using an application package based on the 2.5-dimensional PIC-code 'Karat' [48]. The geometry of the simulated region roughly corresponded to the experimentally investigated vircator. A pulse with a voltage of 900 kV and a duration of 40 ns was applied to the diode gap the shape of which corresponded to the previously recorded experimental oscillogram. The current in the diode reached ~35 kA. A typical instantaneous phase portrait of a particle collective is shown in Fig. 3.33. It follows from the figure that a VC is formed in the system, the oscillations of which are a source of microwave radiation. It is seen that the electrons coming from the diode are reflected from the space charge of the VC and oscillate in the potential well formed by the cathode, grid and the VC. Calculations have shown that the instant of VC formation coincides with the beginning of microwave radiation generation, and the moment of VC destruction – with the moment of its termination.

Figure 3.34 shows a typical calculated 'oscillogram' of the power of the output radiation with a 12 mm gap between the cathode and the grid. The calculated maximum of the peak radiation power corresponds to a gap of ~12–14 mm, which fits the experimental data (see below). Also, the obtained values of peak power (~400–500 MW) were close to them. The closeness of the calculated and experimental results suggests that this computer model is suitable for preliminary optimization of the system.

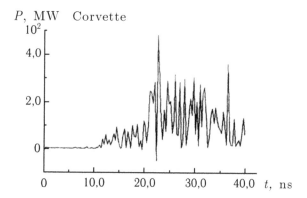

Fig. 3.34. Calculated 'oscillogram' of microwave power.

In the first series of experiments, the maximum microwave energy was recorded by a calorimeter for a cathode with a diameter of 72 mm, consisting of 8 emitters. The dependence of the measured energy on the value of the cathode–anode gap had a rather narrow peak, which agrees with the experimental data obtained by other laboratories (see, for example, Ref. [49]). With a 10 mm gap, the energy of a single microwave pulse was 2.65 J. In this case, the peak power estimated with a semiconductor detector reached 520 MW. The synchronized oscillograms of the cathode current and microwave pulse are shown in Fig. 3.35. Hence, taking into account the propagation time of the microwave radiation from the VC to the detector, the starting current of the vircator was determined, which was 19 kA. This value roughly corresponds to the value of the passing current.

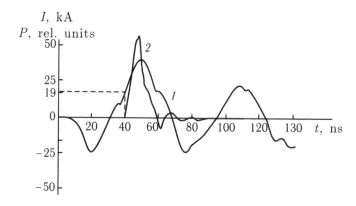

Fig. 3.35. The synchronized oscillograms of the cathode current pulse (*1*) and the microwave pulse (*2*).

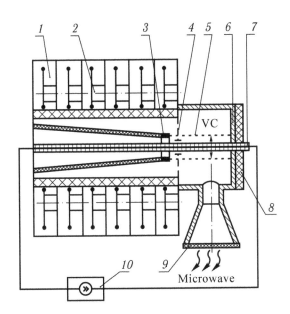

Fig. 3.36. A vircator based on a two-block LIA (when working on the second half-wave voltage): *1* – block inductors; *2* – current-carrying thread; *3* – cathode; *4* – grid anode; *5* – beam; *6* – collector of electrons; *7* – contact; *8* – insulator; *9* – window for outputting radiation; *10* – current source.

Thus, the authors [45] were the first to realize a powerful vircator based on a 'non-iron' linear induction accelerator on radial forming lines. As a result of experimental optimization of the vircator, pulses of microwave radiation with a peak power of more than 500 MW, duration 40 and 18 ns at a current of 35 kA, were obtained.

The experiments carried out by far do not exhaust the possibilities of using LIA for experiments with generators based on a virtual cathode, both in the mode of the vircator and its modifications, and in the mode of the reflex triode. In Fig. 3.36 is a version of the vircator, built on the basis of a two-block LIA. The peculiarity of this generator is the presence of an axial current-carrying thread to create an additional azimuthal magnetic field, which makes it possible to ensure equality of the electron oscillation frequencies in the potential well 'cathode–virtual cathode' and cyclotron rotation of virtual cathode electrons in the total azimuthal field. Note that the low-inductive galvanic isolation of the current-carrying wire from the electrodes of the accelerator diode is constructively possible only in LIA.

3.4.3. Hybrid microwave generator based on the vircator + travelling wave tube (virtode) system

One of the interesting ideas in terms of the development of microwave generators with a virtual cathode is the idea of hybrid generators [50], in which the modulated transit electron current enters a slowing-down system tuned to the mode of a backward-wave tube (BWT). The microwave wave excited in the BWT is returned back to the VC (feedback is performed), and the radiation output is made near the VC. Thus, this hybrid generator, called the virtode, is a vircator+BWT system.

A hybrid microwave generator of the virtode type was implemented in Ref. [51]. The scheme for the construction of the virtode is based on the fact that a modulated electron beam enters the slowing-down system and, in contrast to the conventional running wave tube (RWT), the pumping of the system with a good Q-factor occurs with this beam and the microwave radiation of the vircator. The power source for the virtode was a two-block linear induction accelerator I-3000, described in Ref. [52] (see 1.3.4). The electron beam parameters in this experiment were as follows: electron energy 2.4 MeV; beam current 12 kA; pulse duration 20 ns.

The electrodynamic structure of the RWT was an open resonator in the form of a section of a corrugated waveguide of circular cross section with an external diameter of 67 mm, a corrugation period of 16 mm, a depth of corrugation of 7 mm and a number of periods equal to 18. The profile of the corrugation was in the form of two conjugate semicircles. The electrodynamic structure was placed in a solenoid, creating a magnetic field with a value of up to 3 kG. Figure 3.37 *a* represents the configuration of a relativistic RWT, similar to that studied in Ref. [52]. When it was optimized by varying the geometric parameters of the generator, the diode impedance and the configuration of the magnetic field, it was possible to get the microwave generation mode with an efficiency of more than 10% at a frequency of 10 GHz. In this case, as the results of measurements of the spatiotemporal structure of the radiation extracted into the far zone showed, it had a narrowband and coherent character. The optimal RWT configuration in this paper was used as a reference.

Figure 3.37 *b* shows the configuration of a vircator with a smooth drift tube with a diameter that coincides with the internal diameter of the corrugated RWT structure. The diode region of the vircator was separated from the drift tube by a metal grid with a geometric

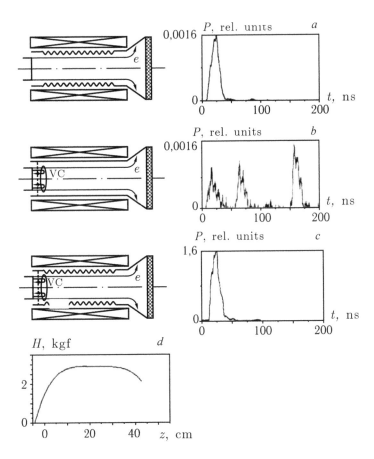

Fig. 3.37. The investigated configurations and normalized oscillograms of the envelopes of microwave pulses: *a*) RWT; *b*) the vircator; *c*) virtode; *d*) the spatial profile of the magnetic field combined with the configurations.

transparency of 90%, made of tantalum wire 0.1 mm in diameter. The width of the cathode–grid gap was 60 mm. This vircator was also used by the authors of [52] for comparison.

Figure 3.37 *c* shows the geometry of the virtode on the basis of the hybrid vircator + RWT system. We emphasize that both the RWT, and the vircator, and the virtode were realized in the mode of microwave generators, that is, unlike the microwave amplifiers, an external signal was not applied to the input of these devices. Therefore, their comparison in terms of the generated power is methodologically justified. The configuration with a smooth waveguide was also investigated, but, unlike the vircator, without a grid. All the listed configurations were investigated under identical regimes of high-voltage feeding of the accelerator and the same magnetic field (Fig. 3.37 *d*).

Typical normalized oscillograms of the microwave pulses are shown in Fig. 3.37 *d* from the corresponding configurations, the peak power density of the RWT being taken as one for convenience of comparison. The power of the vircator radiation turned out to be much smaller than the RWT power, however, the power of the virtode considerably exceeds the total power of the RWT and the vircator. In a configuration with a smooth waveguide without a grid of microwave radiation, there was no fixed. Thus, the virtode is the most efficient of the investigated generators.

Attention is drawn to the fact that the vircator generates three consecutive microwave pulses, the third of which is the highest in power. The presence of several pulses is explained by the fact that in the mode of load mismatch in the LIA, a sequence of decaying high-voltage echo pulses is formed on the basis of lines with distributed parameters. Such a multi-pulse operating mode was previously observed in the relativistic RWT in the I-3000 accelerator [52]. The third impulse in the vircator has the greatest power. Consequently, the vircator most effectively radiates at power voltages of 100–500 kV, and at the ultrarelativistic electron energy its efficiency drops sharply. The latter circumstance does not hold in the case of a virtode in which the VC plays the role of only a beam modulator.

Thus, when a RWT, similar to that of the anode grid investigated in Ref. [52], can be used, the generation efficiency can reach 16% due to the output of the microwave oscillator to the virtode regime. We note that the arbitrary placement of the grid in the RWT does not always lead to an increase in the generation efficiency. To do this, it is necessary to select the position of the grid so that, first, the VC can be formed in the drift tube and, secondly, the frequency of the VC oscillations is equal to the natural frequency of the RWT generation.

3.5. Microwave generator of Cherenkov type

The 'non-iron' LIA I-3000, which forms electron beams with an electron energy of more than 2 MeV and a current of more than 10 kA, was used to study the Cherenkov microwave generator. The fact is that, in addition to LIA, there are practically no high-current electron accelerators with an output voltage of more than 1.5 MeV. (As noted in Chapter 1, there are elements in the design of the

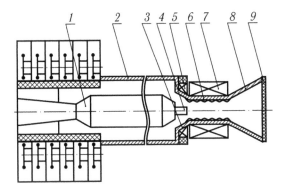

Fig. 3.38. Microwave generator of the Cherenkov type on the basis of I-3000: *1* – cathode holder; *2* – transmission line; *3* –anode; *4* – cathode; *5* – correcting solenoid; *6* – electrodynamic structure; *7* – magnetic field solenoid; *8* – a horn; *9* – window for output of microwave radiation.

high-current electron accelerators (HCEA) on which the total output voltage is loaded, and in LIA such elements can be eliminated.) The scheme of the Cherenkov-type microwave radiation generator is shown in Fig. 3.38 [52]. The electron beam injected by the cathode, passing through the electrodynamic structure, interacts with the spatial harmonics of its field, generating at the output a powerful pulse of microwave. To output radiation from the generator, a conical horn and a vacuum window made of polyethylene or teflon are used. The electron beam is formed in a coaxial diode with magnetic insulation. The magnetic system consists of two solenoids – the main and correcting. With the help of special calculations followed by experimental refinement, the diode system was designed in which the profiles of the cathode and conical anode coincide with the magnetic field line. With a magnetic field strength in the system from 4 to 20 kG, this configuration of the electrodes and lines of force provides a magnetic isolation of the diode on the first (with a positive polarity of the cathode electrode) and the formation of an electron beam on the second half-wave of the accelerating voltage. The diameter of the electron beam print in the cylindrical drift region at a distance of 200 mm from the cathode was 35 mm and practically coincided with the diameter of the hollow cylindrical cathode. Thickness of a print ≈1.5 mm, the displacement relative to the generator axis is insignificant (≈1 mm). Figure 3.39 shows an oscillogram of a current pulse with an amplitude of 15 kA, recorded in the experiment. The boundary energy of the beam electrons depended on the charging voltage of the PVG of I-3000 and could vary between 2.4 and 3.5

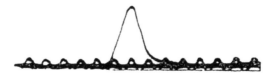

Fig. 3.39. Oscillogram of the current pulse in the I-3000 generator. Markers 100 MHz.

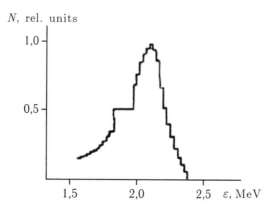

Fig. 3.40. The electron beam spectrogram.

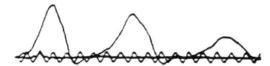

Fig. 3.41. The oscillogram of consecutive current pulses formed with increasing diode gap. Markers 100 MHz.

MeV. Figure 3.40 shows a spectrogram of the electron beam obtained with the help of a spectrometer with a toroidal dispersing field [53]. As can be seen from the figure, the boundary energy of the electrons reached 2.4 MeV. As the distance between the edge of the cathode and the entrance to the electrodynamic structure increases, the I-3000 accelerator forms two electron current pulses that follow one after another in 40 ns. Their amplitudes were 12 and 10 kA, the duration at the base of 20 ns (Fig. 3.41).

The experiments used an electrodynamic structure, which is a sinusoidally corrugated cylindrical waveguide with cylindrical inlet and outlet stainless steel sections. The inner surface was polished to provide electrical strength. The cycle of calculations of the interaction of an electron beam with an electrodynamic structure, performed by the method of determining the dispersion characteristics of a corrugated waveguide with allowance for the electron beam [54, 55] and the gain factor [56], showed that the oscillator operates as a traveling-wave tube with a feedback distributed along the length of the structure. The starting field excitation current of an electromagnetic field with spatial characteristics similar to the E_{01} mode of the cylindrical waveguide was about 8 kA. The generation frequency was close to 9.5 GHz. The electrodynamic structure was placed in a stainless steel tube, inside of which, as well as in the accelerator, the vacuum was at least 10^{-5} Torr.

The typical dimensions of some elements of the microwave generator: cathode diameter is 30–35 mm; the average diameter of the electrodynamic structure is 59 mm; corrugation period is 16 mm; optimum depth of corrugation 6 mm; diameter of vacuum cylindrical volume is 76 mm; length of the horn for output of microwave radiation into the atmosphere is 850 mm; its aperture is 300 mm.

To measure the radiation pattern of different polarization, twenty semiconductor detectors were sufficient in one generator pulse [57]. The detectors were located in the wave zone of the horn antenna at a distance of 9 m from the output window of the installation. The spectral composition of the microwave radiation was estimated using a set of band tunable resonant filters. The microwave radiation was visualized either using a panel of neon lights, or by recording the glow of air in the vicinity of the exit window of the installation horn.

Figure 3.42 *a* illustrates the characteristic oscillogram of the microwave pulse. The duration of this pulse (about 20 ns on the base) practically coincides with the duration of the current pulse of the generated electron beam, which indicates a high coefficient of amplification of the electrodynamic structure and qualitatively confirms the correspondence of the obtained data to the calculation results. The integration of power along the radiation pattern with allowance for radial and angular polarization gives the value of the radiation output to the atmosphere of ~3 GW. An increase in

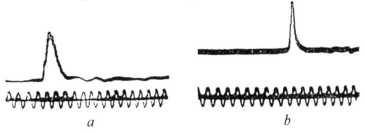

a   *b*

Fig. 3.42. Oscillogram of the pulse of microwave radiation (*a*); shortened pulse of microwave radiation (*b*). Markers 100 MHz.

the charging voltage of the I-3000 pulsed voltage generator and, correspondingly, an increase in the power of the electron beam, lead to a shortening of the pulse duration along the base to 10 ns (Fig. 3.42 *b*). In such modes, breakdowns were detected not only at the output window of the horn, but also at the output of the electrodynamic structure. Apparently, we can assume that the shortening of the pulse duration is due precisely to these phenomena and, possibly, to the reflection of radiation from the plasma formed during breakdown. In this case, the plasma density reaches values sufficient for complete shielding of the microwave radiation, within 10 ns from the beginning of the pulse.

The oscillogram of the microwave pulses generated by the device with an increased distance between the edge of the cathode and the entrance to the electrodynamic structure, when the accelerator forms two electron current pulses corresponding to the second and fourth half-waves of the voltage, is shown in Fig. 3.43. The amplitude of the first microwave pulse is 15% higher than the amplitude of the second one. The pulse duration at the base is 20 and 15 ns. The time between pulses corresponds to the time between the second and fourth half-waves of the accelerating voltage and is 45 ns.

Measurements have shown that practically all the energy of microwave radiation is in the frequency band 9.5–10 GHz, which corresponds to a wavelength of 3.16–3 cm and is in satisfactory agreement with the calculated data.

Additional data on the degree of monochromaticity of the generated microwave radiation were obtained by investigating a luminous ring discharge with a periodic structure arising from the reflection of a microwave wave by a plane metallic mirror perpendicular to the radiation direction. A photograph of the integral glow of the discharge with a distance between the output window and a mirror of 60 cm is shown in Fig. 3.44. The characteristic diameter of the luminous region is 20 cm. The analysis of the microwave breakdown image allows one to determine the wavelength of the microwave radiation from the distance between the ring discharges.

Fig. 3.43. The oscillogram of consecutive pulses of microwave radiation. Markers 100 MHz.

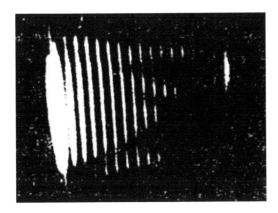

Fig. 3.44. Glow of a microwave discharge in a standing wave in the air.

The corresponding measurements lead to a value of $\lambda \sim 3.0-3.2$ cm, which agrees with the measurement data by resonant filters.

The axial symmetry of the discharge, its azimuthal homogeneity, as well as the appearance of an interference pattern in the addition of diametrically opposite parts of the wave field when the wave is reflected at an angle of 45° indicate not only the monochromaticity, but also the coherence of the generated radiation.

The authors note that the combination of high (in tens of kiloamperes) current with the possibility of accelerating electrons to an energy of several MeV opens the prospect of creating super-power microwave generators, limited only by the electric strength of the electrodynamic systems. In the modernized accelerator LIA-10 M, it was supposed to obtain a peak power of $1.25 \cdot 10^{12}$ W at a beam current of the order of 50 kA, a particle energy of 25 MeV, and a pulse duration of 25 ns. In the case of successful development of electrodynamic systems with such power parameters, it was supposed to achieve a microwave pulse power of the order of 100 GW.

Additional possibilities of using the 'non-iron' LIAs arise when the two inductor units are connected in parallel. In this case, the input and output of one of the modules are inverted relative to the second, and the cathode electrode is connected to their common electrode (Fig. 3.45). An electron beam propagates between the two coaxial anodes. Such an anode configuration is of interest in terms of investigating diffraction type generators.

The multimodular structure of accelerators on the blocks of inductors with radial lines makes it possible to form several electron beams with different energies (Fig. 3.46). Investigation of their

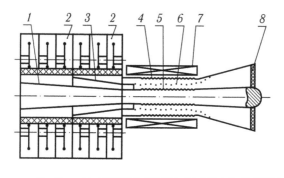

Fig. 3.45. The diffraction generator based on LIA (when operating on the second half-wave voltage): *1* – anode holder; *2* – block of inductors; *3* – cathode; *4* – external electrodynamic structure; *5* – internal electrodynamic structure; *6* – beam; *7* – solenoid; *8* – coaxial output window.

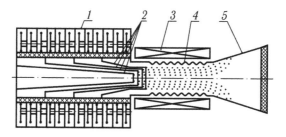

Fig. 3.46. A multi-beam microwave generator based on four LIA units: *1* – inductor block; *2* – cathodes; *3* – the solenoid; *4* – electrodynamic system; *5* – horn.

interaction in a common resonant system is of interest in application to electron-wave generators. The design of the interaction space in the form of a set of coaxial cylinders shielding electron beams from each other makes it possible to study methods of coherent addition of radiation fluxes formed in a set of coaxial electrodynamic structures. Note that in this case, at least two pairs of series-connected inductor units are required (Fig. 3.47). The latest proposals concern the use of resonant structures with a large diameter and wavelength, which unites them with multiwave Cherenkov generators. But at the same time it makes it possible to substantially increase the total current and, correspondingly, the power of the electron beam interacting with the vacuum electrodynamic structure.

We also note that the possibilities given by the LIA to increase the energy of particle acceleration are promising for mastering the millimeter and submillimeter.

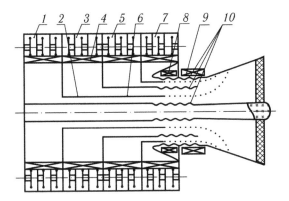

Fig. 3.47. The scheme of a multi-beam microwave generator based on two coaxial electrodynamic structures with phased addition of powers: *1, 3, 5, 7* – inductor blocks; *2* – cathode holder; *4* – solenoid; *6* – cathode; *8, 9* – solenoids; *10* – electrodynamic structures.

3.6. Antenna-amplifier

Potential applications of sources of high-power microwave pulses in information and telecommunication systems, such as radar systems for surveying, sensing the atmosphere and the Earth's surface, systems for remote space communications, are still held back by insufficient progress in solving the problem of controllability [58]. In these applications, it is necessary to provide the possibility of varying as many of the characteristics of the output radiation as possible (power, frequency and phase for monochromatic signals, the spectrum for multifrequency signals and ultrashort pulses, and the directivity of the output microwave beam). Hence it is necessary to develop amplifying devices, while the bulk of research in the field of relativistic high-frequency electronics related to different types of generators. Despite the progress made here, amplifiers potentially provide more opportunities. This and a greater range of capacities, and a broad frequency band, and the ability to control the phase. At the same time, up to the present time, narrowband amplifiers, such as the relativistic klystron or the two-section relativistic traveling-wave tube with regenerative amplification in the first section, have had the greatest development in terms of successful experimental studies of super-power microwave amplifiers. Broadband amplifiers at a high power level have not yet been mastered at all. As for phase control, up to now, in almost all experimental studies with relativistic amplifiers, the duration of the electron beam pulse is shorter than the duration of the input microwave signal. The consequence of this is an uncontrolled phase shift at the voltage pulse front.

In addition to the development of amplifiers per se, for many information and telecommunications applications the compactness of the microwave source and its ability to operate at a high pulse repetition rate is crucial. For amplifiers on high-current electron beams, the obstacle to achieving the necessary compactness is not only the size of the accelerator, but also the existence of cumbersome systems for input of the microwave signal, the transmission lines between the output of the amplifier and the radiating antenna, transformers of the types of waves providing a field structure suitable for excitation of the antenna, as well as the considerable size of the antennas commonly used. Therefore, at present, there is an increased interest in the study of such configurations,which are not associated with the use of large-volume installations and in which the interaction space is integrated with compact sources of impulse voltages.

Below are the results of studies carried out at the INP in the development of an antenna-amplifier based on a compact module of a linear induction accelerator. Their final goal was an experimental demonstration of amplification with wide possibilities of controlling the parameters of the output microwave pulse with a power level of the order of tens of megawatts and a duration of the order of a dozen nanoseconds in the 3 cm range of wavelengths.

3.6.1. The concept and elements of the antenna-amplifier

The concept of the antenna-amplifier, proposed by A.S. Shlapakovsky, includes the idea of creating a compact controlled source of powerful microwave radiation by combining an electronic accelerator and an electrodynamic system of interaction with the input system of an external amplified microwave signal and a radiating antenna. This possibility exists, in particular, if: 1) the radiating antenna is a dielectric rod antenna of the surface wave; 2) amplification is provided by Cherenkov interaction of a coaxial electron beam with the operating mode of the antenna; 3) the LIA module is the source of the accelerating voltage for beam generation.

Running wave antennas. The antennas of the running wave realize the mode of axial radiation. They are performed on the basis of slowing-down systems capable of supporting surface waves. Excitation of runing wave antennas is realized from one end, and the running wave mode is provided by proper choice of the parameters of the slowing-down system or, very rarely, by applying matching loads at the opposite end. With a change in frequency, the phase velocity in the slowing-down system can change, as well as the effectiveness of the action and the quality of matching of the exciter. Usually, the antennas of a running wave have a operating frequency band, measured by percentages, much less often – tens of percent. A unique property of such antennas is the small dimensions of the cross section of the radiating system. This allows the running wave antennas to be placed on the smooth surface of the bodies of flying objects [59]. Dielectric rod antennas (Fig. 3.48 *a*) apply at frequencies from 2 GHz and higher. They are dielectric rods *1* (sometimes tubes) of a circular or rectangular cross section with a length of several wavelengths excited by a piece of a circular or rectangular metal waveguide *2*. In the dielectric rod, the lowest hybrid slowed wave HE_{11} is used.

Schematic diagram of the antenna-amplifier. The circuit of the antenna-amplifier is shown in Fig. 3.48 *b*. In the absence of a high-frequency field, the electrons move rectilinearly along the longitudinal axis *z* at a certain constant velocity v_0 (that is, the beam electrons are completely magnetized by a strong focusing magnetic field, which is applied along the interaction space). An electromagnetic wave propagates only in the positive direction of the *z* axis and has E_z component of the electric field strength [60]. If the waveguide, wave, and electron beam parameters are chosen in such a way that the synchronism condition is met, then electrons with a

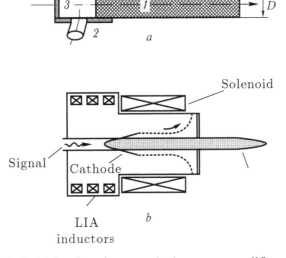

Fig. 3.48. Dielectric rod antenna (*a*); antenna-amplifier concept (*b*).

certain injection phase will pass into the decelerating half-wave of the wave at the maximum of the electric field (in this case they give the wave their kinetic energy), and in the accelerating half-period at the minimum [61]. The described interaction corresponds to the induced Cherenkov radiation. A separate particle moving at a velocity v_0 greater than the speed of light in a medium (v_l – speed of light in the slowing medium) emits Cherenkov radiation and the induced Cherenkov radiation is determined by the existence of electrons in a wave grouping mechanism [62].

If a coaxial relativistic beam passes along the surface of the rod, it can amplify the signal that excites the antenna. In this case, the signal that excites the antenna is at the same time the input signal of the amplifier – the running wave tube (RWT). After interaction with the electromagnetic wave, the magnetized electron beam diverges, moving along the lines of force of the magnetic field, and the amplified signal is radiated by the antenna to the space. The antenna is led out through a vacuum window. Such a hybrid configuration does not require the presence of a transmission line between the amplifier and the antenna or a wave type transformer providing a field structure suitable for excitation of an antenna [63, 64].

The concept of an antenna-amplifier can be realized if a coaxial beam is generated in a linear induction accelerator module with a hollow cathode holder (Fig. 3.48 *b*). Due to the physical principles of LIA operation, its cathode holder is on the outside of the ground potential and can be connected to an external microwave source. In such a case, the hollow cathode holder can simultaneously serve as a waveguide for excitation of the antenna. The LIA module itself is a compact device and is able to operate at a high pulse repetition rate. All this means that the antenna-amplifier is potentially a compact source of high-power microwave pulses with a high repetition frequency and with the ability to control the power, frequency spectrum, phase, and also the radiation pattern of the output microwave beam. An essential physical feature of such a system is that the operating mode of the rod antenna (its main mode HE_{11}) is azimuthally asymmetric, which is not typical for traditional RWTs operating on symmetric TM-modes.

When designing an amplifier antenna, one of the main problems is the probability of plasma occurrence for the following reasons: 1) when a dielectric rod is placed inside the cathode, the high voltage applied to the cathode can lead to a surface breakdown; 2) the plasma can arise as a result of such processes as electron bombardment and accumulation of a charge that appear when the beam propagates along a dielectric rod. Therefore, the initial studies of the antenna-amplifier [61] were connected with the study of possible plasma phenomena and included: 1) theoretical consideration of the slowing-down system with a plasma layer near the rod surface to determine the degree of plasma effect on the characteristics of the mode HE_{11}; 2) experiments on the generation and transport of a beam in a guiding magnetic field with an internal rod for estimating the plasma density under various conditions. The purpose of research – to prove that the concept is realizable, i.e., the plasma can not prevent the Cherenkov interaction between the electron beam and the mode HE_{11} of the dielectric rod.

3.6.2. Calculation of the electrodynamic properties of a waveguide with a dielectric rod and plasma

Let the dielectric rod of radius a be in a circular metallic waveguide of radius b, and the plasma layer of outer radius r_p is adjacent to the surface of the dielectric. In analytical calculations it will be assumed that the density distribution in the plasma layer is uniform and the dielectric permittivity tensor will be used for the model cold collisionless fully magnetized plasma neglecting the motion of the ions for which the diagonal elements are equal to zero, $\varepsilon_{rr} = \varepsilon_{\varphi\varphi} = 1$ and $\varepsilon_{zz} = 1 - \omega_p^2/\omega^2$ (where ω is the circular frequency of the wave, ω_p is the electron plasma frequency).

In the case of azimuthally non-symmetric waves in which all the perturbations are proportional to $\exp[i(l\varphi + kz - \omega t)]$ (where l is the azimuthal index, k is the longitudinal wave number), the presence of a dielectric leads to coupling of E- and H-type waves, as a result of which the eigenmodes have all six components of the microwave field. Therefore, in order to derive the dispersion equation, it is necessary to 'sew together' the solutions of the wave equations for both E_z- and H_z-components in different regions: the dielectric, the plasma layer, and the vacuum gap.

The system is placed in an infinite axial magnetic field. In this case, the magnetic field restricts the motion of electrons so that they move only in the direction of the z axis and therefore can not interact with modes that have only transverse components of the electric field. This means that in this case the waveguide TE-modes are not distorted due to the presence of plasma. In TM-modes there is an electric field component directed along a constant magnetic field. Therefore, in the presence of a plasma, the waveguide TM-modes are distorted:

$$\Delta_\perp E_z + p^2 E_z = 0, \quad r < a,$$
$$\Delta_\perp E_z + \kappa^2 E_z = 0, \quad a < r < r_p,$$
$$\Delta_\perp E_z - q^2 E_z = 0, \quad r_p < r < b, \qquad (3.10)$$
$$\Delta_\perp H_z + p^2 H_z = 0, \quad r < a,$$
$$\Delta_\perp H_z - q^2 H_z = 0, \quad a < r < b,$$

Here

$$\Delta_\perp = \frac{1}{r}\frac{d}{dr}\left(r\frac{d}{dr}\right) - \frac{l^2}{r^2}$$

is the transverse part of the Laplace operator; p, q and k are the transverse wave numbers in the dielectric, vacuum and plasma, respectively:

$$p^2 = \varepsilon\frac{\omega^2}{c^2} - k^2, \quad -q^2 = \frac{\omega^2}{c^2} - k^2, \quad k^2 = -q^2\left(1 - \frac{\omega_i^2}{\omega^2}\right), \qquad (3.11)$$

where ε is the permittivity of the rod material; c is the speed of light.

Only slow waves for which $q^2 > 0$ are analytically considered, so that the fields in the vacuum gap decrease from the boundary of the plasma layer to the conducting wall. In the plasma region, the field distribution has a bulk character ($k^2 > 0$) if the frequency of the wave is less than the plasma frequency. In the region of a dielectric, the wave is volumetric ($p^2 > 0$) if its phase velocity exceeds the speed of light in the medium. Thus, in the system there exist plasma modes in pure form the field of which decreases on both sides of the plasma ($k^2 > 0$, $p^2 < 0$), waveguide modes in pure form, the field of which is 'pressed' to the surface of the dielectric rod ($k^2 < 0$; $p^2 > 0$), and hybrid waves, both bulk in the plasma and in the dielectric ($k^2 > 0$; $p^2 > 0$).

In Fig. 3.49 it was demonstrated that the phase velocity changes in the presence of plasma in the system. This is due to the fact that the plasma is a region of charged particles that change the permittivity so that it becomes less than unity, which leads to an increase in the phase velocity. The transition from curve *1* to curve *3* corresponds to a change in the plasma density by approximately an order of magnitude, so that for $b = 2$ cm the density varies in

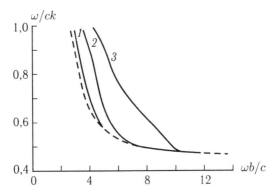

Fig. 3.49. Dispersion of the phase velocity of the waveguide mode HE_{11} in the region of slow waves ($\varepsilon = 5$; $a/b = 0.3$; $r_p/b = 0.5$) at $\omega_p b/c = 4$ (*1*), 8 (*2*) and 12 (*3*). The dashed line corresponds to the absence of a plasma ($\omega_p = 0$).

order of magnitude from $\sim 10^{12}$ to $\sim 10^{13}$ cm^{-3}. It can be seen from the figure that for all three curves, a significant change in the phase velocity in comparison with the absence of a plasma (dashed line) occurs only in the frequency range $\omega < \omega_p$, i.e., in the region of hybrid waves. It should also be noted that for curve *1* (the case of the smallest plasma frequency), an appreciable deviation from the dotted curve begins even at $\omega > \omega_p$, while curve *2* in this range practically merges with the dotted curve, and for curve *3*, in which the deviation is the sharpest, it starts only with a certain frequency less than ω_p. This is due to the high concentration of the microwave field in the dielectric slow wave at high frequencies: for influencing the dispersion one requires not only a significant change in the environment characteristics (ε_{zz} of the plasma compared to vacuum), but also sufficient penetration of the field through the entire thickness of the plasma layer. At a plasma density of 10^{12} cm^{-3}, the dispersion of the system varies little, that is, for the frequencies of the three-centimeter range it is permissible to have a plasma density not exceeding 10^{12} cm^{-3}. For larger values the phase velocity of the mode HE_{11} will vary greatly and may exceed the speed of light, and hence the Cherenkov interaction with the electron beam becomes impossible.

Figure 3.50 shows the dependence of the phase velocity of the HE_{11} mode on the plasma frequency for cases of different thicknesses of the plasma layer at a value of the normalized frequency $\omega b/c = 4$, which at $b = 2$ cm corresponds to a 3 cm range of wavelengths. Naturally, the thinner the layer, the weaker this

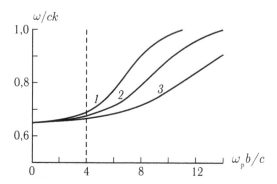

Fig. 3.50. Dependence of the phase velocity of the waveguide mode HE₁₁ at a fixed frequency on the plasma frequency for various values of the outer radius of the plasma layer: $r_p/b = 0.5$ (*1*), 0.4 (*2*) and 0.35 (*3*) ($\varepsilon = 5$; $a/b = 0.3$). The vertical dashed line corresponds to $\omega_p = \omega$.

dependence and the greater the plasma frequency at which the wave becomes fast. The figure shows that even for very thick layers between the dielectric and the beam (curve *1*), the phase velocity is changed little, if the plasma density is less than or about 10^{12} cm^{-3}. At a density of the order of 10^{13} cm^{-3} the Cherenkov interaction with the waveguide mode HE₁₁ is completely excluded.

At a later stage, it was necessary to determine the intensity of the Cherenkov interaction through a quantitative characteristic – the coupling resistance, and also the power flux for the azimuth-asymmetric modes of the rod antenna in the presence of plasma. In addition, it was required to consider the dispersion and electrodynamic properties of plasma modes for various parameters.

The coupling resistance for the radius corresponding to the boundary of the plasma layer is given by

$$R_c = \frac{\left|E_z(r_p)\right|^2}{2k^2 P},$$ (3.12)

Wherein

$$P = \frac{c}{8\pi} \int_S \text{Re}\left(\left[\mathbf{EH}^*\right]_z\right) dS,$$ (3.13)

where P is the power; S is the cross section of the waveguide.

Finally, the coupling resistance in the GHS system has the form

$$R_c[\text{ohm}] = \frac{60|E_z(r_p)|^2}{k^2 \int\limits_0^b \text{Re}(E_r H_\varphi^* - E_\varphi H_r^*) r dr}.$$
(3.14)

From the analysis of the data presented in [65], we can establish the following:

1) the maximum power flow falls on the dielectric and decreases when the waveguide is removed to the wall. However, at high plasma densities, a redistribution of power from the region of the dielectric to the vacuum region is observed;

2) as the plasma frequency increases, the profile of the power density transferred in the plasma varies from the surface one to the volumetric one;

3) in the case of a hybrid mode, the power transferred in vacuum sharply increases with increasing plasma frequency;

4) the coupling resistance increases with increasing plasma density and reaches a maximum at $\omega_p/\omega \approx 1.65$, which is approximately 6 times greater than the value for a system without a plasma, and then decreases;

5) at a low plasma density, the mode is similar to the HE_{11} mode for a system without a plasma (about 80% of the power is transmitted through a dielectric rod). At high plasma density, the mode becomes similar to the TE_{11} mode of the coaxial waveguide formed by the outer surface of the plasma and the metal wall (about 80% of the power is transmitted through the vacuum);

6) the plasma modes do not affect the interaction of the waveguide asymmetric modes with the electron beam, despite the relatively high coupling resistance, since they have a phase velocity substantially lower than the hybrid modes.

Here we briefly presented the results of Ref. [65] devoted to an analytic study of the propagation characteristics and the microwave field pattern for slow azimuthally asymmetric waves in a system that is a dielectric rod with an adjacent plasma layer in a circular waveguide in a strong longitudinal magnetic field. It is found that the plasma density of 10^{12} cm^{-3} does not significantly affect the interaction of the electron beam and the microwave signal. At a plasma density of more than 10^{13} cm^{-3} the coupling resistance of the plasma modes decreases, since the beam is completely shielded from the dielectric, and no amplification occurs. Thus, in order to realize the concept of an antenna-amplifier, an experimental study

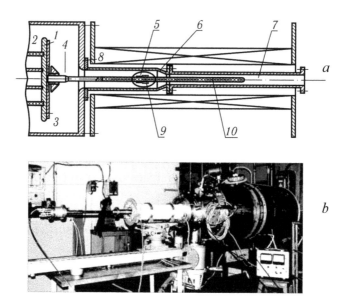

Fig. 3.51. Scheme of the experiment (a cathode is singled out by a circle) (*a*): *1* – a vacuum chamber of the LIA; *2* – insulator; *3* – high-voltage flange; *4* – cathode holder; *5* – the diode; *6* – buffer section; *7* – drift region; *8* – solenoid; *9* – cathode; *10* – dielectric rod. Installation of LIA with a solenoid and a Faraday cylinder (*b*).

of plasma appearance in the formation and transport of an electron beam is an actual task.

3.6.3. Generation and transportation of an electron beam in an antenna-amplifier

In experiments with the transportation of a coaxial electron beam with a dielectric rod and subsequent experiments on microwave amplification, a linear induction accelerator (voltage up to 400 kV, current up to 2 kA) was used in the antenna-amplifier. The appearance of the installation is shown in Fig. 3.51. The accelerating voltage induced by the LIA was applied to the cathode. The guiding magnetic field was created by a pulsed solenoid. To measure the current of the electron beam, a Faraday cylinder, mounted at the end of the anode, was used.

The purpose of experiments with the generation and transport of a beam in a guiding magnetic field encompassing a dielectric rod in the absence of an external microwave signal consisted in studying the region of plasma formation on the dielectric surface of the rod and in estimating its density.

The experimental scheme (Fig. 3.51) included a vacuum chamber LIA *1* with an insulator *2*, a high-voltage flange *3* and a cathode holder *4*, a coaxial magnetically insulated diode *5* (anode diameters of 60 mm, a 20 mm cathode), a conical buffer section *6* (26 mm length) and a drift region *7* (diameter 40 mm). The figure shows the alignment element of the cathode holder on the high-voltage flange. The axis of the magnetic field was aligned with bolted connections between the flange of the solenoid *8* and the vacuum chamber. The edge of the cathode *9* was located in the region of a homogeneous magnetic field (52 mm before entering the drift region). Dielectric rod *10* (diameter 12 mm) was inserted into the cathode holder. In the experiments, rods of various lengths were used (the maximum distance from the edge of the cathode to the end of the rod was 240 mm) from three different materials (plexiglas, polyethylene, and quartz).

The accelerating voltage of the LIA was measured by a capacitive voltage divider located on the inner cylindrical surface of the vacuum chamber. The current in the LIA in the flanges of the vacuum chamber was measured. The current of the electron beam in the drift tube was measured by a moving Faraday cylinder. The magnetic field for all series of experiments was 2.6 T.

The formation of plasma can occur in two cases: 1) when a dielectric rod is placed inside the cathode and a high voltage is applied to the cathode, which can lead to a surface breakdown; 2) as a result of processes arising because of the propagation of the beam along the dielectric rod. First, the possibility of the appearance of a plasma inside the cathode was considered. The construction of the first cathode is shown in Fig. 3.52 *a*. It can be seen that the region in which the rod contacts the inner surface of the cathode is fairly close to the edge of the cathode, where a high voltage is applied, which causes the formation of a very dense plasma. As a consequence,

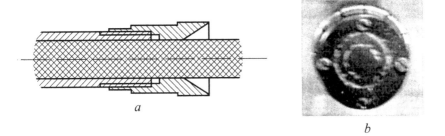

a

b

Fig. 3.52. The cathode scheme (*a*) and the electron beam autograph on copper foil (*b*).

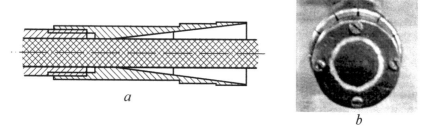

Fig. 3.53. The scheme of the modified cathode (*a*) and the autograph of the electron beam on copper foil (*b*)

two concentric circles with virtually identical brightness exist on the electron beam autographs (Fig. 3.52 *b*). The outer circle is the result of the action of the main beam (it corresponds to the diameter of the cathode), and the inner circle is the trace of electrons from the plasma on the surface of the rod (corresponds to the diameter of the rod). Similar prints were obtained with long and short plexiglass rods. Autographs of the electron beam show that the plasma density on the surface of the dielectric rod is as high as the density of the plasma formed by the explosive emission on the metal surface of the cathode. Such a plasma is able to exclude the interaction of microwave fields of a three-centimeter range with an electron beam, which is unacceptable.

If we change the design of the cathode as shown in Fig. 3.53, then the formation of plasma on the surface of the rod can be avoided. According to the autograph on copper foil it is evident that in this case the 'shot' of the installation was allegedly produced without a dielectric rod.

However, in the case of a long dielectric rod (350 mm), a new trace appears inside the main autograph of the beam. It is not as bright as in the first case. Consequently, the plasma formed on the surface of the rod has a low density. Since the internal trace occurs only in the case of a long rod, it can be concluded that the plasma is formed in the buffer region between the electron beam and the dielectric. After doing the 'shots' for all used materials (plexiglass, quartz, caprolon, polyethylene), surface breakdown signs appeared on the surface of the dielectric rod.

The main goal of further experiments was to obtain data on the actual density of the plasma formed on the surface of the dielectric.

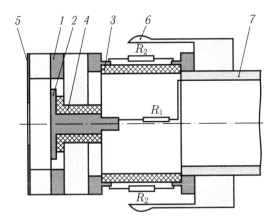

Fig. 3.54. The scheme of the compound Faraday cylinder: *1* – external collector; *2* – internal collector; *3* – insulator of external collector; *4* – insulator of internal collector; *5* – diaphragm (target); *6* – collet; *7* – grounded rod.

3.6.4. Experimental studies of plasma formation

The diagnostics of the plasma density consisted in measuring the current carried inside the generated coaxial beam. Its source is the plasma on the surface of the rod. Assuming that this current is the electron saturation current, the plasma density n can be estimated from the formula $j \approx nev_t$ (where j is the measured current density, v_y is the thermal velocity). If we assume that the electron temperature is several eV (which is typical for a surface discharge plasma), then the value of the saturation current for the plasma density will be 10^{12} cm^{-3}, whence $j \sim 20$ A/cm^2. In experiments it is necessary to measure the current through a known surface and compare the current density with this value.

The experiments used a Faraday compound cylinder with two collectors, capable of measuring the current of the main electron beam and the current inside the beam (Fig. 3.54). The main beam current is output to the external collector *1*. The internal collector *2* is made so that its diameter can be changed (the maximum diameter should be smaller than the diameter of the beam). Additional elements of the external collector and the diaphragm of the foil *5* protect the insulator *4* internal collector from the plasma formed by the beam. In addition, using diaphragms with different hole diameters, one can obtain the dependence of the current on the internal collector on the hole size, i.e., find the radial profile of the plasma density. The

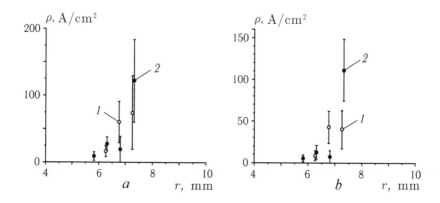

Fig. 3.55. Radial current density profile for diameters of a quartz rod of 12 mm (*1*) and 10 mm (*2*) at a magnetic field of 2.6 T. The voltage of the LIA: *a*) ~280 kV (beam current ~1.1 kA); *b*) ~250 kV (beam current ~0.9 kA).

collet 6 makes contact with the anode, and the registration cables pass through the hollow rod 7.

The use of diaphragms of different diameters made it possible to determine the radial distribution of the plasma density in the system. In the experiments we used diaphragms with a diameter of 11 to 15 mm in steps of 1 mm. For the collection of statistics, a series of 10 'shots' was conducted. To reduce the formation of plasma from the electron beam at the diaphragm, molybdenum targets were used.

In the experiment, two quartz rods with diameters of 10 and 12 mm were used. The current taken from the inner and outer collectors of the Faraday cylinder was obtained for different values of the accelerating voltage and the magnetic field. The radial profile of the current density is shown in Fig. 3.55.

It is seen from the presented graphs that the current density is higher with a small gap between the rod and the beam and increases with increasing voltage and beam current. In addition, the profile has a maximum near the surface of the rod. A further increase in the current density can be caused by the internal peripheral electrons of the main beam. Since the value of the current density does not exceed 20–40 A/cm², which corresponds to a density of the order of 10^{12} cm⁻³, it can be hoped that the formation of a plasma will not affect the operation of the amplifier antenna.

3.6.5. Installation for a demonstration experiment with an antenna-amplifier

The demo installation includes the following elements (Fig. 3.56):

1) a pulse magnetron MI316 frequency tunable from 8.6 to 9.6 GHz with an output power from 100 to 200 kW;

2) a two-frequency microwave compressor with adjustable peak power for demonstrating control over the frequency and power of the antenna-amplifier (Fig. 3.57) [66–68];

3) compact module LIA (voltage 250–340 kV, current 0.9–2 kA, pulse duration ~150 ns at the base, module diameter ~70 cm, module length together with power supply ~100 cm);

4) antenna-amplifier (cathode, anode, dielectric rod) with the length of the interaction range, varying within 20–30 cm (which should correspond to the amplification from 13–15 to 20 dB) due to the change in the length of the quiding magnetic field;

5) a pulse solenoid with a power supply (length 60 cm, maximum induction of the magnetic field up to 3.6 T).

Control of the experimental installation of the antenna-amplifier is carried out according to a scheme consisting of two units (Fig. 3.58): a) the control unit of the LIA, which allows changing the voltage of the charge of the forming line for the variation of the electron beam energy, the operating frequency, and also allows operation in a single and pulse periodic modes; b) a master pulse generator (MPG), which synchronizes the operation of the above elements. It has in its

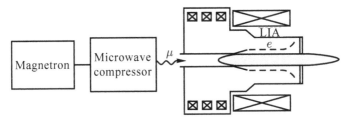

Fig. 3.56. Installation for demonstration of amplification.

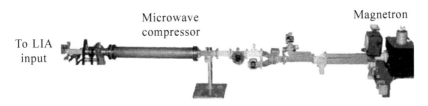

Fig. 3.57. The appearance of the microwave pulse generator based on a magnetron and a microwave compressor.

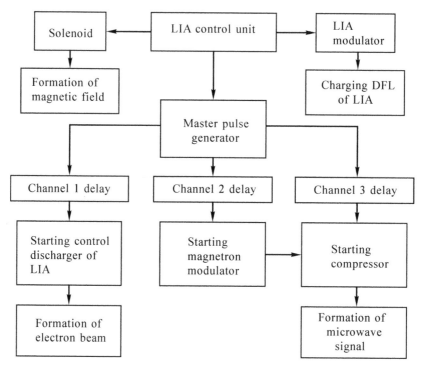

Fig. 3.58. Block diagram of the installation control system.

composition adjustable electronic delay lines (minimum value of 17 ns) for sequential activation of elements.

In the course of the work, the components of the control system were synchronized. The control unit of the LIA provides a control signal to the power supply system of the solenoid, which forms a magnetic field. To trigger the accelerator at the maximum of the magnetic field, after 8.8 ms, the LIA modulator is started, which charges the DFL for 30 μs. In addition, the control unit applies a sync pulse to the master pulse generator, which has in its composition artificial delay lines of three channels. The first channel starts a starter discharger, which switches a multichannel spark switch. A multichannel discharger commutates DFL. An electron beam with a duration of 120 ns is formed. The second channel starts the modulator of a pulsed magnetron with a duration of 1 μs. The output pulse of microwave radiation from the magnetron is accumulated in the resonator of the compressor, which is triggered by the pulse of the third channel and generates a microwave signal with a power ~ 1 MW and a duration of 4 ns. The moments of sequential activation of the elements are shown in Fig. 3.59.

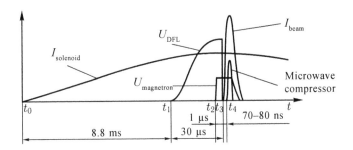

Fig. 3.59. Moments of switching on the solenoid of the magnetic field (t_0), the modulator of the LIA charging the DFL (t_1), the modulator of the pulsed magnetron (t_2), the starting discharger of the LIA (t_3) and the microwave discharger (t_4).

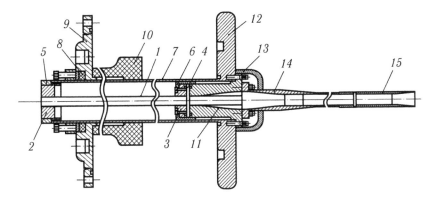

Fig. 3.60. Transmission system inside LIA: *1* – rectangular waveguide; *2, 3* – waveguide flanges; *4* – vacuum window; *5, 6* – adjusting nuts; *7* – central electrode; *8* – the sealant; *9* – ground flange of LIA; *10* – base for the impulse converter; *11* – converter of a rectangular waveguide into a circular waveguide; *12* – high-voltage flange; *13* – the screen; *14* – cathode holder; *15* – cathode

Consider the coupling tract of an external microwave generator and an amplifier. Experiments with the amplification of the microwave signal were conducted using a hollow cathode holder which made it possible to install a dielectric rod inside it. The cathode holder is also a part of the system for transmitting an external signal of a 3 cm range. The transmission system is shown in Fig. 3.60. Inside the LIA there was an electrode *7*, a standard rectangular waveguide of the 3 cm range ($23 \cdot 10$ mm^2) *1* and a converter *11*. The flange *2* and the adjusting nut *5* located on the external grounded surface provided contact of the central electrode with the flange of the LIA *9*. On the opposite side, the flange of the waveguide *3* with the vacuum window *4* was installed with the help of the nut *6* in the converter *11*, which

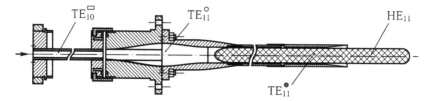

Fig. 3.61. A horn (cathode) exciting the HE_{11} mode of the rod antenna.

in turn was fixed by the high-voltage flange of the LIA *12* on the vacuum side. The cathode holder *14* was connected to a converter. Screen *13* shielded the bolted connections. The inner surface of the cathode holder (and cathode *15*) was a waveguide transition from the waveguide to the dielectric rod exciting the HE_{11} mode. The cathode circuit with an installed dielectric rod is shown in Fig. 3.61.

3.6.6. Demonstration of microwave amplification

The corresponding experiments were carried out in the single-pulse regime. The amplitude of the accelerating voltage of the LIA varied in the range 280–380 kV at a current of 1–2 kA. The experiments were carried out using quartz rods (measured in the three-centimeter range of $\varepsilon \approx 3.8$) with a diameter of 16 mm equal to the inner diameter of the cathode holder. A stainless steel cathode played the role of a matching conical horn for exciting the HE_{11} mode. The emitting edge of the cathode was in the region of a uniform magnetic field. The diameter of the edge, which determines the distance between the beam and the rod, was 23 mm, the diameter of the drift tube was 39 mm. The magnitude of the guiding magnetic field in most experiments was ~2.7 T. The receiving antenna for recording the amplified microwave signal was located on the axis of the system at a distance of ~5 m from the output window. In the microwave input channel, a directional coupler with a detector was installed to register the reflected signals. The same detector also detected a direct signal of the pulse generated by the compressor. The scheme of the experiment differed from the original proposed amplifier circuit in that the quartz rod was entirely in the vacuum volume, and the radiation into the free space was provided by the output conical horn (opening diameter 60 mm), and not by the dielectric surface wave antenna. This configuration of the experiment made it possible to demonstrate the principal possibility of amplification.

In the course of the experiments it was established that the number of microwave signals from the compressor that passed along the rod depends on the length of the rod protruding beyond the end of the cathode. When the end of the rod is located at a distance of 2–2.5 mm from the plane of the cathode edge, almost 100% of the signals are transmitted. As the rod extends beyond the plane, the number of transmitted signals drops. Even with the length of the protruding part being ~30 mm signals practically do not pass. At the same time, an increase of 5–15% in the amplitude of the current pulse is observed at a fixed accelerating voltage. The decrease in the amplitude of the transmitted microwave signal can be related to the appearance of a plasma discharge. Indeed, as the results of the numerical simulation of the amplifier antenna [69] show, in the case of a solid rod, there is a narrow region on its surface inside the beam in which a strong tangential electric field is concentrated. According to the calculations, the maximum field is at a distance ~3 mm from the plane of the cathode edge and reaches ~60 kV/ cm. Such a field leads to a breakdown along the surface of the dielectric and the appearance of a plasma in the region between the cathode edge and the rod, which prevents the transmission of the input signal. This was confirmed by experiments, originally carried out with a continuous long rod. They did not record any microwave signal from the receiving antenna if the delay between the input of the input signal and the beginning of the voltage pulse of the LIA exceeded a certain value.

Estimates made according to increasing beam current, give the values of the electron density in the generated plasma of the order of 10^{12} cm^{-3}. To eliminate the discharge, it was suggested to use a dielectric rod with a gap of 20–25 mm in the region of the end of the cathode (Fig, 3.62). In fact, two rods were used. One rod, inserted into the cathode and cathode holder, provided passage of the input microwave pulse (for an empty waveguide 16 mm in diameter, the critical frequency exceeds the signal frequency). The end of this rod was located at a distance of ~2 mm from the plane of the edge of the cathode inside the cathode. The interaction of the electron beam with the synchronous wave occurred on the section of the second rod. The gap between the rods varied from 2.5 to 9 cm. With a change in the gap, the length of the interaction space varied from ~34 to ~27.5 cm, respectively (if the coordinate of the exit boundary of the interaction region is determined by the deviation of the magnetic field line by 0.5 mm radius of the beam).

Figure 3.63 shows the oscillograms of the incident and transmitted microwave signals in the absence of beam current and with a

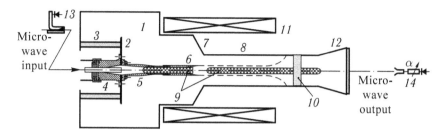

Fig. 3.62. Scheme and geometry of the experiment: *1* – vacuum chamber of the LIA; *2* – high-voltage flange; *3* – insulator of the LIA; *4* – waveguide converter (from rectangular to round) with a radio-transparent vacuum window inside the high-voltage electrode of the LIA; *5* – cathode holder; *6* – edge cathode; *7* – anode; *8* – drift tube; *9* – quartz rods; *10* – sleeve-holder of plexiglass; *11* – solenoid; *12* – conical horn with vacuum window; *13* – detector of the input and reflected microwave signal; *14* – a receiving antenna with an attenuator and a detector for recording the transmitted (amplified) microwave signal. Coaxial electron beam is shown as a dotted line.

operating LIA with a rod with a break of 25 mm. Studies of the amplification of microwave signals in the antenna-amplifier were performed using as an input signal source an industrial pulsed magnetron with a pulse duration of ~1 μs and an output power of ~40 kW, which was limited to the electrical strength of the waveguide. This choice of source is due to the fact that, unlike the compressor, the magnetron allows you to easily adjust the frequency of the output signal in a wide range. In this case, this range was more than 700 MHz (from 8700 to 9500 MHz). In the course of the experiments it was found that there is no clearly pronounced amplification of the pulses from the magnetron in this frequency range. However, there is generation of microwave radiation due to the rupture of the rod, which leads to the appearance of a feedback sufficient for its occurrence. The voltage range at which lasing was observed was ~360–400 kV at a beam current of ~1.6–2 kA. The estimated power of generated pulses of about 40 ns duration was not less than 1 Megawatt. The mode of generation of microwave radiation was absent if an absorbent tip was adhered to the end of the rod. The lack of amplification of the microsecond signal is possibly due to the low power of the magnetron. Figure 3.64 shows the oscillograms of typical envelopes of generated microwave pulses at different voltages and currents.

Further experiments were carried out when the microwave amplifier signal was applied to the input of the device. The power

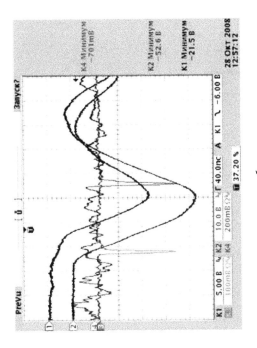

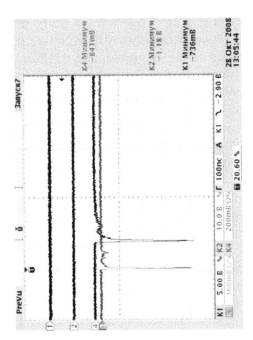

Fig. 3.63. The oscillograms of the incoming (*3*) and transmitted microwave antenna-amplifier (*4*): *a*) without an electron beam; *b*) with an electron beam at a voltage on the cathode *U* = 320 kV (*2*) and beam current *I* = 1.42 kA (*1*).

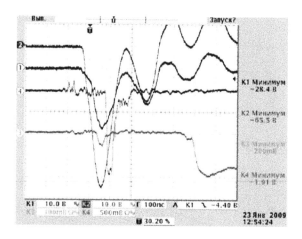

Fig. 3.64. The oscillograms of the voltage, the total current of the LIA and the input microwave pulse of the industrial magnetron (*3*) and the generated microwave pulse (*4*) at U = 400 kV and I = 1.88 kA.

coming from the magnetron to the compressor inlet was 32 to 38 kW. At the same time, the compressor gain reached 16–16.2 dB, that is, at the compressor output, the power of nanosecond microwave pulses was 1.25–1.55 MW. A study of the possibility of amplifying nanosecond microwave pulses was performed at the operating frequencies of the compressor of 9158 and 9386 MHz. At the lower frequency, the amplification was not detected. At 9386 MHz, amplification was achieved in several experiments, reaching a maximum value of 12.5 dB. Typical oscillograms of amplified signals are shown in Fig. 3.65. It can be seen that the duration of the microwave pulse arriving at the receiving antenna coincides with the duration of the compressor signal.The time shift between the input microwave signal and the signal from the receiving antenna has always remained the same, regardless of the delay of the input signal relative to the start of the voltage pulse of the LIA. The signal power from the receiving antenna was determined from the calibration curve of the microwave detector and attenuator attenuation. The gain value was calculated as the ratio of the measured power input to the receiving antenna to the power input to the receiving antenna when the LIA was switched off, for a given output power of the microwave compressor. Thus, for the oscillogram shown in Fig. 3.65, the gain factor determined in this way is $G \approx 11.5$ dB. The error in measuring the gain was estimated to be less than 2 dB. The output power was about 22 MW. The operating voltage range of the antenna-amplifier was in the range of ~300–380 kV with beam

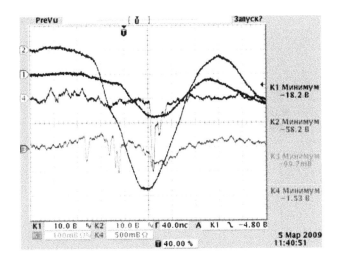

Fig. 3.65. The oscillograms of the voltage, the total current of the LIA and the input microwave pulse of the industrial magnetron (*3*) and the amplified microwave pulse (*4*) at $U = 340$ kV and $I = 1.05$ kA. Gain 12.5 dB.

current ~1.1−1.6 kA. The maximum gain took place with the length of the rod in the interaction space of about 36 cm. At that, the gain per unit length of the rod reached ~0.35 dB/cm. With a rod length in the interaction space of about 42 cm, the gain did not exceed 8 dB. Thus, the linear gain in this case was 0.2 dB/cm. Signals of maximum power were obtained by induction of a magnetic field of 2 T. Because of problems with synchronization and alignment of the rod with respect to the electron beam in the first experiment, it was extremely difficult to detect the dependence of the gain on the voltage, the length of the interaction space, and the magnetic field.

For all the oscillograms shown in Figures 3.63−3.65, it is characteristic that the incident microwave signal is ahead of the current pulse (voltage) by 84−85 ns. This is due to: 1) a time shift of 107−108 ns between the incident and transmitted microwave pulses due to the different lengths of the measuring cables; 2) the time delay of the transmitted microwave pulse relative to the current pulse (voltage) of the LIA (about 23 ns) associated with the run of the electromagnetic wave from the cathode to the receiving antenna.

Figure 3.66 shows a sample of the results obtained for roughly the same voltage and current values, which illustrates the dependence of the gain on the input power. It can be seen that the maximum gain (~12.5 dB) corresponds to the minimum power of the compressor. A decrease in the gain with increasing input power indicates that

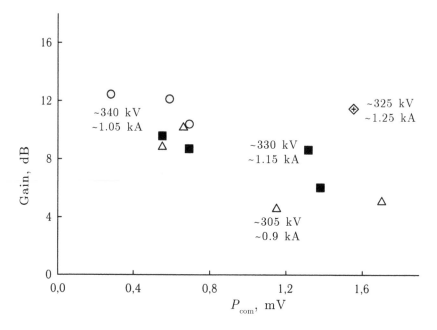

Fig. 3.66. The experimental dependence of the gain on the compressor power for different values of the voltage and current of the LIA at the time of the microwave signal entering the interaction region. The gap between the quartz rods is 9 cm. The value of the guiding magnetic field for the point indicated by the diamond is 1.35 T, and for all others 2.7 T.

the amplifier operates in a nonlinear mode, and the length of the interaction region is clearly greater than the optimal one. This is also indicated by the fact that in the case when the gap between the rods was 2.5 cm, the maximum fixed gain was ~7.7 dB, i.e., at a larger (by 6.5 cm) length of the interaction region, a smaller gain was obtained. This agrees with the results of numerical simulation [70], from which it follows that the length of the interaction region is of the order of 27–28 cm for the input power ~0.2 MW already exceeds the optimal value. In order to determine the output power of the amplifier antenna from the compressor gain and power measured in the experiment, it is necessary to take into account the losses during the passage of the microwave signal from the compressor output to the input of the interaction space. In this case, the most significant, obviously, are the reflection losses from the region of the quartz rod rupture. Thus, the demonstration of amplification in these experiments was possible due to a certain decrease in the output power. It is significant that the reflection losses are different in the absence and in the presence of an electron beam. Therefore, for the output power, only an estimate based on the data of the 'cold' measurements is possible. Such an estimate gives the value of the maximum output power ~16 MW [71]. Note that this result (it corresponds to the point indicated in Fig. 3.66 by the diamond, i.e., the gain value of ~11.5 dB at the compressor power ~1.55 MW) was obtained with a guiding magnetic field value that is half that in other cases. A detailed study of the dependence of the gain on the magnetic field was not carried out in this work.

As a result of the theoretical and experimental work carried out:

1) the propagation characteristics and the microwave field patterns for slow azimuthally-asymmetric waves in a system that is a dielectric rod with an adjacent plasma layer in a circular waveguide in a strong longitudinal magnetic field are investigated. It is found that a plasma with a density of about 10^{12} cm^{-3} does not significantly affect the interaction of the electron beam and microwave signal. With increasing plasma density above 10^{13} cm^{-3} the coupling resistance of the plasma modes decreases because the beam is completely shielded by the dielectric and amplification does not occur;

2) The experimental results of plasma formation revealed that the plasma density in the system does not exceed 10^{12} cm^{-3}. Consequently, even in the presence of a plasma layer on the surface of the rod, the concept of the antenna-amplifier remains realizable. The causes of plasma formation associated with surface breakdown of the dielectric rod inside the cathode (high voltage applied to the cathode) and the propagation processes of the beam along the dielectric rod are elucidated;

3) it was found out that among all the materials of the dielectric rod (plexiglass, polyethylene, quartz) used in the experiment, quartz is the most preferable, since it has the highest melting point and is subject to a smaller surface breakdown;

4) experimentally demonstrated the efficiency of the antenna-amplifier. An amplification of 12.5 dB was obtained at a power output level of 22 MW in the three-centimeter wavelength band at a frequency of 9388 MHz. At the same time, the electronic efficiency of the device was ~5%. Obviously, the length of the interaction space in the experiments was greater than optimal. Further investigation is needed of the dependence of the gain and output power on the length of the homogeneous region of the guiding magnetic field, and also on the magnitude of this field.

In conclusion, we note that with a lower accelerating voltage, it is possible to work with a solid dielectric rod without the formation of a plasma on its surface, since in this case the longitudinal component of the electric field within the beam near the cathode decreases. Therefore, the configuration of an antenna-amplifier with a rod of a larger diameter or of a material with a greater dielectric constant, in which, at given frequencies, the synchronism of the HE_{11} wave with a beam of lower energy is worthy of attention.

3.7. Super-reltron

In this section, we describe the original device of relativistic high-frequency electronics, similar to the relativistic klystron, which was proposed by R.B. Miller of Titan Advanced Innovative Technologies of Albuquerque (USA) [72]. This device consists of sequentially located one after another an electron gun, a modulating resonator (buncher), a tubular high-voltage anode electrode, an output section and an electron collector (Fig. 3.67). The electron gun of the super-reltron is a system of two electrodes: an explosive-emission cathode and a grid anode. The cathode is under ground potential. The anode is made of a metal grid and is installed at the input of the buncher. The grid anode and the buncher are supplied with a voltage with an amplitude of 150–200 kV of positive polarity. The distance between the cathode and the anode is chosen from the condition for obtaining the necessary operating current. The cathode is made of metal or graphite and operates in the mode of explosive electron emission. An electron beam is formed in the longitudinal electric field between the cathode and the grid anode under the action of an electric field of high intensity. Passing the grid anode, the electrons enter the buncher, in which they are modulated in speed. A metal drift tube between the resonator and the output section screens the drift space (grouping space) from external constant and alternating electric fields. When electrons are transported in this region, the velocity modulation of the electron beam is transformed into density modulation. In the space between the buncher and the tubular high-voltage electrode the electrons are accelerated to the total energy of the power source (a pulsed voltage of positive polarity 500–700 kV is fed from the power supply to the tubular high-voltage electrode). The resulting bunches of electrons enter the output resonator. There their kinetic energy is transformed into the energy of microwave oscillations. The output resonator is equipped with a waveguide output of microwave power. Passed through the output resonator, the electrons are deposited on the collector, scattering the remaining kinetic energy. For the supply

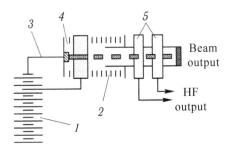

Fig. 3.67. The super-reltron circuit: *1* – high-voltage generator; *2* – accelerating gap; *3* – electron injector; *4* – modulating resonator; *5* – output section.

of superreltelons pulse voltage generators are used, the total output voltage of the power source being applied to the tubular high-voltage anode electrode, and the partial output voltage (20–30% of the total voltage) is fed to the buncher. The described device generated pulses of power ~600 MW at 1 GHz with a duration of 0.5 to 1 μs with a 40% efficiency. The injection voltage was 250 kV, the current was 1.35 kA, and the voltage at the accelerating gap was 850 kV. A similar device, but with a frequency of 3 GHz at an injection voltage of 200 kV, a current of 1 kA and an accelerating voltage of 750 kV, generated an output power of 350 MW with an efficiency of 37%. One of the variants of the super-reltron had a device for mechanical tuning of the radiation frequency within 5% due to a change in the size of the modulating resonator. As shown by the results of experiments, the super-reltron is a powerful and effective device of relativistic high-frequency electronics.

At the same time, an improved super-reltron design [73] based on the use of LIA can be proposed (Fig. 3.68). This design eliminates the shortage of devices associated with a single mode of operation when using power supplies such as pulse voltage generators. One-shot operation of the PVG is associated with a slow process of charging capacitors and using multiple spark dischargers. In addition, such a power source has a low efficiency and large weight dimensions, its operation is accompanied by a high level of electromagnetic interference. Periodic revision of the dischargers is necessary because of the erosion of their electrodes under the action of high discharge currents. It should also be noted that when using the PVG, the voltage on the elements of the super-reltron does not come simultaneously. First there is voltage on the buncher, an electron beam is formed and only through the time of the voltage wave run through the remaining

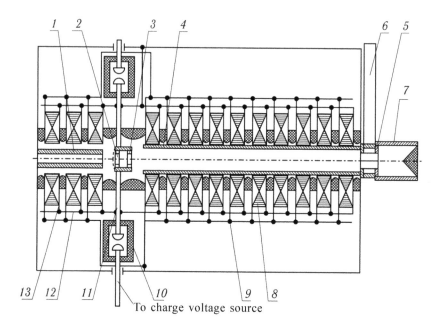

To charge voltage source

Fig. 3.68. Scheme of the super-reltron with LIA as a power source: *1* – cathode; *2* – anode; *3* – buncher resonator; *4* – anode electrode; *5* – output resonator; *6* – waveguide for output of microwave radiation; *7* – collector; *8* – induction system; *9, 13* – magnetizing turns of inductors; *10* – multichannel discharger; *11* – grounded flange; *12* – FL electrodes.

set of capacitors voltage appears on the tubular high-voltage anode electrode. Thus, in the initial time interval the device does not work, which reduces the overall efficiency of the device.

The proposed super-reltron design allows increasing the repetition rate and the overall efficiency of the device and significantly reducing its weight-and-size parameters by constructively combining the elements of the power source and the super-reltron. This is achieved by using the original layout as the power source of a linear induction accelerator. The proposed super-reltron (Fig. 3.68) contains an explosive emission cathode *1*, a grid anode *2* located at the entrance end of the buncher *3* facing the cathode *1*, a buncher *3*, a tubular high-voltage anode electrode *4*, an output resonator *5* with a microwave output waveguide *6* and the collector *7*. This design is almost the same as the prototype device. The difference lies in the fact that these elements are placed on the axis of the induction system of LIA. The accelerator comprises an induction system *8*, forming line *9* – and a multichannel spark discharger *10*. It differs from the known constructions in that the induction system

8 is made of two sets of ferromagnetic cores (for example, 3 and 10 inductors) located on opposite sides of the grounded flange *11*, and the strip forming line *9* is made of two sets of electrodes *12*. Magnetizing turns *13* of the first set of ferromagnetic cores of the induction system *8* are connected to the electrodes of the first set of the forming line, and the turns of the second set of the ferromagnetic cores to the electrodes of the FL second set. The sets of forming lines can consist of ground and potential electrodes, forming a single strip forming line, as well as from ground, potential and free electrodes, forming a strip double forming line. (The strip FLs were discussed in detail in section 1.2.3.) The arrangement of the electrodes *12* and their connection to the magnetizing coils of the cores, the electrodes of the multichannel discharger *10* and the source of the charging voltage for the SFL and DFL are carried out in the same way as in the case of conventional LIA developed at the TPU. It is also possible to use coaxial FLs (for example, as in CLIA, see section 4.8).

A source with an output voltage of the order of 60 kV is used to charge the LIA forming lines (it is not shown in Figure 3.68). Forming lines *9* are charged from the source when voltage is applied to the potential plates. When the multichannel discharger *10* is triggered, the potential SFL electrodes are closed to ground and through the magnetizing turns *13* of the ferromagnetic cores of the induction system *8*, the discharge current of the forming line begins to flow with a duration equal to the time of the double path of the wave along the line. In the case of using DFL, the central (free) electrode of the forming line is closed to the ground. The vortex electric field induced by the first set of cores is summed by the cathode *1* and the vortex electric field of the second ferromagnetic core set is summed by a coaxial high-voltage anode *4*.

In the case of a double forming line, the voltage values at the cathode *1* and the coaxial high voltage anode *4* are approximately equal to the charge voltage of the forming line multiplied by the number of inductors of the first and second sets, respectively. When applying a single forming line, the voltages on the cathode and the tubular high-voltage anode electrode are half as large. Thus, unlike the prototype, in the super-reltron, the voltage on the cathode and the tubular high-voltage anode electrode appears simultaneously, which makes it possible to use for the generation of microwave radiation a current pulse of full duration.

The distance between the cathode *1* and the grounded anode *2* located on the flange *11*, is selected from the condition for the formation of an explosive-emission cathode plasma and the formation of a current of the required value determined according to the three-halves power law, i.e., depends on the magnitude of the voltage, cathode area and cathode-anode distance. The parameters of the elements of the super-reltron for a wavelength range of 10 cm can be approximately the following. The outer diameter of the cathode is 30 mm, and the length is 100 mm. The end of the cathode is located at a distance of 33 mm from the grid anode. An unmodulated electron beam formed in the cathode–anode gap with a current of 750 A and an electron energy of 180 keV enters the buncher and excites the microwave field in the resonator. The microwave field of the buncher produces a high-speed modulation of the electron beam. The buncher has an outer diameter of 76.4 mm, a length of 40 mm, an internal hole of 36–40 mm in diameter and is excited at a frequency of 3000 MHz. Tubular high-voltage anode electrode has a length of 300 mm and is located at a distance of 150 mm from the output end of the buncher. In the space of drift between the buncher *3* and the tubular high-voltage anode electrode *4*, the electrons increase their energy to 600 keV. In this region, the electron beam begins to be modulated in density. Inside the tubular high-voltage anode electrode *4*, electron bunches are formed which enter the output resonator *5* at a frequency equal to the frequency of the buncher *3*. The output cavity *5* with the microwave output waveguide *6* and the collector *7* located outside the body of LIA. The output resonator is made of a piece of a standard rectangular waveguide with a hole of 36–40 mm in diameter for the passage of the electron beam. The microwave output is made of a standard waveguide with a cross section of 72 × 34 mm^2. The electron collector is a metal tube with a cone-shaped graphite absorber in its centre.

The pulse repetition rate is determined by the power of the charging voltage source and the frequency characteristics of the multichannel spark discharger *10*. The expected level of output power of the super-reltron is ~180 MW. In principle, the LIA intended for feeding the relativistic super-reltron can have one common forming line that is discharged in parallel into two sets of ferromagnetic cores. However, the voltages induced by the first and second voltage sets be interdependent. So, for example, when the impedance of the diode gap changes, a voltage variation occurs between both the cathode and the anode, and on the high-voltage anode electrode.

So, the application for feeding the super-reltron of the LIA of the original design allows for a pulse-periodic operation with a significant reduction in the overall dimensions of the device as a whole.

References

1. Sprangle F., Coffey T., Usp. Fiz. Nauk, 1985. Vol. 146. No. 2. P. 303−316.
2. Didenko A.N., et al., DAN SSSR. 1981. V. 256. P. 1106.
3. Vasilyev V.V., et al., Zh. Teor. Fiz. 1983, No. 1. 149, 150.
4. Didenko A.N.,et al., In: Relativistic high-frequency electronics, Gorky, IPF of the USSR Academy of Sciences, 1981. P. 22−35.
5. McDermott D.W., et al., Phys. Rev. Lett. 1978. V. 41, No. 20. P. 1368−1371.
6. Orzechowski T.J., Phys. Rev. Lett. 1986. V. 57, No. 17. P. 2710−2713.
7. Orzechowski T.J., et al., Phys. Rev. Lett. 1985. V. 54, No. 9. P. 889−892.
8. Pasour J.A., et al., Phys. Rev. Lett. 1984. V. 53, No. 18. P. 1728−1731.
9. Pasour J.A., et al., Proc. SPIE. 1984. V. 453. P. 328.
10. Briggs R., Induction accelerators and free-electron lasers at LLNL, Phys. Part. Acceler. Cornell Summer School. 1987. V. 2. P. 2182−2207.
11. Kovalenko V.F., Introduction to electronics of ultrahigh frequencies. 2 nd ed., Moscow, Sov. radio, 1955..
12. Lebedev I.V., Technique and Devices of Microwave. 2 nd ed. In 2 vols. V. 2, Moscow Vysshaya shkola, 1972.
13. Sandalov A.N., et al., in: Proc. of Pulsed RF sources for linear colliders workshop. 1994. P. 134−145.
14. Sandalov A.N., Terebilov A.N., Radio Engineering and Electron Physics. 1983. V. 27, No. 9. P. 90−97.
15. Sandalov A.N., et al., in: Proc. of 'Beams 92. Washington. 1992. V. 3. P. 1673−1678.
16. Sandalov A.N., et al., Investigation of multicavity relativistic klystron with TW output section, EUROEM'94. Bordeaux. 1994. P. 123.
17. Sandalov A.N., Mikheev VV, Stogov AA Experimental investigation of Multiple-beam Relativistic Device. Preprint / Physics Dept. of MSU, No.10. MSU Press, 1984.
18. Tomsky O.N., Furman E.G., Prib. Tekh. Eksper., 1991, No. 5. Pp. 136−138.
19. Bugaev S.P., et al., Izv. VUZ, Fizika, 1968, No. 1. Pp. 145−147.
20. Bugaev S.P., et al., DAN SSSR. 1971. Vol. 196, No. 2. Pp. 324−326.
21. Bugaev S.P., et al., Prib. Tekh. Eksper., 1970, No. 6. Pp. 15−17.
22. Aiello N., et al., High power microwave generation at high repetition rates. Proc. Beams Meeting. 1993. Washington.
23. Friedman M., et al., in: Proc. SPIE Intense microwave and particle beams. Bellingham. 1991. P. 1407.
24. Levine J.S., Harteneck B.D., Appl. Phys. Lett. 1994. V. 65, No. 17. P. 2133−2135.
25. Sincerny P., et al., High average power modulator and accelerator technology developments at Physics International, Preprint, NATO Series. V. G34. Part A. 1993.
26. Aver'yanov V.I., et al., Zh. Teor. Fiz., 1987. Vol. 57, No. 6. Pp. 1213−1216.
27. Kazacha V.I., et al., Collective methods of acceleration, Dubna, 1982. P. 108−110.
28. Rusin F.S., Bogomolov G.D., in: Electronics of Large Capacities, Moscow, Nauka, 1968. Vol. 5. P. 45−58.
29. Trubetskov D.I., Khramov A.E., Lectures on microwave electronics for physicists.

In 2 vols. T. 2, Moscow, Fizmatlit, 2004.

30. Alikayev V.V., et al., in: Vacuum microwave electronics, Nizhnyi Novgorod, I|PF RAS, 2002. P. 71–76.

31. Thumm M., State of the art of high power gyro-devices and free-electron masers, Karlsruhe: Forschungszentrum, 2007.

32. Zaitsev N.I., et al., Pis'ma Zh. Teor. Fiz., 1982. Vol. 8. No. 15. P. 911–914.

33. Didenko A.N., et al., in: Intern. Top. Conf. On Electron Beams Research and Technology. Albuquerque. 1975. V. 2. P. 424–429.

34. Didenko A.N., et al., in: Proc. of 3 Inf. Top. Conf. On High Power Electron and Beam. Novosibirsk. 1979. V. 2. P. 683–691.

35. Didenko A.N., et al., in: Plasma Electronics. Coll.. scientific works, Kiev, Naukova dumka, 1989. P. 112–131.

36. Didenko A.N., et al., DAN SSSR. 1989. V. 309, No. 5. Pp. 1117–1120.

37. Vintizenko I.I., Furman E.G., in: 12 Int. Conf. on High Power Particle Beams, Tel-Aviv: Program and Abstract, 1998. P. 107.

38. Vasil'ev V.V., et al., Relativistic microwave generator, Patent of the USSR for invention No. 1830228.

39. Vintizenko I.I., Furman E.G., in: 12 Int. Conf. on High Power Particle Beams, Tel-Aviv: Program and Abstract, 1998. P. 333.

40. Dubinov A.E., et al., Inzh.-Fiz. Zhurnal, 1998. V. 71, No. 5. Pp. 899–901.

41. Zherlitsyn A.G., et al., Pis'ma Zh. Teor. Fiz., 1990. Vol. 16. Issue. 11. P. 69–72.

42. Zhdanov V.S., et al., Proc. doc. XXV Zvenigorod Conf. on Plasma physics and CF, Zvenigorod, 1998. P. 169.

43. Azarkevich E.I., et al., Teplofiz. Vys. |Temper. 1994. V. 32, No. 1. Pp. 127–132.

44. Pavlovsky A.I., In: Relativistic high-frequency electronics, Gorky: IAP RAS, 1992. Issue 7. P. 81–103.

45. Selemir V.D., et al., Zh. Teor. Fiz., 2001. V. 71, No. 11. Pp. 66–72.

46. Pavlovsky A.I., et al., DAN SSSR. 1980. Vol. 250, No. 5. Pp. 1118–1122.

47. Gerasimov A.I., et al., Prib. Tekh. Eksper. 1998, No. 1. Pp. 96–101.

48. Tarakanov V.P., II User's Manual for Code Karat, Springfield, VA: Berkeley Research Associate Inc., 1992.

49. Yatsuzuka M., et al., Inst. Plasma Phys. Annu. Rev. Nagoya Univ. 1990. P. 96.

50. Gadetsky N.P., et al., Fiz. Plazmy, 1993. V. 19, No. 4. Pp. 530–537.

51. Selemir V.D., et al., Pis'ma Zh. Teor. Fiz., 2001. Vol. 27. Issue. 14. P. 25–29.

52. Pavlovsky A.I., et al., in: Relativistic high-frequency electronics, Nizhnyi Novgorod, IAP RAS, 1992. Issue. 7. P. 81.

53. Minashkin N.V., et al., Method for determination of the spectrum of a pulsed electron beam, Authors. certificate for invention №1681658.

54. Kurilko V.I., et al., Zh. Teor. Fiz., 1978. Vol. 49, No. 12. P. 2569–2575.

55. Swegle J., et al., Phys. Fluids. 1985. V. 28, No. 9. P. 2882–2894.

56. Rukhadze A.A., et al., Physics of high-current relativistic electron beams, Moscow: Atomizdat, 1980.

57. Raizer M.D., and Tsopp L.E., Radiotekh. elektron., 1975. Vol. 20, No. 8. Pp. 1691–1693.

58. Didenko A.N., Yushkov Yu.G., Powerful microwave nanosecond pulses, Moscow, Energoatomizdat, 1984..

59. Sazonov D.M., Antennas and microwave devices, Moscow, Vysshaya shkola, 1988..

60. Weinstein L.A., Electromagnetic waves. 2 nd ed., Moscow, Radio i svyaz', 1988.

61. Trubetskov D.I., Khramov A.E. Lectures on microwave electronics for physicists. In 2 vols., V. 1, Moscow, Fizmatlit, 2003.

62. Nikolsky V.N., Electrodynamics and propagation of radio waves. Moscow, Nauka, 1973.

63. Voskoboinik M.F., Chernikov A.I., Microwave technology and devices, Moscow, Radio i svyaz', 1982..

64. Shlapakovski A.S., Pis'ma Zh. Teor. Fiz., 1999. Vol. 25. No. 7. P. 43–50.

65. Shlapakovski A.S., SPIE Proc. Intense Microwave Pulses VI. 1999. V. 3702. P. 108–113.

66. Shlapakovski A.S., Krasnitsky M.Yu., Fiz. Plazmy, 2008. V. 36, No. 1. Pp. 34–46.

67. Artemenko S.N., et al., Frequency agile high-power resonant microwave compressor. Int. Conference EUROEM 2008. Lausanne, Switzerland, 2008. P. 45.

68. Artemenko S.N., et al., in: Proc. of 15 Int. Symposium on High-current Electronics. Tomsk, 2008. P. 395–398.

69. Avgustinovich V.A., et al., Prib. Tekh. Eksper., 2008, No. 3. Pp. 93-96.

70. Shlapakovski A.S., et al., in: 14th Int. Symposium on High-Current Electronics. Tomsk, 2006. P. 417–420.

71. Avgustinovich V.A., et al., Izv. VUZ, Fizika, 2009, No. 11/2. Pp. 307–310.

72. Miller R.B., et al., IEEE Transactions on Plasma Science. 1992. V. 20, No. 3. P. 332–343.

73. Vintizenko I.I., Fomenko G.P., Super-reltron, The patent of the Russian Federation for the invention No. 2239255. BI. 2004, No. 30.

4

Relativistic magnetrons powered by linear induction accelerators

Introduction

The magnetron was invented by A. Hall [1] in 1921 as a power converter and a power adjustment device. In 1928, two Japanese professors, Yagi and Okabe, discovered that a magnetron, in which the anode is cut into two or more segments, can generate high frequencies. The multiresonator magnetron was created in 1936 by Soviet engineers N.F. Alekseev and D.E. Malyarov. In 1940 this device described in the press [2]. Around the same time, H. Booth and D. Randall developed a similar design in the UK. The successful use of these devices during the Second World War for radar stations caused a rapid development of theoretical and experimental research, as well as the emergence of work on the practical use of magnetrons [3–6].

If the anode of the coaxial diode with magnetic insulation (see Fig. 1.25) is produced in the form of an azimuthal periodic structure, then in the Cherenkov phase matching conditions (proximity of the drift velocity of electrons β_e to the azimuthal wave phase velocity β_p):

$$\beta_e = \frac{cE_0}{H} \approx \beta_p = \frac{V_p}{c}, \qquad (4.1)$$

the energy of electrons can be converted into electromagnetic radiation with high efficiency. The essential difference between the magnetrons and the O-type Cherenkov devices is that the translational motion of electrons is a drift in crossed electrostatic (E_0) and magnetostatic (H) fields. The drift of electrons to the anode, during which they receive energy from the electrostatic field and transmit it to the electromagnetic field, occurs under the action of a synchronous wave, and the kinetic energy of the electrons does not change significantly. The theory of processes in the resonator block of the magnetron and their calculations go beyond the present monograph. For the material further described, it is essential that during the operation of the magnetron a running electromagnetic wave occurs and is synchronized with the motion of electrons between the cathode and the anode. The best experimental samples of magnetrons achieve efficiency of up to 82%.

The use of high-current accelerators for powering magnetrons opens new prospects for these devices. In contrast to O-type generators, the interaction of a high-current relativistic electron beam with high-frequency fields occurs directly in the diode, that is, in the region of creation of the electron beam. In this case, the limitations associated with ensuring a high energy of electrons with a small velocity spread and mastering of high currents are removed.

Such advantages of devices with crossed fields, such as the frequency and phase stability of generated microwave radiation, high efficiency, small weight and dimensions and cost, low level of side oscillations and harmonics, became the basis for intensive research of relativistic magnetrons (RM) [7]. The first experiments with relativistic magnetrons, which yielded notable results (the Massachusetts Institute of Technology in the USA, the Institute of Physics and Technology of the USSR Academy of Sciences (Gorky) and the Tomsk Polytechnic University in the USSR), allowed to obtain power levels from hundreds of megawatts to several gigawatts with an efficiency of 10–30%.

Theoretical and experimental studies have shown that relativistic magnetron generators are a high-current version of a conventional magnetron [8]. Relativistic voltages are needed to excite explosive electron emission [9] and to obtain high currents. The principle of action is identical to the classical device. However, determining the range of operating conditions and calculating the output parameters require the taking into accouny of relativistic factors in the corresponding formulas. It is known that the conditions for neglecting relativistic corrections are reduced to the requirement of smallness of the voltages measured in units of $m_0 c^2 / e$ and the smallness of all the geometric dimensions of the magnetron, determined in units of $n\lambda/2\pi$ (where m_0 is the electron rest mass, n is the number of type of oscillations, λ is the wavelength). Despite the fact that in magnetron-type devices the conversion of non-kinetic energy of electrons occurs, and potential relativism is not so tightly connected with the cathode-anode voltage, a relativistic factor must be taken into account for a number of devices with a small slowing-down of the electromagnetic wave. Indeed, theoretical studies performed in the late 70s of the last century show the influence of the relativistic dependence of the electron mass on its velocity on the microwave generation processes. The averaging method proposed independently by P.L. Kapitsa [10] and V.E. Nechaev [11] for solving non-relativistic equations of motion of electrons in the interaction space of the device, was used as a basis for developing an elementary theory of the relativistic magnetron of planar geometry. However, in the cylindrical geometry of the magnetron, there is an essential feature of the processes of interaction of electrons with electromagnetic fields. It is connected with the fact that here the condition of synchronism of the electromagnetic wave and the electron beam is exactly satisfied only at a certain radius, while in the case of a planar magnetron, the electrons are in phase with the wave along the entire height of the spoke. The impossibility of exact synchronism in the entire interaction space leads to the appearance of an additional azimuthal drift of electrons, a curvature of the trajectories of the motion of electrons in the spokes, and a decrease in the efficiency [8].

To date, a rich experimental experience has been accumulated and a large number of papers devoted to RM experimental research have been published. Data on the input and output characteristics of the devices are shown in Table 4.1, on the basis of which it is possible to determine the parameters of relativistic magnetrons and compare them with classical instruments.

A typical construction of a relativistic magnetron generator is shown in Fig. 4.1. The cathode *2* is mounted coaxially to the anode block *1*. The region between the cathode and the anode forms the interaction space of the electrons with the microwave waves of the resonator system. The anode unit consists of identical resonators separated by lamellae and connected with the interaction space by the coupling slots. Usually, the cathode is connected to the high-voltage electrode *5* by a supply source of negative polarity, and the anode is grounded. Elements of RMs are placed into the vacuum chamber *4*. To limit the loss of current from the interaction space, vacuum chambers (drift tubes) of large diameter are used [12]. Using such devices allows approximately 20% increase in output power. In classical devices, end screens are used for this purpose on the

Table 4.1 Parameters of relativistic magnetrons

Parameter	Classical magnetron	Relativistic magnetron
Type of cathode	Thermoemission	Explosive emission
Voltage	Less than 50 kV	100–1500 kV
Current	Less than 0.1 kA	3–100 kA
Voltage pulse duration	1–20 µs or continuous operation	from 30 ns to 1.5 µs
Steepness of the voltage pulse front	up to 100 kV / µs	up to 100 kV/ns
The duration of the microwave pulse	Corresponds to the duration of the voltage pulse	from 20 ns to 1.2 µs
Power	Less than 10 MW	100–10 000 MW
Radiation wavelength	0.8–60 cm	3–30 cm
Efficiency	50–82%	Less than 30%

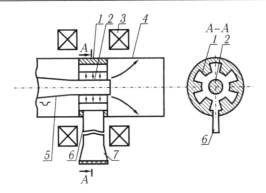

Fig. 4.1. Typical RM construction: *1* – anode block; *2* – cathode; *3* – magnetic system; *4* – vacuum chamber; *5* – cathode holder; *6* – waveguide output of power; *7* – antenna.

cathode, but for relativistic magnetrons they are not used to avoid 'parasitic' electron emission and breakdown.

Electrons are emitted from the cathode surface when an electric field strength is created between the cathode and the anode, exceeding the critical value for excitation of explosive electron emission accompanied by evaporation of the material and formation of a cathode plasma. The axial magnetic field formed by the magnetic system *3* causes the electrons to move along the cycloidal trajectories and prevents their rectilinear movement to the surface of the anode. The microwave energy from the resonators of the anode block is emitted into the free space by means of the waveguide output of the power *6* and the antenna *7*.

As sources of power for relativistic magnetrons, high-current electron accelerators containing pulse voltage generators discharged through the forming line to a magnetron diode were initially used. Despite the fact that the voltage pulses had a nanosecond duration, the shortcomings of the relativistic magnetrons were associated with the destruction of the anode blocks for several hundred pulses. This process is caused by the following factors: 1) high specific power of the anode current, leading to evaporation, erosion, mechanical deformation of elements under the action of thermal shock; 2) lack in matching of the internal resistance of the forming line of the HCEA (2–24 Ohm) with the impedance of the relativistic magnetron (40–100 Ohm); 3) the appearance of repeated pulses due to incomplete matching of the power source and the relativistic magnetron and additional release of energy in a magnetron diode. When a relativistic magnetron operates in a pulse–periodic mode, when a large number of pulses are collected during short time intervals, it becomes particularly necessary to perform thermal calculations on the surface of the resonator system to determine the parameters of electron beams that allow a long operation of the device. Thermal processes on the surface of the anode blocks of the RM are discussed in one of the paragraphs of this chapter.

The element base of the HCEA on the basis of pulse voltage generators allows them to operate exclusively in a single mode or, in extreme cases, with a frequency of one hertz. The experiments with a relativistic magnetron performed for the first time in the TPU in a pulse-periodic regime (a packet of three pulses with a repetition rate of 160 Hz was generated) stimulated the development of research in this direction. Linear induction accelerators were used in these and subsequent experiments.

The technical problem of achieving both high peak and high average power is that in the frequency mode it can: 1) release material from surfaces that increases the pressure in the system and causes breakdown in strong electric fields (~100 kV/cm) by the pulse sequence; 2) the emission of electrons from the surface of the cathode in the breakdown of a monolayer (gas, oil, water, etc.) is stopped (hampered) before it recondenses to the next impulse. Prior to the beginning of the studies, it was assumed that the formation of plasma due to evaporation of the monolayer occurs faster than in the case of explosive emission of metallic microedges. The issue of vacuum poisoning can be solved by applying the best high-vacuum equipment. However, these limitations can only be detected when operating at high peak power (>100 MW) and high frequency (>100 Hz).

It is the pulse-periodic mode of RM operation that seems most promising in terms of practical application of devices for creating compact radiating systems with high output parameters for pulse power, efficiency, pulse repetition rate. The following directions are singled out in which the use of microwave sources with a high average power is especially topical: microwave power supply systems for linear resonant electron accelerators with a high rate of acceleration; radar, including non-linear; research on the electromagnetic compatibility of radioelectronic equipment; sterilization.

The appearance of consumers of high-power microwave radiation determined the list of requirements for such complexes:

1) high reproducibility of output pulses in amplitude and shape;

2) the voltage range is not more than 500 kV to ensure satisfactory X-radiation protection at low material costs;

3) small weight dimensions;

4) 'convenient' for the user type of the emitted wave;

5) the possibility of frequency tuning;

6) low level of side oscillations and harmonics;

7) durability of the elements;

8) the maximum efficiency of converting the primary storage energy into electromagnetic radiation;

9) high repetition rate;

10) the duration of the microwave pulse, close to the duration of the power pulse.

The last three parameters determine the average power of the installation, which is its most important characteristic: $P_{av} = P_{im} f \tau$ (where P_{im} is the pulse power of the RM, f is the pulse repetition frequency, and τ is the pulse duration).

To obtain a high average power of electromagnetic radiation, it is necessary to realize a high frequency of repetition of pulses. This task is of a complex nature and is related to the development of: 1) power supplies for specific parameters (voltage, current, pulse duration) most fully conforming to RM requirements; 2) magnetic and vacuum systems permitting long-term operation of RM with a high repetition rate; 3) directly pulse–periodic relativistic magnetron generators with increased durability of the elements.

4.1. Relativistic magnetrons powered by LIA with multichannel dischargers

The first studies of relativistic magnetrons in the pulse-periodic mode of operation were carried out in Russia (TPU) in 1986 [13]. The appearance of the installation is shown in Fig. 4.2. The scheme of the experiment is presented in Fig. 4.3. As a power source, a LIA with 12 ferromagnetic cores and a 24-channel spark discharger switching 8 parallel DFLs was used. In the nominal mode, the charging voltage of the lines was 40–42 kV, which provided a voltage on the cathode of ~400 kV at a load of 130 Ohm. The accelerator operated at a repetition rate of up to 50 Hz in a continuous mode and up to 160 Hz in a burst mode of 5–10 pulses. The pulse repetition rate in the continuous mode was determined by the speed of the gas circulation of the working volume of the discharger. In the burst mode, the frequency was limited by the time characteristics of the charging circuit of the DFL. We note that the LIA has an internal resistance that is in good agreement with the impedance of the relativistic magnetron. In the formulation of research, this circumstance was the key. In the experiments, a 6-cavity magnetron of a 10-cm wavelength band was used. The magnetic field was formed by two coils forming

Fig. 4.2. Appearance of a pulse-periodic relativistic magnetron with a power source – LIA with a multichannel discharger (1986).

a Helmholtz pair. The power supply circuit of the magnetic system operated in the single-pulse mode, however, a large current pulse duration (~100 ms) made it possible to realize a packet of three 80 ns magnetron voltage pulses with a repetition rate of 160 Hz at an almost constant magnetic field.

In order to eliminate the 'parasitic' reverse current, which leads to the breakdown of the radial high-voltage insulator *8* of the electron gun, a protective screen *9* was installed on the cathode-holder. The outer diameter of the screen was chosen so that the magnetic field lines outgoing from the interaction space of the magnetron *11*, crossed its surface. Such a design intercepted the reverse flow of electrons from the RM interaction space and solved the problem of breakdown of the insulator, which allowed for a long series of experiments.

The maximum microwave radiation power of a relativistic magnetron reached 360 MW, which corresponds to an electron efficiency of ~30%. Since there was no voltage sensor in the experimental scheme, estimates of the accelerator voltage were made on the basis of measurements of the synchronous magnetic field under the assumption $\beta_p \approx \beta_e \approx 0.34$. They amounted to ~300 kV. A characteristic feature of the work of a relativistic magnetron during feeding from a linear induction accelerator in this experiment was the growth of the working current in the region of synchronous magnetic fields. This phenomenon, typical of high-efficiency low-voltage magnetrons, was noted earlier only for inverted relativistic magnetrons. Such a mode of current resonance is possible at comparable values of the static electric field strengths and electric components of the high-frequency field, and also at low current losses due to the 'parasitic' emission of the cathode-holder surface and the end current magnitude due to the relatively low cathode–anode voltage. A high electronic efficiency was achieved precisely due to these factors.

In the experiments, a good repeatability of the amplitude and shape of the recorded pulses was noted. Analogous pulses were obtained also with a single operating mode of the accelerator. Attention was drawn to the close coincidence of the duration of the current pulses and the microwave signal, which indicates a high efficiency of using the accelerator energy and a short oscillation time not exceeding several nanoseconds.

Thus, the first experiment demonstrated the possibility of the relativistic magnetron operating when fed from the LIA in a pulse-

Fig. 4.3. Appearance of pulse–periodic RM with a section of LIA located on a mobile platform.

periodic mode with a repetition rate of at least 160 Hz. Subsequently, several RMs were fabricated using LIA with multichannel dischargers to conduct experiments on the addition of microwave power from two magnetron outputs, and also on the addition of power from two magnetrons fed from two LIAs [14].

The result of the first studies was the development in 1988–1989 of a pulse–periodic relativistic magnetron placed on a mobile platform (Fig. 4.3) [15]. The experimental scheme of the installation is shown in Fig. 2.8. The output parameters of the relativistic magnetron reached 200 MW, which corresponds to an electron efficiency of ~20%. In addition, the pulse-periodic relativistic magnetron was equipped with a harmonic filter for non-linear radar detection of radioelectronic devices and a high-directivity pyramidal antenna of microwave radiation.

The systems described above were used to study the effect of high-power electromagnetic radiation on various semiconductor elements, radioelectronic devices, and biological objects.

Synchronization of relativistic magnetrons. By the mid-80s of the last century, the registered capacities of RM with HCEA reached the limit levels for the electrical strength of resonators and radiation output devices. An increase in the volume of the resonators V leads to mode competition, since the number of modes is proportional to $(V/\lambda)^3$. Therefore, in order to further increase the power (up to 10 GW and higher) research on the creation of sources on the basis of phase-locked (synchronized) relativistic generators began in various scientific centres. It was assumed that the most natural way to ensure the coherent operation of many powerful generating devices (auto-generators) is their synchronization due to mutual coupling. With the spatial formation of radiation, this should have led to an increase in the density of the power flux, which is proportional to the square of the number of sources.

The beginning of research of the processes of mutual synchronization of several magnetrons can be attributed to the 40-ies of the last century. When creating multi-generator microwave systems, regardless of the specific implementation of the device and the range of operating frequencies, a number of general requirements were advanced. These include the high efficiency of summation of power, high quality of the output signal spectrum, and also the most important condition is the stability of the operating mode. The stability of the operating mode is understood as the ability of the system to retain the specified characteristics (primarily phase ones) of the coherent mode with admissible deviations of the amplitude and frequency characteristics of the partial oscillators, with changes in loads, the quality of the elements of the communication circuits (channels), the parameters of the power sources, etc.

The first reports on the experimental study of powerful autogenerator systems began to appear in the late 80's and early 90's of the last century [16, 17]. Researchers from the United States managed to synchronize two relativistic magnetrons of a 10-cm wavelength band. The magnetrons were fed from a common high-current electron accelerator through a branched magnetically isolated transmission line and connected to each other by a waveguide with a length multiple of the operating wavelength. They worked initially at 2.5 GHz for the π-type and at a frequency of 3.8 GHz for the 2π-type and had a frequency detuning of less than 50 MHz. Their power is about ~1.5 GW. In the experiment, the power flux density in the far zone was doubled when each magnetron was operating on its emitting antenna. Phase synchronization was observed only 5 ns after the start of generation. Later on, the same researchers attempted to synchronize modules, including four to seven relativistic magnetrons [16]. In the same place, various ways of combining generators into modules were considered. To construct such sources, it was proposed to use magnetrons with power outputs from several resonators. In [17], a study was reported of the degree of coherence of magnetron oscillations by means of phase measurements. In the experiments the stability of the synchronous mode was not satisfactory: the synchronization band did not exceed 50 MHz, the coherent mode was observed only in half of all pulses. This, apparently, was the result of non-optimal interactions of generating devices connected with each other by different waveguide communication channels. Probably, therefore, interesting studies were curtailed.

More optimal interaction of the generators was realized in experiments conducted at the TPU [18, 19]. In them, mutual synchronization of two relativistic magnetrons was carried out. In contrast to the American studies, in this case the stability of the synchronous regime was ensured by the inclusion of a dissipative element in the waveguide channel, a common load, which provided the so-called resistive coupling [20]. Such a connection is optimal from the point of view of the stability of synchronous sinphase, antiphase or close to them modes. A distinctive feature of communication channels with dissipative elements is a wide range of adjustment of phase parameters. This leads to low criticality of coherent systems in tuning, and with strong links – a wide synchronization band, which is very important from a practical point of view.

The experiments were carried out at the research facility (Fig. 4.4), which allowed to study various variants of the RM connection, including the mode of summation or subtraction of powers in the total load. Two RMs of a 10-cm wavelength range with identical sector-type resonators were used. Two LIAs were used as power supplies of magnetrons. Accelerators were fed from the general power system. Synchronization of LIA was carried out by a common start-up discharger. The spread in the operation of the accelerators did not exceed 5 ns. The RMs had one or two power outputs from the resonators of the anode block and operated at 3 GHz. The RMs were connected to each other and the common load (antenna A_2) by the waveguide. The total length of the communication line was selected in the course of various experiments. The second RM outputs were loaded with matched antennas A_1 and A_3. Approximately half the power of each magnetron was output through these antennas. The control of microwave radiation from the side antennas made it possible to easily diagnose the operation of the magnetrons both during their joint operation and in an autonomous mode.

At the total electrical length of the communication channel, approximately equal to $2n\pi$ (the length of the waveguide communication lines was 7–8 wavelengths), summation of the powers in the total load A_2 occurred in the symmetrical system. The total power in this load was 1.9 times higher than the power of an individual magnetron and reached 190 MW (Fig. 4.5).

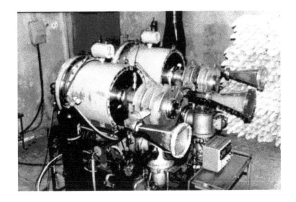

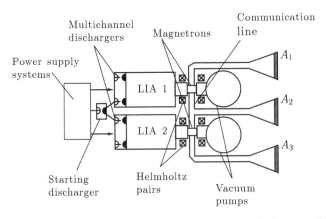

Fig. 4.4. Appearance and installation scheme for experiments with phase synchronization of relativistic magnetrons: A_1, A_2, A_3 – output antennas.

With the total electrical length of the communication channel $(2n+1)\pi$, the anti-phase mode was stable in the system, with subtraction of the oscillations in the total load A_2. The radiation level through this antenna was ~25% of the power of an individual magnetron. A sufficiently high level of radiation was due to the difference between the partial the powers of the magnetrons and the detuning of their own frequencies [21]. In the case of RMs with two outputs, radiation in the antiphase mode was output through the side antennas and, when the phase shift was compensated, was summed in space. It should be noted that it was not possible to determine the limiting permissible frequency detuning of magnetrons (synchronization bands). At the same time, in some experiments, stable synchronization took place for detuning of the order of ~60 MHz.

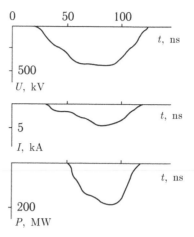

Fig. 4.5. Forms of pulses of voltage, current and microwave signal.

4.2. The basic elements of the pulse–periodic relativistic magnetron

A new stage in the development of pulse–periodic RMs is associated with the emergence of LIA on magnetic elements (early 90s of the last century) [22]. An example of such sources are the two options of the microwave generators based on LIA 04/4000 [23, 24] and LIA 04/6 [25]. The design, principle of operation and the element base of such accelerators were described in Ch. 2. Since LIAs on magnetic elements allow to significantly increase both the repetition rate of pulses and the number of pulses in a continuous series, additional requirements arise for the elements of the relativistic magnetron, the magnetic and vacuum systems. This was the subject of research, which will be discussed below.

Conditionally, it is possible to distinguish the following basic elements of the considered installations (the corresponding block diagram is shown in Fig. 4.6):

1) a linear induction accelerator consisting of a high-voltage generator and a power source. It ensures the operation of the installation with a repetition rate of up to 320 Hz with a high repeatability of the amplitude–time characteristics of the pulses (instability less than 5%);

2) high-voltage generator of the LIA, consisting of a magnetic pulse generator, forming lines and induction system. It is intended

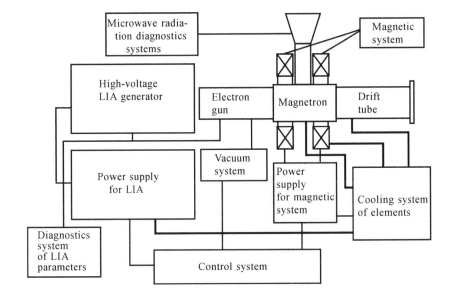

Fig. 4.6. Block diagram of a pulse–periodic relativistic magnetron.

for the formation of high-voltage pulses at the central electrode (the cathode holder of the relativistic magnetron);

3) the power supply of the LIA with the stabilization of the voltage level of the charging pulses of the LIA and the energy recuperation;

4) the control system of the installation, which determines the pulse repetition rate, the number of pulses in the packet, the interval of pause between packets, the automatic or manual start of the packet, sets the amplitude of the charging pulse voltage, the magnitude of the magnetic field, is provided with locks that exclude incorrect operation of the installation;

5) a vacuum system designed to produce a vacuum of no higher than $2 \cdot 10^{-5}$ Torr in a vacuum chamber. It has additional channels for pumping the anode block from the side of the drift tube and the waveguide output of power;

6) a system for diagnostics of the parameters of LIA, consisting of capacitive voltage dividers and Rogowski coils;

7) a system for diagnostics of the microwave radiation parameters, intended for measuring the amplitude, energy and frequency of pulses. It includes detectors, directional couplers, tunable attenuators, bandpass filters, etc.;

8) a relativistic magnetron containing a graphite cathode of special geometry for reducing the loss current and a water-cooled anode block with a microwave output device;

9) a magnetic system that is a Helmholtz pair and is powered by a constant current source. It creates in the interaction space of the magnetron a uniform magnetic field with induction up to 0.55 T and ensures continuous operation of the unit. The magnetic system is made of a hollow copper tube with a central hole for the flow of cooling water;

10) a water cooling system that establishes the required thermal conditions of the anode block, drift tubes, magnetic field coils, semiconductor elements of the LIA power supply and the magnetic power supply system.

4.3. Construction and thermal calculation of the anode block

Experimental studies carried out at the Tomsk Polytechnic University and other organizations showed that for a stable operation of the device its slowing-down system should have a small number of resonators. Therefore, 6- and 8-cavity anode blocks of sector type were chosen for the pulse-periodic regime. Differing in simplicity of design, the slowing-down structure of this type nevertheless satisfactorily adapts to the operating c0onditions with the power source (LIA). A characteristic feature of LIA on magnetic elements is the change in the amplitude of the voltage during the pulse. At the initial time, the accelerator operates at idle, since the anode current of the relativistic magnetron is small. As the oscillations develop, the current increases, the impedance of the load decreases, which leads to a decrease in the amplitude of the voltage. In these conditions, for the stable operation of the magnetron, it is necessary to maximize the separation of the operating mode of oscillations from 'parasitic' ones with respect to the phase velocity. With an appropriate choice of the geometry of the anode block, this can be achieved. In addition, a very important condition for relativistic magnetrons is the possibility of operating at low values of the magnetic field, such that electromagnetic systems can be used with direct current. The fulfillment of this condition is necessary when working with a high repetition rate of pulses. All this imposes significant restrictions on the choice of the geometry of the resonator system.

One of the most stringent requirements for a pulse-periodic magnetron is the durability of the cathode and the anode block both in terms of the total number of pulses and the number of pulses in a separate series with high stability of the generated frequency. These parameters of the magnetron are associated with the processes of destruction of the anode surface and a change in the emissivity of the cathode material.

As a result of calculations and recommendations, the following geometric dimensions of the resonator systems of magnetrons were chosen: the inner radius of the anode block is $R = 2.15$ cm; depth of resonators $D_r = 4.30$ cm; number of resonators $N = 6$; angle of opening of resonators $2\Theta = 40°$; cathode radius $r_c = 1.0–1.2$ cm; the length of the anode block $L = 7.2$ cm ($N = 8$; $2\Theta = 22.5°$; other dimensions coincide). The slowing-down of the electromagnetic wave for the operating π-mode of the oscillations is $\beta_p^{N=6} \sim 0.45$ and $\beta_p^{N=8} = 0.366$, and the required the intensity of the magnetic field is equal to 3–5 kG for the level of the output voltages of the LIA of 300–400 kV. Here $\beta_p = 4\pi R/\lambda N \approx \beta_e = cE_0/H$ (where λ is the wavelength of the oscillations), i.e., the phase velocity of the electromagnetic wave for operating π-oscillation mode with the number $N/2$ approximately equal on the drift velocity of electrons in crossed electrostatic (E_0) and magnetostatic (H) fields.

The relativistic magnetron with an 8-cavity anode block was designed to increase the efficiency of the installation due to the growth of the electronic efficiency. The excitation of the operating π-mode of oscillations in a given magnetron occurs at higher synchronous magnetic fields. In this case, the radius of cyclotron rotation of electrons decreases, which should lead to an increase in the electronic efficiency of the device in accordance with the estimated formula

$$\eta_{el} = 1 - \frac{2r_c}{d}\gamma_p, \tag{4.2}$$

where r_c is the radius of cyclotron rotation of the electrons; $\gamma_p = (1-\beta_p^2)^{-1/2}$.

For a correct comparison of the results of the experiments, both anode blocks had the same geometric dimensions (cathode and anode radii, internal radius of the resonators), which ensured the same load of the power source in the initial part of the pulse. Table 4.2 shows the results of calculating the wavelengths and the slowing-down rates

Table 4.2. Results of calculations

N = 8		N = 6	
3π/4-mode of oscillations		2π/3-mode of oscillations	
λ, cm	9.305	λ, cm	10,093
β	0.448	β	0.669
β_{-1}	0.29	β_{-1}	0.335
β_{+1}	0.132	β_{+1}	0.167
λ, cm	9.68 *	λ, cm	*10.4*
π-mode of oscillation		π-the mode of oscillation	
λ, cm	9.223	λ cm	9.77
β	0.366	β	0.461
β_{-1}	0.366	β_{-1}	0.461
β_{+1}	0.122	β_{+1}	0.154
λ, cm	*9.43*	λ, cm	*10.6*

* Italics - measurements of P2-78.

of the electromagnetic wave of the operating mode of oscillations closest to the competing $(N/2 - 1)$ and its (-1) and $(+1)$-harmonics, and also the results of 'cold' measurements at a low power level using a microwave panoramic meter P2-78 (in italics).

Some difference in the calculated and measured wavelengths of oscillations is caused by the presence of a waveguide output of power, which was not taken into account in the calculations. As can be seen from the table, the separation of the modes by the value of the phase velocity for the 6-resonator relativistic magnetron microwave generator is $\beta_\pi/\beta_{2\pi/3} \approx 27\%$, and for its (-1)-harmonics is equal to $\beta\pi/\beta_{2\pi/3}^{-1} \approx 31\%$. Separation of the types of oscillation for the 8-cavity anode block is less than for the 6-cavity anode block, by about a third. It is $\beta_\pi/\beta_{3\pi/4} \approx 21\%$ and $\beta_\pi/\beta_{3\pi/4}^{-1} \approx 24\%$. It also follows from the table that the value of the slowing-down of the operating type of oscillations of the 8-cavity anode block in comparison with the 6-cavity anode block is approximately 26% lower. Thus, from this block we can expect a higher efficiency of work. On the other hand, a small amount of slowing-down necessitates the operation of the magnetron at higher values of the synchronous magnetic field.

4.3.1. Thermal processes in a relativistic magnetron

The thermal calculation presented below can be used to analyze thermal processes not only on the surface of anode blocks of relativistic magnetrons, but also on the surface of collectors of O-type devices.

The anode block of the magnetron is heated by several sources: radiation from the cathode surface, electron bombardment and the circulation of high-frequency currents along the surface of the resonators. For approximate calculations, the heating of lamellae due to radiation and microwave currents can be neglected because of their small influence [26]. The increase in the temperature of the resonator system leads to the deformation of the lamellae and, consequently, to a change in the wavelength. In addition, as the temperature rises, the active losses increase, that is, the intrinsic Q-factors of the oscillating system and the circuit efficiency decrease.

In the RMs, the power density of the electron beam deposited on the surface of the anode block becomes so significant that it is possible to develop evaporation, erosion, and mechanical deformation processes under the action of thermal shock. At the same time, the pulse duration is so short that the power dissipated at the anode does not have time to be directed outward, but even spread over the entire thickness of the anode, causing instantaneous pulse heating. The surface layer of the metal is heated during each pulse, and cools during pauses between the pulses. Expansion and contraction of the surface layer with respect to the neighbouring unheated layer causes internal stresses, under the influence of which inhomogeneities are formed. With a large number of impulses, inhomogeneities unite and microcracks appear. Microcracks prevent the process of heat transfer from the surface layer into the interior of the metal. Therefore, temperature of the layer increases with each following pulse, which leads to melting and evaporation of the material. Thus, the main channels for dissipating the energy of an electron beam from the surface layer are thermal conductivity, the propagation of thermomechanical stresses, phase transitions, and evaporation of the material. Depending on the specific density of the energy released and on the pulse duration, it is possible to develop one or other of these processes.

Experimental studies of relativistic magnetrons with the use of the high-current electron accelerators in the single-pulse mode showed a short longevity of the anode blocks (the number of pulses often did not exceed one hundred). In this case, fractures were subjected to separate parts of the lamellae in the region of sharp edges. At the same time, with the pulse–periodic regime, the lifetime of the anode block must be at least 10^6 impulses. Therefore, it is necessary to evaluate the limiting modes of RM operation, which do not lead to destruction of the anode blocks, depending on the electron beam energy flux density, repetition frequency and pulse duration, electron beam parameters, and suggest measures to increase the longevity of the anode device blocks. To do this, it is required to determine the areas of deposition of electrons, the depth of their penetration and the uniformity of heating, to consider the process of heat dissipation during the supply pulse and in a pause between pulses.

Areas of electron deposition on surface of the anode block. In both classical and relativistic magnetrons, electrons move along trajectories close to cycloidal ones and fall on the surface of the anode at different angles, determined by the ratio of the radial and tangential velocity components near the anode (tan $\alpha = r'/\phi'$) and the phase of cyclotron rotation [26]. The areas of electron deposition are most easily determined from the traces of erosion of lamellae obtained in the high power modes of operation of a relativistic magnetron. Such regimes were implemented in experiments on the HCEA Tonus-1 installation. At the same time, the erosion of the angles of the lamellae was fairly uniform along the length of the anode block. This fact indicates the presence of areas that are strained in terms of heat generation. Indeed, heat fluxes from the electrons incident on the cylindrical and lateral surfaces of the lamellae are summed at the lamella angles. We note that at the edges of the lamellae, the static and high-frequency electric fields increase and the anode current density increases. To reduce this effect, a radius of 0.5 mm was used to round the edges of the lamellae of the anode blocks of relativistic magnetrons, which amounts to ~3% of the width of the lamella. A further increase in the radius of the lamellae leads to a redistribution of the amplitudes of the spatial harmonics of various modes of oscillation in the interaction space and, as a consequence, to a decrease in the electronic efficiency [27].

Thermal regime of the anode block of the relativistic magnetron For pulsed heating with a duty cycle of more than 10^3, the rise in surface temperature over time is described by expression

$$T(t,\tau) = T_{av}(t) + T_{im}(\tau), \tag{4.3}$$

where $T_{av}(t)$ is the average value of the temperature of the surface heated by the pulse packet; $T_{im}(\tau)$ is the temperature jump due to heating by a single pulse.

To determine the impulse component of the surface temperature, it is necessary to take into account the specific features of the RM operation. First, for an impulse the duration of which is from several hundredths to several tenths of a microsecond, an energy with a density of the order of 10^7–10^8 W/cm² is released on the working surface of the RM anode, warming up the latter. Secondly, the thermal state of the anode is affected not only by the power density of the incident electron beam, but also by the energy of the electrons with which they reach its surface. In the RM, the electron energy reaches hundreds of keV, so that the electrons penetrate into the anode material to a depth comparable to the depth of penetration of heat, thereby reducing the temperature of its surface. This phenomenon is taken into account by means of a correction factor $G\left(\dfrac{d_{el}}{2\sqrt{a\tau}}\right)$, called the penetration function, which depends on the ratio of the depths of penetration of electrons (d_{el}) and thermal field ($\sqrt{a\tau}$) inside the anode material [28] (where a is the thermal diffusivity coefficient and τ is the pulse duration).

The greatest idealization in solving this thermal problem is that the power density of the electron beam is taken to be uniformly distributed over the working surface of the anode. Thus, the impulse component of the temperature is calculated by the formula for the surface thermal shock taking into account the penetration depth of the electrons:

$$T_{im}(\tau) = \frac{2\,P_0\sqrt{a\,\tau}}{\lambda_m\sqrt{\pi}}\,G\left(\frac{d_{el}}{2\sqrt{a\tau}}\right). \tag{4.4}$$

Here P_0 is the surface density of the pulsed power; λ_m is the coefficient of thermal conductivity of the anode material.

The thermal shock is the cause of significant temperature stresses on the working surface of the anode. Fatigue in the metal causes micro- and macrocracks to form, which increase the thermal resistance of the anode surface, and with further development can lead to its complete destruction. To avoid premature destruction of the anode by an electronic flux, the amplitude of the impulse temperature on the surface should not exceed a safe value ΔT_b:

$$T_{im}(\tau) < \Delta T_b. \tag{4.5}$$

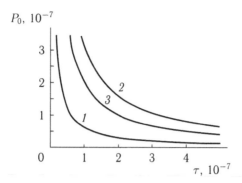

Fig. 4.7. Dependences of the electron beam power density on the pulse duration at $T_{im}(\tau) = \Delta T_b$: *1* – copper; *2* – molybdenum, *3* – stainless steel.

The value of ΔT_b is calculated by the formula [29]

$$\Delta T_b = \frac{2\sigma_T}{\alpha_t E_m}, \tag{4.6}$$

where σ_T is the yield strength; α_t is the coefficient of linear expansion; E_m is the modulus of elasticity of the material.

Figure 4.7 shows the dependence of the surface density of the pulsed power of the electron beam on the pulse duration for copper, molybdenum and stainless steel under the condition $T_{im}(\tau) = \Delta T_b$. The safe temperature values calculated using formula (4.6) are 74, 485 and 192°C, respectively. When operating with beams whose parameters are higher than the curves shown in Fig. 4.9, the surface of the anode block of the RM is destroyed in a single-pulse mode.

The component $T_{av}(t)$ in the formula (4.3) can be found by solving the problem of heating an unbounded plate of finite thickness where one of the surfaces is heated by a flux with a power equal to the average pulse power of the electron beam and the other one is forced to cool according to Newton's law. Its solution for the heated surface has the form

$$T_{av}(t) = T_0 + \left[1 + \frac{1}{Bi} - \sum_{n=1}^{\infty} A_n \exp\left(-\mu_n^2 \frac{at_p}{h^2}\right)\right] \frac{P_0 h}{v \lambda_m}, \tag{4.7}$$

where T_0 is the initial temperature of the anode; t_p is the duration of a packet of pulses; $Bi = \dfrac{\alpha}{\lambda} h$ is the Biot criterion, depending on the thickness of the anode and cooling conditions; α is the heat transfer coefficient (for water cooling $\alpha \sim 0.05$–1 W/cm$^2 \cdot$ deg, for the air $\alpha \sim$

10^{-3}–10^{-2} W/cm^2 · deg); λ_m is the coefficient of thermal conductivity of the anode material; h is the thickness of the plate; v – the off-duty factor of pulses; $A_n = \dfrac{2(\mu_n^2 + \mathrm{Bi}^2)}{\mu_n^2(\mu_n^2 + \mathrm{Bi}^2 + \mathrm{Bi})}$; μ_n are the roots of the characteristic equation ctg $\mu_n = \mu_n/\mathrm{Bi}$. As follows from expression (4.7), $T_{av}(t)$ is determined by the average value of the power of the electron flux P_0/v and depends on the heat removal rate from the anode surface. If $0.1 < \mathrm{Bi} < 100$, then the heat dissipation rate is defined by both the transfer of heat inside the material by means of thermal conductivity, and by the transfer of heat from the surface of the material to the cooling liquid by heat transfer. At $\mathrm{Bi} < 0.1$, the heat removal intensity is determined only by transfer.

Thus, the equation for calculating the temperature of the working surface of the anode at a high off-duty ratio will have the form

$$T(t,\tau) = T_0 + \left[1 + \frac{1}{\mathrm{Bi}} - \sum_{n=1}^{\infty} A_n \exp\left(-\mu_n^2 \frac{a t_p}{h^2}\right)\right] +$$
$$+ \frac{P_0 h}{v \lambda_m} + \frac{2 P_0 \sqrt{a \tau}}{\lambda_m \sqrt{\pi}} G\left(\frac{d}{2\sqrt{a\tau}}\right). \tag{4.8}$$

It is obvious that heating in the pulse-periodic mode is aggravated by the presence of a constant component $T_{av}(t)$, with respect to which temperature jumps occur. Reduction of this component is possible due to the intensification of cooling, reducing the thickness of the anode at the sites of deposition of electrons. In this case, the temperature above which the anode surface may be destructed can be determined with sufficient accuracy by the melting point T_{melt} [29]:

$$[T] = \frac{1}{3} T_{melt}. \tag{4.9}$$

For the materials under consideration, the temperature calculated according to formula (4.9) is: for copper 360°C, for molybdenum 875°C, for stainless steel 467–475°C. It is lower than the temperatures recommended for the use of these materials during continuous heating (copper – 500°C [31], molybdenum – 1700°C [31], steel – 600°C [32]).

Since the impulse component of the temperature, limited by the condition (4.9), does not affect the durability of the anode, it is obvious that the bases A contribution to the disruption of the surface is made by the constant component of the temperature $T_{av}(t)$, which, as mentioned above, depends on the heat removal intensity. In this

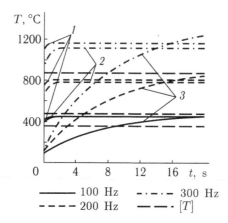

—————— 100 Hz   — · — · — 300 Hz
— — — — 200 Hz   — — · — [T]

Fig. 4.8. Dependence of the surface temperature of the anode on the duration of the pulse packet at $\tau = 10^{-7}$ s and $P_0 = 0.5 \cdot 10^7$ W/cm² : *1* – copper; *2* – molybdenum; *3* – stainless steel.

case, at $\alpha = 0.15$ W/cm² · deg, the Biot criterion calculated for copper, molybdenum and steel have values of 0.01, 0.03 and 0.3, respectively. Thus, the fraction of heat removed from the cooling surface from the heat supplied through the wall is several tens of times higher for steel than for copper or molybdenum.

Figure 4.8 shows the dependence of the surface temperature of the anode on the duration of the pulse packet with characteristic values for RM $\tau = 10^{-7}$ s and $P_0 = 0.5 \cdot 10^7$ W/cm² for pulse repetition frequencies of 100, 200 and 300 Hz [33]. From the data given it follows that copper can not be used for pulse–periodic RMs, and the use of molybdenum is limited by a repetition rate of slightly more than 200 Hz. Based on the calculations carried out, and also taking into account the high processability, availability and low cost of stainless steel, its use for the production of anode RM blocks can be considered a fairly successful solution.

The above calculations of the thermal state of the surface of the anode blocks of the RM are approximate. It is known that in a relativistic magnetron a considerable part of the anode current does not reach the surface of the resonator system due to drift in the axial direction. This leads to a decrease in the efficiency of the device but, at the same time, to the weakening of the heat load of the anode. On the other hand, a number of simplifications were introduced in the calculations, such as the rectangular shape of the pulse, the uniform distribution of the power flow, and additional heating sources were not taken into account. Similarly, the depth penetration function of electrons was not determined. Nevertheless, the calculations are in satisfactory agreement with the results of experimental studies of the operating life of the anode blocks and allow us to work out the following recommendations for the construction of the RM anode blocks. Preference should be given to the RM anodes with sector-form resonators having a large lateral surface in comparison with resonators of the 'slot' or 'slot–hole' type. The edges of the lamellae need to be rounded to reduce the strength of the electric fields. The RM anode blocks should be produced from materials whose mechanical properties allow them to withstand sufficiently large jumps in the pulse temperature. The melting point of the material must be high enough to ensure that the maximum permissible operating temperatures can be reached. The thermophysical properties of the material should ensure the value of the Biot criterion in the range 0.1–100. It is necessary to cool the anode block lamellae. Cooling should be carried as close as possible to the most heat-loaded parts of the anode. Therefore, the thickness of the lamellae in the region of electron beam deposition should be as small as possible. The material should be accessible and have suitable technological effectiveness. All of the above requirements are best satisfied by stainless steel, although it is not a traditional material for microwave devices.

For devices operating in the pulse-periodic mode, it is permissible to use anode blocks with a low slowing-down of the electromagnetic wave (despite the reduction of the electronic efficiency). This leads to the appearance of high values of the energy of cyclotron rotation of the electrons. As a result, electrons penetrate deeper into the metal, reducing the temperature of pulsed heating.

4.3.2. The construction of the anode block of the relativistic magnetron

Based on the results of thermal calculations, the anode blocks of relativistic magnetrons (Fig. 4.9) are made of stainless steel. All permanent joints are made by argon-arc welding. The anode block is a cylinder 86 mm high with an external diameter of 116 mm. A technological boss is welded to the lateral surface of the anode block intended, firstly, for an accurate connection to the anode unit of the microwave energy output device, and secondly to ensure, as far as possible, deformation-free welding of the anode block to the output device. The internal cavity of the anode block in cross

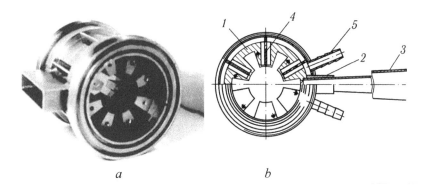

a   *b*

Fig. 4.9. Appearance of the anode block without a water cooling jacket and power output (*a*); drawing of the anode block of the 6-resonator pulse-periodic RM (*b*): *1* – anode block; *2* – boss; *3* – waveguide output of power; *4* – diaphragm; *5* – nozzle.

section consists of a central anode hole with a diameter of 43 mm and six sector-shaped resonators with an external diameter of 86 mm and angular dimensions of 20°. The height of the working part of the anode block is 72 mm. The inner surface of the anode block is manufactured by electroerosion technology, which allows obtaining the necessary accuracy and quality of the product.

In a series of six anode blocks, the resonant frequencies of the modes and the values of the loaded Q-factor were distinguished in the range of 0.3%. The most stringent requirements are imposed on the geometric dimensions of the central anode aperture (diameter $43^{+0.03}$ mm) and cavity slots ($7.4^{\pm0.08}$ mm), on the symmetry of the structure (the angle between the axes of the slits is 60 ± 10 min, slit asymmetry 0.05 mm), flatness of lateral lamellar walls (0.02 mm), arithmetic mean deviation of anode hole profiles and resonator walls (0.63 µm). To reduce the edge strength of the electric field in order to increase the durability of the anode block, the radius of the side edges of the lamellae is set equal to 0.5 mm. To remove the microwave energy from the anode block in the rear wall of one of the resonators, a rectangular 10 mm wide window with a height of 72 mm of a total thickness of 2 mm was made.

The cooling system of the anode block consists of internal and external circuits and a regulating device. The inner circuit is formed by rectangular cavities milled in the anode block lamellae from its outer side. The outer circuit is formed by the outer wall of the anode block and the inner surface of the shell. The regulating device consists six special diaphragms, partitioning the inner and outer contours along the radius and ensuring the sequential flow of cooling water through each of the six cavities in the lamellae and regulating the water flow along the radius and height of the cavities. Cooling water is supplied through the nozzle *5*.

The device for the output of microwave energy *3* of length 345 mm consists of a rectangular waveguide of variable cross section and a choke flange welded to it. The internal cross-section of the waveguide smoothly varies from 10×72 mm^2 at the point of docking with the lead-out window of the anode block to 34×72 mm^2 and a length of 312 mm. Pyramidal horns are used to output radiation to free space. Since the emission of radiation from the magnetron is carried out at the lowest wave of a rectangular waveguide, it is not difficult to calculate pyramidal antennas with a given directivity. The antennas are made evacuated. The aperture is covered by a dielectric window made of plexiglass. To prevent surface discharge one must select the appropriate window size. Assuming an operating power level of ~300 MW, the window size was 320×300 mm^2. At the dielectric strength of the outside (air side) surface of ~30 kV/cm and a pulse duration of microwave radiation of ~100–150 ns the output window of the antenna of a predetermined size is capable of tentatively transmitting through without breakdown ~600 MW. The length of the horn was chosen to be 730 mm, which formed the opening angle in the *H*-plane of 25°. Upon excitation by a wave H_{10} ($\lambda \sim 10$ cm) the calculated beam width between zero values is 60°. In 'cold' measurements of the antennas the parameters determined were the effective aperture area, the directivity ratio, the beam width in the *E*- and *H*-planes at different wavelengths corresponding to types of the oscillations of the resonant system. Microwave probes (loops) were placed on the lateral surface of the antenna for recording the output power. There was also a branch pipe for additional vacuum pumping of the interaction space of the magnetron through the communication window, power output and the antenna. To determine the power, vacuumed directional couplers were installed in some experiments between the output waveguide and the antenna.

4.4. Cathode of a pulse-periodic relativistic magnetron

The magnetron diode is structurally attached to the LIA module by means of a cantilever coupling. The mounting location of both consoles is the output flange of the LIA. The distance from the module flange to the centre of the anode block is 740 mm, and to the end flange of the drift tube is 1100 mm. Since the individual

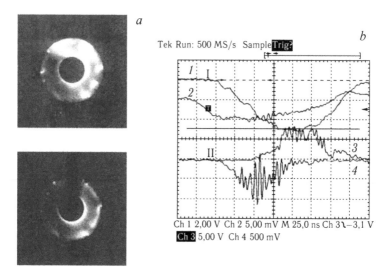

Fig. 4.10. Glow of plasma in the interelectrode gap of the RM at radial displacement of the cathode relative to the axis of the anode block (*a*); number of pulses 20 and 10. Oscillograms of pulses with a radial displacement of the cathode (*b*): *1* – current; *2* – voltage; *3*, *4* – signals of microwave detectors.

elements are rigidly connected through vacuum and oil resistant rubber gaskets (high-voltage insulator of the electron gun of the LIA), it is very difficult to align the cathode and the multi-resonator anode block at such a large length during assembly. This is illustrated by photographs of the interelectrode gap made during the experiments through the vacuum window of the drift tube. Figure 4.10 *a* shows that the volume discharge in the space between the anode and the cathode is non-uniform in the azimuth. One can draw a conclusion about the radial displacement of the cathode from the axis and the decrease in the interelectrode gap at this point. This circumstance substantially worsens the conditions for the settling of electromagnetic oscillations in the resonators of the anode system of the relativistic magnetron and affects the formation of electron spokes in the interaction space. Figure 4.10 *b* shows typical oscillograms of pulses of the total current, voltage and microwave radiation, demonstrating the work of a relativistic magnetron with a radial displacement of the cathode. A distinctive feature here is strong pulse modulation. We also note that an asymmetric discharge in the interaction space leads to local overheating and destruction of the surface of the anode block and the cathode.

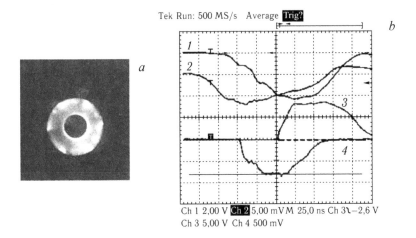

Tek Run: 500 MS/s   Average   Trig?

Ch 1 2,00 V Ch 2 5,00 mV M 25,0 ns Ch 3ᴸ−2,6 V
Ch 3 5,00 V Ch 4 500 mV

Fig. 4.11. The glow of plasma in the interelectrode gap of the RM using the alignment unit (*a*) and the corresponding oscillograms of the pulses (*b*): *1* – current; *2* – voltage; *3*, *4* – signals of microwave detectors.

Figure 4.11 shows a photograph of the interelectrode gap of the RM during the installation of the cathode using an alignment unit and the corresponding oscillograms of the pulses. It can be seen that the discharge has a more uniform azimuth character, and the oscillograms have no modulations. This device demonstrated high reliability, providing good reproducibility of the output parameters of the relativistic magnetron when replacing cathodes and the durability of the anode block.

The thermal regime of the cathode is determined by electrons that are not captured in the generation mode, but fall into the accelerating phases of the microwave wave on the cathode. According to the known data on the heating of the cathode by 'inverse' ('irregular' phase) electrons, 5–10% of the power supplied to the magnetron is expended. However, there is no reliable data on the energy of the incident electrons. If we assume that the energy of the 'inverse' electrons is greater than 20 keV, then the impulse temperature of the cathode surface of stainless steel will reach 500°C. The admissible number of pulses even at the average cathode temperature $T_0 =$ 200 °C exceeds 10^5. In this case, the greatest danger for the stable operation of the magnetron is the loss of the emission capacity of the cathode at a significant number of pulses. Experimental data published recently [34, 35] indicate that in planar and coaxial diodes at voltages close to the conditions of these studies, but with a pulse duration of ~20 ns, the permissible number of pulses for cathodes made of different metals does not exceed 10^3–10^5. Therefore, considerable attention was paid to the choice of the cathode material.

Cathode material. Various laboratories tested cathodes made of stainless steel, graphite, and metal–dielectric cathodes. Cathodes made of stainless steel are traditionally used for relativistic magnetron generators, beginning with the first experiments. Their advantages are high resistance to electron bombardment by incorrectly phased electrons, high heat capacity, low material entrainment. The main disadvantage of such a cathode is low durability in terms of emissivity. As the operation increases, the amplitude of the voltage increases and the current decreases with a simultaneous decrease in the rate of rise of the front of the current. Also noticeable are the shortening of the duration and the appearance of instability in the amplitude of the signals from pulse to pulse.

A cathode is proposed in [36] for planar vacuum diodes, in which the metal–insulator contact surface serves as the emission surface. For such a cathode, the threshold for the formation of explosive electron emission is lower. Due to this, a greater number of emission centres are formed than cathodes made of homogeneous materials. Metal–dielectric cathodes for pulse–periodic RMs were produced as follows. Copper interlayers was interlaced with one or two mica gaskets. The outer diameter of the cathode was machined on a lathe. The surface was etched in nitric acid to remove copper burrs. Such cathodes from the point of view of obtaining high output power and generation stability at large (~10^5) the number of pulses exceed the stainless steel cathodes. However, they have limitations on the pulse repetition rate. Continuous operation of RM with a metal–dielectric cathode is possible with a frequency not exceeding one hertz, due to a drop in the vacuum caused by the large amount of the evaporated cathode material, both due to explosive electron emission and bombardment with incorrectly phased electrons. Operation of the device with such cathodes is possible with a frequency of up to 80 Hz in the mode of short bursts of pulses (no more than 20 pulses).

The best parameters for the criteria of durability, stability and the low threshold for the formation of explosive electron emission were demonstrated by pyrolytic graphite cathodes. From the oscillograms shown in Fig. 4.11 it can be seen that in this case the duration of the microwave pulses is close to the duration of the current pulses. Nevertheless, with prolonged continuous operation of the RM (more than 10^6 pulses), an increase in the amplitude of the voltage pulses and a decrease in the accelerator current were observed. Accordingly, the power of the microwave generation also fell. These facts indicated a decrease in the emission capacity of the cathode. Thus, abrasive treatment of its surface was required.

Effect of the dimensions of the cathode on the output characteristics of the RM. The adjustment of the RM to the maximum output power consisted in determining the outer diameter of the cathode and its length for synchronous magnetic field values. Figure 4.12 shows the dependence of microwave radiation power of a relativistic magnetron on the cathode diameter. The presence of a maximum on the $P(d_c)$ curve is explained by the best matching of the impedances of the linear induction accelerator and the relativistic magnetron, as well as by the physical processes of the interaction of electrons with the microwave waves of the anode block. Further experiments were carried out only with cathodes with a diameter of 19 mm.

The selection of the length of the cathode and the profile of its ends makes it possible to reduce the current leaving the interaction space (the end current of the drift tube). For pulse–periodic magnetrons operating at relatively low currents (3–6 kA), the end current is determined by the electric field strength at the edge of the cathode. In connection with this, it was proposed to use a cathode with a length exceeding the length of the anode block [37]. The end of the cathode is rounded with a radius equal to the radius of the cathode. Due to this, an uniform electron beam of the end current is formed [38]. It is known that the limiting transport current in the drift tube of an uniform beam is less than the limiting transport current of a coaxial beam.

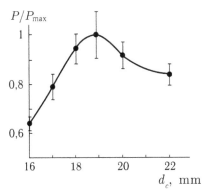

Fig. 4.12. Dependence of the microwave radiation power of RM on the diameter of the cathode; P_{max} = 170 MW.

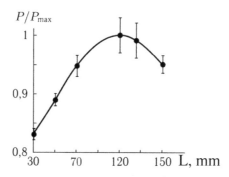

Fig. 4.13. Dependence of the microwave radiation power of RM on the length of the cathode; P_{max} = 200 MW.

Figure 4.13 shows the dependence of the output power of a relativistic magnetron on the length of the cathode. The maximum of the microwave power corresponds to a cathode length of 110 mm. An increase in length above 110 mm caused the end of the cathode to fall into the edge field of the magnetic system and the current losses increased.

4.5. Magnetic system

For the effective work of magnetrons, including relativistic ones, a magnetic field with high circular symmetry and maximum uniformity along the length of the interaction space is necessary; the diameter of this field is larger than or in proportion to its length. When designing, manufacturing and operating magnetrons, it is necessary to minimize fluctuations in the magnitude of the field along the anode block [39]. The production of a uniform magnetic field is based on the principle of superposition: two or more sources of the magnetic field are located relative to a given volume in such a way that the vector sum of the source field strengths is constant in this volume. The resulting magnetic field is uniform with a certain degree of accuracy, depending on the design of the system.

To create a uniform magnetic field in a certain interval (in this case in the operating region of the magnetron), it is necessary to fulfill the conditions for locating and selecting the geometric dimensions of the magnetic coils creating this field (the Helmholtz condition). Helmholtz coils are a system of two coils that create a uniform field in the axial region when the following conditions are met:

a) the distance between the geometric centres of the coils is equal to their average radius: $A = R_{av} = (R_1 + R_2)/2$, where R_1 is the inner radius, R_2 is the outer radius of the coils;

b) the thickness of the coil windings $K = R_2 - R_1$ and its L_c are small compared with A. If the cross section of the winding is chosen so that the condition $31K^2 = 36\ L_c^2$ is satisfied, the magnetic field represented as a series of Legendre polynomials will be uniform up to a term in the expansion containing r^4/A^4, where r is the radius of the interior volume ($0 < r < R_1$). When this condition is satisfied, we have $K/L_c = 1.077$ (ratio of the thickness of the winding to its length). The area in which a uniform magnetic field with an induction of up to 0.5 T is required is determined by the dimensions of the relativistic magnetron: the diameter of the cathode is 18–22 mm; the inner diameter of the anode is 43 mm; the length of the anode is 72 mm.

In our case, several additional requirements are imposed on the geometric parameters of the investigated magnetic system, which are due to its design features. This is primarily a limitation on the outer radius of the magnetic coils R_2, which is associated with the overall dimensions of the installation and is limited by lag studs. The internal radius R_1 was chosen by a compromise between several requirements that are opposite in their functions. On the one hand, a decrease in R_1 leads to the possibility of creating the necessary magnetic field due to additional turns without a sharp increase in the current density in the magnet winding. On the other hand, this violates the first Helmholtz condition, since with the existing design parameters, the value of A (the distance between the geometric centres of the coils) already exceeds the average radius R_{av}. In addition, this causes a decrease in the transverse dimensions of the drift tubes. Consequently, losses are increasing due to the end current of the RM. Further reduction of the distance between the coils is limited by the size of the waveguide for the output of microwave energy, whose width is 80 mm.

Analogous arguments can also be given regarding the length of each magnetic coil. Thus, an increase in L to the goal of adding additional turns and thus obtaining the required field 0.5 T without substantially increasing the current in the windings leads to an increase in the distance between the geometric centres of the coils (A) and, of course, to a violation of the uniformity of the magnetic field.

Within the framework of these limitations, the following dimensions of the magnetic system of the relativistic magnetron were chosen to maximize the configuration of the Helmholtz coils:
– the radii of the coils R_1 = 100 mm and R_2 = 236 mm;
– the width of each coil L_c = 100 mm;
– the distance between the coils is 90 mm.

The calculated axial non-uniformity of the magnetic field in the RM working area did not exceed 3%.

4.5.1. Calculation of the thermal regime of the magnetic system

To operate the relativistic magnetron with an output voltage of LIA ~400 kV, a magnetic field with an induction of 0.4–0.5 T is required, which corresponds to a current density in the conductor j of ~8.4–10.5 A/mm² (preliminary calculations were carried out for copper bar with a cross section of 3.05 × 3.28 mm²). Effective cooling is necessary to remove heat generated in the magnetic system. Therefore, at the design stage, calculations of the thermal state of the coil windings were performed. A numerical simulation method based on the solution of the finite-difference heat conduction equations with internal heat sources was used [39]. The task was to determine the maximum temperatures in the windings in order to select the optimal design of the cooling system for the implementation of the continuous mode of RM operation.

The temperature field in the magnet coils is described by the heat conduction equation of the form

$$\frac{\partial T}{\partial \tau} = a\left(\frac{\partial^2 T}{\partial r^2} + \frac{v}{r}\frac{\partial T}{\partial r} + \frac{\partial^2 T}{\partial x^2}\right) + \frac{q_v}{\rho_{mat}C_p}. \tag{4.10}$$

Here we use the effective thermophysical properties of the coil, considered as a homogeneous body: a the coefficient of thermal diffusivity; ρ_{mat} is the density of the material; C_p is the specific heat capacity of the coil material; v is the coefficient of the shape taken for the cylinder to be unity; q_v is the power of internal heat sources. The field of the arguments is of the form $0 < \tau \le \tau_3$, $R_1 < r < R_2$, $0 < z < L_c$, where τ_{pr} is the predetermined calculation time.

The initial condition is given as follows: for $\tau = 0$ we have $T = T_0$. On the cylindrical surfaces of the coil, the boundary conditions have the form

$$-\lambda_{mat}\left(\frac{\partial T}{\partial r}\right)_i = \alpha_i\left(T_{suri} - T_{med}\right), \quad i = 1,...4, \tag{4.11}$$

where α_i is the heat transfer coefficients on the corresponding surfaces; λ_{mat} is the coefficient of thermal conductivity of the material of the coils; T_{sur} and T_{med} are the temperatures of the coil surfaces and the medium.

We replace the region of continuous variation of the arguments by the grid region, the finite nodes of which go beyond the boundaries of the coil:

$$\{k\Delta\tau; \; r_i = R_1 + (i-1,5)h_r; \; z = (j-1.5)h_z\},$$

where $k\Delta\tau < \tau$; $i = 1, ..., N$; $j = 1, ..., M$; k, i, j – time and space coordinates, respectively; $\Delta\tau$ is the time step; h_r, h_z – steps along the radial and longitudinal coordinates:

$$h_r = \frac{R_2 - R_1}{N-2}; \quad h_z = \frac{L}{M-2}. \tag{4.12}$$

Approximation of equation (4.11) on a rectangular grid according to the explicit scheme will be as follows:

$$T_{i,j}^{k+1} = T_{i,j}^k + p_r\left[T_{i+1,j}^k - 2T_{i,j}^k + T_{i-1,j}^k + \frac{vh_r}{2r_i}(T_{i+1,j}^k - T_{i-1}^r)\right]$$
$$+ p_z(T_{i,j+1}^k - 2T_{i,j}^k + T_{i,j-1}^k) + W, \tag{4.13}$$

where $p_r = a\Delta\tau/h_r^2$; $p_z = a\Delta\tau/h_z^2$; $W = q_v\Delta\tau/(C_p\rho_m)$.

In solving the resulting difference equation, the sweep method [40] was used, which ensures the stability of numerical calculations when the parameters of the problem vary widely.

A numerical realization of the solution of equation (4.10) with respect to the implicit difference scheme was conducted. Comparison of the results of the calculation by two methods gave the confidence that there are no errors.

Figure 4.14 shows the results of calculations for the following versions of the design of the coils of the magnetic system:
– water cooling of the outer surface (curve *1*; $j_{req} = 8.4$ A/mm²);
– an outer water cooling and a cooling channel inside the disc of each winding (curve *2*; $j_{req} = 9.46$ A/mm²);

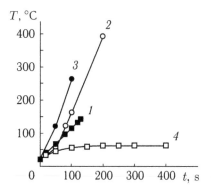

Fig. 4.14. The results of calculating the temperature regime of the winding of a magnetic system made of a copper bar with water cooling of the outer surface: *1* – without the use of additional disc cooling channels inside each winding; *2* – one channel; *3* – two channels; *4* – two channels and reducing current density to $0.7\,j_{req}$.

– external water cooling and two disc cooling channels inside each winding (curve *3*; $j_{req} = 11.5$ A/mm²).

The use of internal cooling channels reduces the copper filling factor of the coil volume, which explains the increase in the required current density.

Calculations showed that none of the above options yields positive results and the stationary thermal mode of operation of the magnetic system is possible only with a significant reduction in the current density in the conductor. For example (Fig. 4.14, curve *4*), with coils for the two cooling channels the stationary regime is feasible if the current density is lowered to $0.7\,j_{req}$. For other variants, an even lower current density is required [41].

4.5.2. The design of the magnetic system of the RM

The final construction consists of the coils of a copper tube of square cross section 8.5 × 8.5 mm² with a central hole 5 mm in diameter for the flow of cooling water. In order to create a magnetic field with an induction of 0.4 T in each coil, it is necessary to lay 196 turns and pass a current of 490 A (current density 10.1 A/mm²). The width of the coils was increased to 140 mm, which led to an increase in the non-uniformity of the magnetic field in the operating zone of the magnetron to 5%. As a result, each winding consists of seven flat-helical sections (Fig. 4.15), connected in series and according to the current and in parallel to the cooling liquid. The length of the tube in each section is 28 m. The cooling circuit of the magnetic system

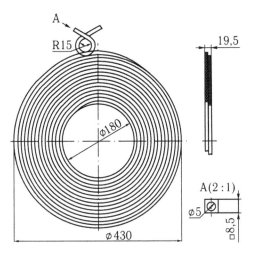

Fig. 4.15. The drawing of the winding section of the magnetic system made of copper tube of square cross-section with a central hole for cooling water.

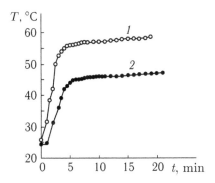

Fig. 4.16. The temperature regime of the windings of the magnetic system at a current: *1* – 500 A; *2* – 400 A.

is made according to a closed circuit. Demineralized water is used as a coolant. By adjusting the water flow, it is possible to achieve the necessary heat dissipation and to provide a steady-state operating mode also at high current density values.

Figure 4.16 are graphs of the change in the temperature of the cooling water at the outlet of the central section of the winding when operating in a continuous mode at a current of 500 A (j_{req} = 10.3 A/mm²; curve *1*) and 400 A (j_{req} = 8.24 A/mm²; curve *2*). The water flow rate was 0.32 l/s at a water inlet pressure of 2.5 atm. As follows from the figure, the magnetic system ensures the continuous operation of the relativistic magnetron during an

Table 4.3. Power supply parameters of a magnetic system

Maximum rectified voltage	90 V
Maximum current	500 A
Limits of current control	50–500 A
The maximum power consumed from the network is 380/220 V, 50 Hz	52 kW
Pulsations of the current of a magnetic system	less than 0.1%

arbitrarily long time interval. The slow rise in water temperature is explained by its heating in the closed circuit.

The power supply of the magnetic system is made according to the scheme of a transformer bridge rectifier. It has the parameters specified in Table 4.3.

Thus, the calculated and fabricated magnetic system ensures the continuous operation of relativistic magnetrons powere by the sections of LIA 04/4000 and LIA 04/6.

The appearance of the coils of the magnetic system of a pulse-periodic relativistic magnetron is shown in Fig. 4.17.

4.6. Vacuum system

The vacuum system was developed on the basis of the following requirements:

– the vacuum in the chamber should be better than $5 \cdot 10^{-5}$ Torr, in the cathode–anode gap of the RM better than $1 \cdot 10^{-4}$ Torr;

Fig. 4.17. Appearance of the magnetic system.

– the partial pressure of vapours of organic materials should not exceed 10% of the vacuum pressure in the chamber;

– the vacuum system must provide a high pumping speed;

– materials with a low operating temperature (less than 100–150°C) should be excluded from heated areas;

– it is necessary to use means of high vacuum evacuation, allowing to receive an oil-free vacuum, or the installation should be equipped with a freezing nitrogen trap.

The developed vacuum system is schematically shown in Fig. 4.18. Its evacuated volume consists of a vacuum chamber *1* with a system of nozzles, an anode unit with a microwave output device *3*, an antenna *4*, a drift tube (collector) of the end current *5* and an output device *6*. An additional path *9* is used to pump the antenna, as well as the cathode–anode gap of the magnetron through the microwave output device. The figure shows the location of the measuring sensors *2*. The pumped volume is mainly made of stainless steel. Permanent joints are made by argon-arc welding, and the

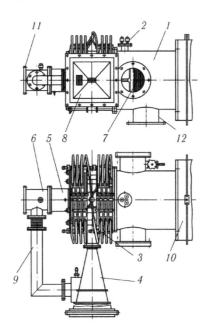

Fig. 4.18. The scheme of the vacuum system of pulse–periodic RM: *1* – vacuum chamber with a system of nozzles; *2* – measuring sensors; *3* – anode block of a magnetron with a microwave output device; *4* – antennas; *5* – end current collector; *6* – output device; *7* – high-voltage insulator of the electron gun of LIA; *8* – vacuum window of the antenna; *9* – additional pumping path; *10* – case of LIA module; *11* – connecting flange ODP-500 l/s; *12* – connecting flange ODP-2000 l/s.

split joints are sealed by gaskets from oil-resistant vacuum rubber, simple vacuum rubber and teflon. The high-voltage insulator 7 of the electron gun and the exit window of the antenna 8 are made of plexiglass.

The intrinsic gas load, consisting of the flows of gas evolution of structural materials and equipment, as well as leakage of gas flow from the outside should, according to the estimates in [42], be have in an untrained vacuum system the value $Q_{int} = (1-1.1) \cdot 10^{-2}$ l · Torr/s. When the magnetron operates, especially in the pulse-periodic mode, a significant additional gas load is possible in the vacuum system, due to thermal desorption of the heated cathode and desorption from the surfaces of magnetron electrodes bombarded by X-rays and electrons. It is rather difficult to accurately estimate the amount of additional gas evaporation. Assuming that the magnetron cathode, which is a stainless steel cylinder 80 mm long and 20 mm in diameter, heated to 500°C and the fractional contribution of the two components of further gassing are approximately the same, it is possible to obtain the numerical value of Q_{add} and the magnitude of the total gas evolution in the system $Q_{\Sigma} = Q_{int} + Q_{add}$ as a function of time after the establishment of the quasi-stationary mode of RM operation.

On the basis of these data, the necessary rates of pumping out the chamber and the magnetron were determined and a decision was made to produce it from two sides. The throughput of the magnetron can be increased by means of holes in the end caps of the anode block. For this purpose, 6 holes were placed in each cover, located in the centre of the resonators. The maximum area of one hole, which can be tolerated without reducing the output power level and maintaining the frequency characteristics of the microwave radiation, is ~90 mm². The second option for increasing the throughput of the magnetron is the use of end caps made of metal mesh with high geometric transparency (70–80%). In this case, a good contact of the mesh with the ends of the resonators is achieved by soldering or by using profiled end caps.

High-vacuum aggregates include oil–vapour diffusion (ODP) or turbomolecular pumps (TMP). The only advantage of TMP is that they contaminate the pumped volume with organic substances at a lower rate than ODP. This is important if the pumped volume does not contain organic materials or other pollution sources, since its own gas load is more than 80% determined by the desorption of gases and vapours from the surfaces of organic materials facing the inside of the pumped volume. Consequently, there is a reason to believe that the 'purity' of the vacuum should not significantly depend on the type of high-vacuum pumps. At the same time, the oil–vapour diffusion pumps should better cope with variable (impulse) gas loads and evacuation of hydrogen, which is an essential component of additional gas evolution.

The high-vacuum evacuation systems consists of two parallel branches: the main pumping of the system through a vacuum chamber and additional evacuation through the output device. The main branch is based on a high-vacuum oil–vapour diffusion pump with a capacity of 2000 l/s, a nitrogen trap and a mechanical two rotary pump. At the inlet of the unit, the pumping speed is 710 l/s. The additional branch consists of a high-vacuum ODP with a capacity of 500 l/s, a nitrogen trap and a mechanical rotary pump. The pumping speed at the inlet of this unit is ~200 l/s. An additional vacuum tract is used for pumping out the antenna and the interaction space of the RM through the waveguide output of power and the coupling window in the resonator. In experiments, the vacuum in the chamber before the start of the packet was:

1) when using a single ODP with a capacity of 2000 l/s – $6 \cdot 10^{-5}$ Torr;

2) two ODPs (2000 and 500 l/s) – $1.8 \cdot 10^{-5}$ Torr;

3) two ODPs (2000 and 500 l/s) with nitrogen vacuum traps – $6 \cdot 10^{-6}$ Torr.

In the vacuum chamber there is a system for diagnostics of the parameters of the accelerator: in the end flanges of the vacuum chamber there are two Rogowski coils for measuring current, and a capacitive voltage divider is installed on the inner surface of the chamber's shell. The Rogowski coils and the voltage divider are calibrated when a resistive load with a known resistance is connected to the output of the LIA.

The end current collector has a cooling water jacket to reduce the temperature of pulsed heating by the electron beam, and hence to maintain the gas evolution at a natural level.

4.7. Relativistic magnetrons powered by LIA on magnetic elements

At different stages, experiments with RM were carried out at two facilities: on the basis of LIA 04/4000 and LIA 04/6 [43–46]. The

Fig. 4.19. Appearance of a pulse-periodic relativistic magnetron with a power source – LIA on magnetic elements (2000).

appearance of one of them is shown in Fig. 4.19. The differences between the two units are as follows.

The power supply for LIA 04/6 has higher output parameters with smaller weight dimensions. The magnetic system is made of a hollow copper tube of a larger cross-section, 13×13 mm^2, with an internal hole 6 mm in diameter. It was possible to reduce ohmic losses and increase the magnetic field. In the same dimensions of the magnetic system and the power source 50 V and 1000 A, the maximum induction of the magnetic field reached 0.55 T.

4.7.1. Scheme and methodology of experimental research

The scheme of experimental studies is shown in Fig. 4.20. The process of studies of relativistic magnetrons can be conditionally divided into two stages. The first step is to set the device to the maximum output parameters in the single mode or at a low pulse repetition rate. The second stage is the investigation of the parameters of a relativistic magnetron at a high pulse repetition rate.

Two vacuum waveguide directional couplers *7* with a coupling coefficient –40 dB, connected in series with each other, were used to measure the power of the microwave radiation of the relativistic magnetron. The couplers were placed between the output of the magnetron and the radiating horn antenna *6*. The channels of each of the branch power couplers incorporated waveguide plate attenuators *8*. The outputs of each of the attenuators were connected to the lamp or semiconductor microwave detectors *9*, signals from which

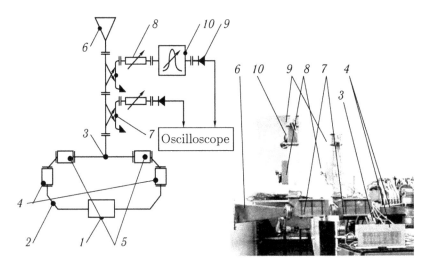

Fig. 4.20. Structural diagram and appearance of the RM measurement system.

along coaxial cables entered the inputs of oscilloscopes. Thus, the microwave power of a relativistic magnetron was determined as the average between the powers measured by the detectors, taking into account the coupling coefficient of the vacuum directional couplers and the waveguide attenuators. It was calculated in accordance with the expression

$$P_{RM} = \frac{(\alpha_{C1} + \alpha_{A1})P_{d1} + (\alpha_{C2} + \alpha_{A2})P_{d2}}{2}, \qquad (4.14)$$

where P_{RM} is the power of the relativistic magnetron; α_{C1}, α_{C2} are the coupling coefficient of vacuum directional couplers; α_{A1}, α_{A2} are the attenuation of the waveguide attenuators; P_{d1}, P_{d2} are the powers registered by the first and second detectors, respectively.

The scheme for measuring the spectrum of microwave radiation is slightly different from the scheme used to measure power. In this case, a tunable rejection waveguide filter *10* was included between one of the attenuators *8* and the corresponding microwave detector *9*. The second detector was used to record the reference signal, which made it possible to significantly reduce the influence on the spectral characteristics of the radiation of the amplitude spread between the pulses. The procedure for measuring the radiation spectrum was as follows. From the shot to the shot, the resonance frequency of the filter *10* changed discretely. At each frequency, the signal powers from each of the detectors were measured. The current spectrum was calculated from the ratio of the power of the signal passed through the filter at a given resonant frequency to the signal power from the output of another detector (reference signal) at a given fixed time:

$$S(t_i, \Delta t) = \frac{P_{\text{filt}}(t_i, \Delta t)}{P_0(t_i, \Delta t)}. \tag{4.15}$$

In spectral measurements of the short-pulse microwave radiation with the help of a narrow-band resonance filter it is necessary to take into account that, in addition to the forced oscillatory process, natural oscillations are excited in it. The duration of the transient process is comparable to the filter time constant τ_k, which is proportional to the quality factor of the filter Q: $\tau_k = 2Q/\omega$. This makes it difficult or impossible to record the spectral parameters during time intervals less than τ_k. A decrease in the Q factor broadens the analysis bandwidth and reduces the resolution of the filter, which does not allow us to evaluate the spectrum of narrow-band processes. The band of the filter used in the region of the magnetron generation frequency (2840 MHz) is 30 MHz, which corresponds to $Q \approx 90$ and $\tau_k \approx 10$ ns.

The experiments were carried out at different charging voltage of the primary storage of the accelerator U_{C0} in the range from 500 to 900 V, which allowed within wide limits to vary the output voltage and current of the LIA. Measurements of the microwave power, the total current and the voltage of the accelerator were carried out in the magnetic field induction range of 0.24–0.55 T. In the process of measuring the power of the spectral composition of the microwave radiation, 10–50 shots were performed for each magnetic field value. Such a volume of experimental data due to automated statistical processing results allowed to significantly reduce the error and improve the accuracy of measurements.

4.7.2. Parameters of relativistic magnetrons

The results of measurements of the output characteristics of pulse–periodic relativistic magnetrons are summarized in Table 4.4.

From the table it follows that the higher output power of the RM was achieved both by increasing the number of anode block resonators (increasing the slowing-down of the electromagnetic wave of the working mode of oscillations) and by increasing the output parameters of the power source. This process was accompanied by a slight spreading of the spectrum and a decrease in the stability of the microwave pulses due to an increase in the spread of the power

Table 4.4. Results of measurements of the output characteristics of pulse–periodic relativistic magnetrons

Options	Based on LIA 04/4000	Based on LIA 04/6
Radiation power, MW	200 ($N = 6$)	300 ($N = 6$) 350 ($N = 8$)
Radiation frequency, MHz	2840	2840 ($N = 6$) 3030 ($N = 8$)
Radiation bandwidth at 3 dB, MHz	40–50	50 ($N = 6$) 60 ($N = 8$)
Voltage of LIA, kV	300	400
Current of LIA, kA	2.6	3.6
Current pulse duration, ns	160	170
The duration of the microwave pulse, ns	120	110
Instability of the microwave power amplitude,%	12	15
Pulse repetition rate, Hz	0.4–320	0.4–200
Average microwave power at the maximum pulse repetition rate, kW	3	4.1

parameters of the magnetron. It should be noted that the value of the pulsed power of the magnetron was repeatedly checked by various methods, including when using the measuring equipment of customers of similar devices.

Figure 4.21 shows the dependence of the power of the relativistic magnetron, its total efficiency, voltage and current of the accelerator on the charge voltage of the primary storage device. The power growth is linear in nature, and the efficiency is constant, which agrees with the theoretical estimates [8]. It can be noted that the increase in the generated power is mainly due to an increase in the anode current (the steepness of the current rise above the corresponding increase in the output voltage).

The stability of the operation of relativistic magnetrons is demonstrated by a set of oscillograms – signals of microwave detectors recorded at different pulse repetition rates (the number of superimposed pulses 100) (Fig. 4.22). It can be seen that the amplitude, duration and shape of pulses in the frequency range of 80–320 Hz practically do not change. This allows one to assume that the relativistic magnetron can work with an even higher frequency.

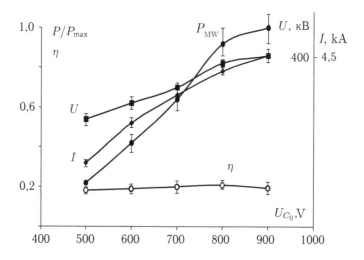

Fig. 4.21. Dependences of the microwave power (P_{MW}) and the full efficiency (η) of the relativistic magnetron, voltage (U) and total current (I) of the LIA 04/6 on the voltage of the primary energy storage device.

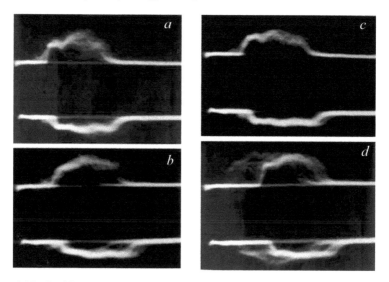

Fig. 4.22. Oscillograms of the signals of microwave detectors at different pulse repetition rates of RM [Hz]: *a*) 80; *b*) 160; *c*) 240; *d*) 320.

Figure 4.23 shows a series of 200 current pulses (upper oscillograms) and signals of a microwave detector at a repetition rate of 80 Hz. The oscillograms are recorded in the peak detection mode (each peak is a separate pulse, the data collection system does not detect the interval between pulses). The instability of the amplitude of the

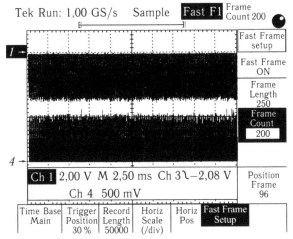

Fig. 4.23. A series of 200 pulses with a repetition rate of 80 Hz. The upper oscillograms are the total current of the LIA, the lower ones are the signals of the microwave detector.

current generated by the LIA is less than 5%, and the instability of the microwave power amplitude does not exceed 12%. When two ODPs are used, the vacuum stabilization (at the level $(2–3.2) \cdot 10^{-5}$ Torr) and the performance characteristics of the device occurred approximately 100–150 pulses after the start of the packet. Eliminating the vacuum traps of the nitrogen traps of ODP led to an increase in the initial vacuum level, but an increase in the evacuation rate made it possible to maintain an acceptable level of vacuum for the stable work of the relativistic magnetron vacuum level $\sim 4 \cdot 10^{-5}$ Torr during the whole packet of pulses. Use of only one ODP with a capacity of 2000 l/s was insufficient: the vacuum dropped to $2 \cdot 10^{-4}$ Torr after approximately 350 pulses with a frequency of 80 Hz and the subsequent disruption of microwave generation.

As demonstrated by long-term operation, the durability of the anode block proved to be extremely high. During the operation at both facilities, about 10^7 impulses were produced and not a single block was destroyed. This is due to the fact that the parameters of the electron beam were selected in accordance with the results of thermal calculations and did not exceed the limiting values. In addition, measures were taken to increase the longevity of the anode blocks by applying cooling and rounding the edges of the lamellae. The longevity of the elements of the relativistic magnetron is influenced by the specific features of the work of the LIA, namely, that the energy of the repeated pulses of the LIA is not scattered in the magnetron diode, but is spent on the magnetization reversal of the cores of the induction system. This feature of LIA, which is most important for RM, predetermined their choice as sources of power for pulse-periodic RMs.

The maximum pulse repetition frequency of a pulse-periodic relativistic magnetron based on LIA 04/6 (200 Hz) was limited by the power of the supply cables to the experimental room. In this mode, the power consumed by the LIA and the magnetic system from the network reached 150 kW. However, the limiting frequency with which the LIA 04/6 accelerator can operate is 400 Hz. It is determined by the duration of the charge–discharge processes of the primary storage (this mode of operation of the accelerator was checked with the use of an ohmic load).

Typically, the pulses with high pulse repetition rates are generated in packets, between which a pause is needed to cool the elements of the relativistic magnetron, semiconductor devices of the power supplies and windings of the accelerator and magnet power supply elements, and to restore the vacuum conditions in the volume. The duration of a pause (in seconds) between series of pulses is estimated by the empirical formula

$$T_{pau} = 0.1N_{im}\left(\frac{1}{f_{cont}} - \frac{1}{f}\right),\qquad(4.16)$$

where N_{im} is the number of pulses in the series; f_{cont} = 8Hz – the repetition rate of the relativistic magnetron pulses in a continuous mode (the regime is not limited by time); f is the pulse repetition frequency.

The pulse-periodic mode of operation of the relativistic magnetron makes it possible to use frequency analyzers of spectrum to record the spectral characteristics of the generated microwave pulses. The results of one of the measurements are shown in Fig. 4.24. The registered bandwidth of the radiation at half-power level was about 50 MHz, which correlates well with the results of spectral measurements by the bandpass filter.

The results of the studies carried out at the Tomsk Polytechnic University on the basis of LIA on magnetic elements were used to construct the pulse-periodic RMs and experimentally implement the modes of operation [45, 45], presented in Table 4.5.

Note that the pulse-periodic relativistic magnetrons are designed as a complex including a generator, a LIA with a power system, a

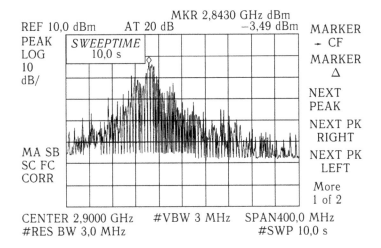

REF 10,0 dBm   AT 20 dB
MKR 2,8430 GHz dBm
−3,49 dBm

PEAK
LOG
10
dB/

SWEEPTIME
10,0 s

MA SB
SC FC
CORR

MARKER
← CF
MARKER
Δ
NEXT
PEAK
NEXT PK
RIGHT
NEXT PK
LEFT
More
1 of 2

CENTER 2,9000 GHz   #VBW 3 MHz   SPAN400,0 MHz
#RES BW 3,0 MHz   #SWP 10,0 s

Fig. 4.24. Frequency spectrum of radiation of a pulse-periodic relativistic magnetron, measured by a frequency spectrum analyzer; P_{max} = 200 MW.

Table 4.5. Modes of operation of pulse-periodic RMs

Operating mode	Parameters of pulse-periodic RMs		
	Power, MW	Pulse repetition rate, Hz	Number of pulses
Continuous	300–350	0.4–8	Not limited
Pulse periodic	300–350	12–80	10^5
Packet	200 300	120–320 120–200	$10^3 - 10^4$

magnetic system with a power source, a vacuum system and a cooling system for the elements.

4.8. Relativistic magnetron of the Physics International Company

The work [47] describes experimental studies of a relativistic magnetron of a 30-cm wavelength band, performed on a linear induction accelerator Compact LIA (CLIA) with the following output characteristics: voltage 750 kV; current 10 kA; pulse duration 60 ns; pulse repetition rate 200 Hz for ≈1 s.

The starting point for these experiments was the results obtained at the TPU by using a pulse-perio0dic linear induction accelerator as a power source for a relativistic magnetron. However, unlike the

LIA of the TPU used in the very first experiment, the CLIA used magnetic elements to switch the forming lines. The research task consisted in obtaining a sequence of 100 pulses with a frequency of 100 Hz with a peak power of each pulse of the order of 1 GW.

4.8.1. Power source of the Compact LIA

The Compact LIA [48] consists of a ten-section accelerator with a cathode holder, which summarizes the voltage applied to the load (Fig. 4.27). Each section consists of an IS and an SFL forming line with water insulation and magnetic commutation. A two-stage magnetic pulse generator charges the SFL from an intermediate energy storage, switched by a thyratron and charged from a common resonant charge block (see Fig. 4.25 at the bottom). Elements of the CLIA allow the compression of pulses, increasing the voltage from 40 to 150 kV, and then, using the sections, to obtain a high voltage output (750 kV). Such a scheme makes it possible to switch primary energy storage with the help of hydrogen thyratrons, and the forming lines by magnetic commutators.

Figure 4.25 shows a block diagram of the CLIA. The DC power supply has the following output characteristics: power 300 kW; the output voltage is 50 kV. It charges capacitors of primary storage up to 40 kV. The energy is extracted from this block and compressed in time by the circuits of the resonance charging unit, switched by thyratrons. The output of the resonance charging unit charges the magnetic pulse generator. Further, the two-stage MPG compresses the energy and increases the voltage, using magnetic commutation and a transformer with a ratio of 2:1. From the output of the MPG ten parallel SFLs with a water dielectric are charged. Then, the forming lines are discharged through the magnetic commutators to the inductors of the induction system of the accelerator. At the output of

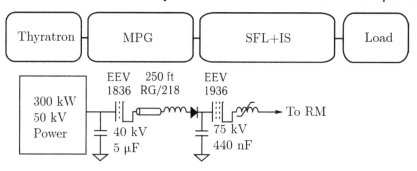

Fig. 4.25. Block diagram of the CLIA.

the accelerator, a pulse with a voltage of 750 kV and a current of 10 kA is formed. When operating on a matched load at a frequency of 250 Hz voltage pulses with an amplitude of 600 kV were obtained.

The authors of [48] note that the power supply–forming lines circuits were designed with a 65% energy reserve relative to the energy required for charging the lines. This increase in energy was done in order to compensate the losses in the elements of the installation, if they are greater than expected. Figure 4.25 schematically shows the power supply–resonant charge unit circuit. The first capacitor is a primary energy storage. It closes through the hydrogen thyratron EEV 1836 to the next capacitive storage, which is charged up to 75 kV per 100 µs. This storage is then discharged through the thyratron EE 1936 to the resonant charge storage. The magnetic commutator shown in the diagram operates as a diode that prevents reverse current through the thyratron. This circuit was initially tested with a resistive load at the output at a frequency of 200 Hz for 5 s at an output voltage of 75 kV. Then it was used to charge an MPG with a frequency of 250 Hz.

Figure 4.26 is a diagram of parts of the LIA: MPG–SFL. The energy from the resonance charging unit charges the first MPG capacitor to 75 kV for 4 µs. This capacitor is discharged through a two-turn magnetic switch to a 2:1 step-up transformer. The output of the latter charges up to 150 kV a water storage unit with a capacitance of 90 pF, which, in turn, discharges through another two-turn magnetic commutator on 10 parallel SFLs.

The forming lines of the SFL are a coaxial structure with water insulation and a wave impedance of 6.8 ohms. The lines are charged up to 150 kV each, forming an output pulse with an amplitude of

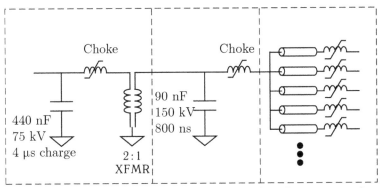

Fig. 4.26. Electrical circuit of elements of the magnetic pulse generator-forming lines CLIA.

75 kV and a duration of 60 ns. Each line has its own magnetic commutator at the output. The outputs of the switches are connected to the magnetizing coils of the inductors of the induction system.

The magnetic switch and transformer cores were made of Allied Metglas 2605CO tape. The cores of MPG switches used a tape with a thickness of 25.4 μm was used, and the cores of the output switches of the SFL – 15.2 μm. The internal diameter of the inductor cores was chosen to be 215.9 mm, the outer diameter 787.4 mm. Their width was 50.8 mm. The actual length of the induction system in the assembly was 1 m, which gives an acceleration rate of 0.75 MV/m. The inductor cores were wound from 2605CO Metglas tape 25.4 μm thick with Mylar insulation between the coils 3.5 μm thick.

The cores of the magnetic elements of the CLIA are returned to their original magnetic state by direct currents from separate power supplies. The first current source is turned on at the point immediately after the second thyratron and demagnetizes the magnetic switch of the first stage of the MPG and the primary winding of the transformer. The second current source is connected at the point between the water storage unit and the magnetic switch of the second stage of the MPG. It serves to return to the initial state of the cores of the transformer, the switch of the second stage of the MPG, the output switches of the SFL, and the induction system of the accelerator. The circuit of the second demagnetizing current source has 11 different grounded channels. The separation of the current between these channels is carried out with the help of thin-walled stainless pipes, into which liquid is filled, providing the necessary resistance. During the CLIA tests, different levels of demagnetizing current were used at different output voltages and pulse repetition rates. If the current was insufficient to restore the magnetic state of the accelerator cores, a slow decrease in the output pulse from pulse to pulse was observed for approximately 50 pulses, after which it completely disappeared.

4.8.2. Relativistic magnetron in the 30 cm range

For experiments with a high repetition rate, a previously manufactured magnetron of a 30 cm wavelength band was modified (the emission frequency was 1.1 GHz at the operating π-mode of oscillations). Its anode block had six resonators with an internal radius of 3.18 cm and a depth of 5.08 cm. The radius of the used cathode 1.27 cm. When operating with power from the HCEA, the magnetron had an output power of 3.6 GW. Note that much lower power was expected, as the CLIA has lower output characteristics. In addition, the matching resistance of the magnetron is ~25 ohms, while the CLIA resistance is ~75 Ohm.

Changes in the anode block of a relativistic magnetron for the frequency operating mode consisted in the application of cooling the anode block lamellae through water channels located 3 mm from the surface. In addition, special attention was paid to the creation of a good electrical contact between the parts of the magnetron, the accelerator and the housing in order to avoid possible current losses. The vacuum system was evacuated using a cryopump to avoid possible oil contamination (vacuum not higher than $4 \cdot 10^{-6}$ Torr). Previous experiments in the single pulse repetition mode show that the peak power of the magnetron increases with decreasing pressure in the system.

As shown in Fig. 4.27, microwave radiation was derived from two opposite resonators, which were connected to the standard waveguide WR650 through quarter-wave transformers. The resonators were collected by an active load in a vacuum. The couplers, designed for high microwave power, with a factor of 80 dB allowed to diagnose the amplitude and frequency of pulses. Signals were recorded in two ways: a crystal detector that detected each pulse, and directly to the high-speed oscilloscope. In this case, one pulse (not necessarily the first one) was selected inside the pulse packet. The only device for

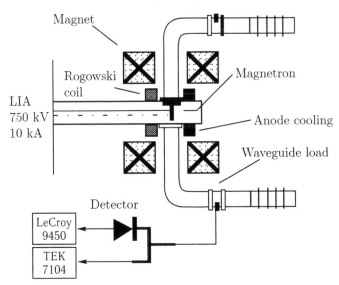

Fig. 4.27. Schematic of experimental studies of the relativistic magnetron.

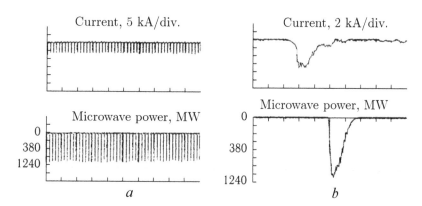

Fig. 4.28. Oscillograms of current and power of microwave radiation of a series of 50 pulses with a frequency of 50 Hz (*a*). Oscillograms of the current and power pulse (*b*)

diagnostics of the parameters of the accelerator was the Rogowski coil, which measures the total current of the relativistic magnetron. The voltage was determined through the current, the charging voltage of the CLIA and the previously measured load characteristic of the accelerator.

Experimental results. At a repetition rate of 10 Hz a packet of microwave pulses of a relativistic magnetron with a power of ~1 GW with a duration of 50 ns and an energy of 44 J each was obtained, yielding 4.4 kW of average power. Figure 4.28 *a* shows the amplitudes of current pulses and microwave radiation (each peak is a separate pulse) for a packet of 50 shots, and Fig. 4.28 *b* shows one pulse from the middle of the packet. A microwave signal from one of the two output waveguides was detected by a crystal detector. The total power was estimated by doubling the detector readings, since it was known from previous experiments that the power in the output waveguides is approximately equal.

As can be seen from Fig. 4.28, all pulses in the series are approximately equal in amplitude, which allows one to assume the possibility of magnetron operating even at a longer duration of the packet. An important feature of the operation of the magnetron with the use of CLIA is that the duration of the microwave pulse (~50 ns) is only slightly less than the pulse length of the accelerator current (~60 ns). This circumstance was first observed in experiments conducted at the TPU. It is noted that when using HCEAs in most experiments with a relativistic magnetron, the typical ratio of the duration of the microwave pulse and the duration of the current pulse is approximately 1/3.

The oscillograms in Fig. 4.28 show that the first few pulses have a larger amplitude, which correlates with the higher amplitudes of the current pulses. This is due to the need for some time to settle a stable energy state in the LIA system. This effect becomes more visible when the pulse repetition rate increases, causing a decrease in peak power, although the average power increases. As shown in Fig. 4.29, at a frequency of 200 Hz peak power decreases to 700 MW, while the average power increases to 6 kW. At 250 Hz, the trend continues (600 MW peak power and 6.3 kW average). The operation of the relativistic magnetron was investigated in the regime of 5 'shots' in the packet at a frequency of 1000 Hz to determine if there is a minimum recovery time between pulses. It follows from Fig. 4.30 that the time interval of one millisecond is sufficient to remove the products of explosive electron emission from the cathode–anode gap of the relativistic magnetron. Using the characteristics of the third pulse in the series, it is estimated that the average power of the magnetron at a repetition rate of 1000 Hz should be ~25 kW at a peak power of a single pulse of 600 MW.

The results of experimental studies of the relativistic magnetron of a 30-cm wavelength band are presented in the summary Table 4.6.

The results of the research carried out at the Physics International Company extend the upper limit of the average and peak power for

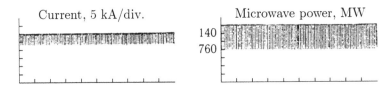

Fig. 4.29. Oscillograms of current and power of microwave radiation from a series of 100 pulses with a frequency of 200 Hz.

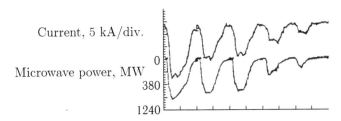

Fig. 4.30. Oscillograms of current pulses and microwave power in a series of 5 pulses with a frequency of 1000 Hz.

Table 4.6. The results of experimental studies of the relativistic magnetron in 30 cm wavelength range

Pulse repetition rate, Hz	Microwave power, MW	Average power, kW	Number of pulses
100	1000	4.4	50
200	700	6.0	100
250	700	6.3	100
1000	600 (third pulse measured)	25	5

relativistic microwave sources. Their authors noted the absence of a problem of gas evolution in a magnetron, which can lead to a drop in the impedance of the cathode-anode gap. When operating from the CLIA, there was no erosion of the anode block lamellae even after several thousand pulses. There was no decrease in the emissivity of the cathode after 1000 pulses. Since the duration of the microwave pulse and the current pulse are approximately the same, there is a reason to hope that the magnetron is capable of generating longer microwave pulses.

It can be assumed that in the relativistic magnetron a higher average microwave power is attainable when operating at a frequency of more than one kilohertz. It should be noted that in the magnetron the emission of a beam, microwave generation, and the deposition of an electron beam occur in the same space. This is its essential difference from linear floating-drift devices, such as the relativistic klystron, the running wave tube, the backward wave tube and the free-electron laser

4.9. Comparison of the specific characteristics of relativistic microwave devices

In conclusion of this chapter we give Table 4.7, which compares the specific weight dimensions of relativistic microwave devices developed in the world. The values of the mass and volume of the experimental setups are taken as approximations based on the author)s visual estimates.

As can be seen from Table 4.7, the relativistic magnetrons when fed from the LIA have rather high specific characteristics in comparison with other generators, which allows us to hope for the possibility of their practical application in the future.

Abbreviations used: P_{im}, f – pulse power and radiation frequency of relativistic microwave devices; F, T_{MW} – repetition frequency and pulse duration; P_{av} – the calculated average radiation power of relativistic microwave devices; M, V – approximate masses and volume of the installations; P_{MW}/M, P_{MW}/V – specific weight and size characteristics of the equipment, including relativistic microwave device, power supply and magnetic system with power supply.

Research centres: IHCE – Institute of High Current Electronics, Siberian Division of the Russian Academy of Sciences; RAS; PIC – Physics International Company, USA; Varian – Varian Associaties, USA; Advanced Technology Group, USA; RFNC-VNIIEF – Russian Federal Nuclear Center, All-Russian Scientific Research Institute of Experimental Physics; AFRL – Air Force Research Laboratory, USA; SNL – Sandia National Laboratory, USA; IE – Institute of Electrophysics, Ural Branch of the Russian Academy of Sciences.

Table 4.7 Characteristics of relativistic microwave generators

Microwave device	Parameters of microwave device					Installation parameters				Reference
	P_{in}, MW	f, GHz	F, Hz	T_{MW}, ns	P_{av}, W	M, kg	P_{MW}/M, W/kg	V, m³	P_{MW}/V, W/m³	
BWT IHCE	700	9–10	200	15–30	900	5000	0.18	10	90	49
RM IRT Corp.Cornell Univ.	200	4.4	1	90	6	20000	0.0003	30	0.5	50
RM AAAI Corp.	700 325	4.63 4.27	2	50	22 12	20000	0.0011 0.0006	30	0.73 0.4	51
RM General Atomics	300	4.4	1	100	10	20000	0.0005	30	0.33	52
RM Rafael, Israel	100	2.5	10	70	17	15,000	0.0001	24	8.5	53
RM TPU	360	2.9	160	80	1550	1500	1	4	388	54

Microwave device	Parameters of microwave device						Installation parameters				Reference
	P_{im}, MW	f, GHz	F, Hz	T_{MW}, ns	P_{av}, W	M, kg	P_{MW}/M, W / kg	V, m³	P_{MW}/V, W/m³		
RM TPU	200	2.9	20	50	70	1500	0.047	4	17.5		
RM PIC	1000 700 600	1.1	100 200 250	50	4400 6000 6300	9000	0.49 0.67 0.7	14	314 430 450		55, 56 57
RM TPU	400	2.84	320	160	6000	3500	1.71	8	750		58
RM CPI (Varian)	60	1.873	10	600	350		No date				59
Virator IHCE	100	2.65	50 20	20–25	350 140	1000	0.035 0.014	20	17.5 7		60, 61
BWT IHCE	500	10	150	5	200	10000	0.02	10			62

Microwave device	Parameters of microwave device					Installation parameters				Reference
	P_{im}, MW	f, GHz	F, Hz	T_{MW}, ns	P_{av}, W	M, kg	P_{MW}/M, W/kg	V, m³	P_{MW}/V, W/m³	
Vircator RFNC-VNIIEF	150	10	10	20	30	2000	0.015	6	5	63
Vircator IHCE	200 350	3	10	20–25	350 140	10000	0.035 0.014	20	17.5 7	64
BWT IHCE	600	10	100	1	60	2000	0.03	3	20	65
BWT IHCE, IE	2200	10	730	0.5	2500	5000	0.5	8	312	65

References

1. Hull A.W., Phys. Rev. 1921. V. 18. P. 31–57.
2. Alekseev N.F., Malyarov D.E., Zh. Teor. Fiz., 1940. V. 10, No. 15. P. 1297–1300.
3. Kovalenko V.F., Introduction to electronics of ultrahigh frequencies., Moscow, Sov. radio, 1955.
4. Magnetrons of centimeter range, Trans. with English. Ed. S.A. Zusmanovsky. In 2 vols. V. 1, Moscow, Sov. radio, 1950.
5. Bychkov C.I., Questions of the theory and practical application of devices of the magnetron type, Moscow, Sov. radio, 1967.
6. Samsonov D.E., Basics of calculation and design of magnetrons, Moscow, Sov. radio, 1974.
7. Bekefi G., Orzechovski T., Phys. Rev. Let. 1976. V. 37, No. 6. P. 379–382.
8. Vintizenko I.I., Novikov S.S., Relativistic magnetron microwave generators, Tomsk: Publishing house of NTL, 2009.
9. Bugaev S.P., et al., The phenomenon of explosive electron emission. Invention. Diploma No. 176, Otkr. Izobr. Prom. Obraztsy, Tov, Znaky, 1976, No. 41. P. 3.
10. Kapitsa P.L., Electronics of high power, Moscow, Publishing House of the USSR Academy of Sciences, 1962.
11. Nechaev V.E., Izv. VUZ, Radiofizika, 1962. Vol. 5, No. 3. P. 534–548.
12. Sulakshin A. With. Restriction of current leakage from the interaction space, Zh. Teor. Fiz., 1983. P. 53, No. 11. P. 2266–2268.
13. Vasiliev V.V., et al., Pis'ma Zh. Teor. Fiz., 1987. Vol. 13. Issue. 12. P. 762–766.
14. Kanaev D.R., et al., *ibid*. 1995. Vol. 21. Issue 20. P. 51–54.
15. Vintizenko I.I., et al., in: Proc. 8th Sympos. High-current electronics, Sverdlovsk, 1990. Part 3. P. 133–135.
16. Levine J.S., et al., J. Appl. Phys. 1991. V. 70, No. 5. P. 2838–2848.
17. Treado T.A., et al., IEEE Trans. on Plasma Science. 1994. V. 22. P. 616–626.
18. Novikov S.S., et al., Proc. SPIE in Intense Microwave Pulses IV, San Diego, 1995. V. 2557. P. 492–498.
19. Vladimirov S.N., et al., Nonlinear oscillations of multifrequency autooscillatory systems, Tomsk, Izd-vo TGU, 1993..
20. Novikov S.S., et al., Proc. SPIE in Intense Microwave Pulses V, San Diego, 1997. V. 3158. P. 271–277.
21. Furman E.G., et al., Prib. Tekh. Eksper., 1993, No. 6. P. 45–55.
22. Butakov F.D., et al., Pis'ma Zh. Teor. Fiz., 2000. Vol. 25. Issue. 13. P. 66–71.
23. Vintizenko I.I., et al., In: Proc. 12 Int. Symposium on High Current Electronics. Tomsk, 2000. V. 2. P. 255–258.
24. Butakov L.D., et al., Prib. Tekh. Eksper., 2001, No. 5. P. 104–110.
25. Samsonov D.E., Basics of calculating and constructing multiresonator magnetrons, Moscow, Sov. radio, 1966.
26. Shlifer E.D., Calculation of multi-cavity magnetrons, Moscow, MEI, 1966.
27. Kovalenko V.F., Elektronnaya tekhnika, Ser. 1. Elektronika SVCh, 1972, No. 1. P. 3–11.
28. Epstein M.S., *ibid*, 1969, No. 1. P. 66–81.
29. Kovalenko V.F., Thermophysical processes and electrovacuum devices, Moscow, Sov. radio, 1975.

30. Properties and application of metals and alloys for electrovacuum devices. Reference guide, ed. R.A. Nilender, Moscow, Energiya, 1973.

31. Anur'ev V.I., Handbook of the designer-machine builder. In 3 vols. V. 1, Moscow, Mashinostroenie, 1982.

32. Vintizenko I.I., Mityushkina V.Yu., Radiotekhnika, 2005, No. 10. P. 74–79.

33. Gunin A.V., et al., Pis'ma Zh. Teor. Fiz., 1999. Vol. 25. Issue 22. P. 84–94.

34. Belomyttsev S.Ya., et al., Zh. Teor. Fiz., 1999. V. 69, No. 6. P. 97–101.

35. Bykov N.M., et al., In: Proc. 10 IEEE Int. Pulsed Power Conf. Albuquerque. 1995. P. 71–74.

36. Vintizenko I.I., Relativistic magnetron, Certificate for utility model No. 13936. B.I. 2000, No. 16.

37. Isakov P.Ya., et al., Prib. Tekh. Eksper., 1988, No. 2. P. 27–29.

38. Mel'nikov Yu.A., Permanent magnets of electrovacuum microwave devices, Moscow, Sov. radio, 1967.

39. Loginov V.S., et al., Izv. VUZ, Elektromekhanika. 1999, No. 4. P. 117–119.

40. Vintizenko I.I., et al., Proc. of the Research Institute of Computer Technologies 'Mathematical Modeling', Khabarovsk, KhGTU Publishing House, 2000. Issue. 10. PP. 80–86.

41. Pipko A.I., et al., Design and calculation of vacuum systems, Moscow, Energia, 1979.

42. Butakov L.D., et al., Pis'ma Zh. Teor. Fiz., 2000. Vol. 25. Issue 13. P. 66–71.

43. Vintizenko I.I., et al., Izv. TPU. 2003. V. 306, No. 1. P. 101–104.

44. Vintizenko I.I., *ibid*, 2006. V. 309, No. 6. P. 47–50.

45. Vintizenko I.I., Izv. VUZ, No. 9, Appendix. Pp. 271–275.

46. Aiello N., In: Proc. 9 Int. Conf. on High Power Particle Beams, Washington, 1992. P. 203–210.

47. Sincerny P. et al. High average power modulator and accelerator technology developments at Physics International, NATO Series, 1993. V. G34. Part A.

48. Rostov V.V., et al., in: Proc. 12 Int. Symposium on High Current Electronics. Tomsk, 2000. V. 2. P. 408–411.

49. Phelps D., et al., in: Proc. 7 Int. Conf. on High-Power Particle Beams, Karlsruhe, 1988. WP112. P. 1347–1352.

50. Spang S.T.,et al., IEEE Trans. on Plasma Science. 1990. V. 18, No. 3. P. 586–593.

51. Phelps D.A., IEEE. Trans. on Plasma Science. 1990. V. 18, No. 3. P. 577–579.

52. Schnitzer I., et al., SPIE. 1995. V. 2843. P. 101–109.

53. Vasil'iev V.V., Pis'ma Zh. Teor. Fiz., 1987. Vol. 13. Issue 12. P. 762–766.

54. Aiello N., et al., In: Proc. 9 Int. Conf. on High Power Particle Beams, Washington, 1992. P. 203–210.

55. Sincerny P., et al., High average power modulator and accelerator developments at Physics International, NATO Series, 1993. V. G34, Part A.

56. Smith R.R., in: Proc. Int. Workshop on High Power Microwave generator and Pulse Schortering, Edinburgh, 1997. P. 1–9.

57. Butakov L.D., et al., Pis'ma Zh. Teor. Fiz., 2000. Vol. 25. Issue. 13. P. 66–71.

58. Treado T.A., in: Proc. 2 Int. Vacuum Electronics Conference, Netherlands, 2001. P. 59, 60.

59. Kitsanov S.A., et al., in: Proc. 12 Int. Symposium on High Current Electronics. Tomsk, 2000, P. 423–428.

60. Korovin C.D., in: Vacuum microwave electronics, Nizhnyi Novgorod, 2002. P. 149–152.

61. Wardrop B.A., Marconi Gec Journal of Technology. 1997. V. 14, No. 3. P. 141–150.

62. Dubinov A.E., et al., Inzh. Fiz. Zhurnal, 1998. V. 71, No. 5. P. 899–901.
63. Korovin S.D., in: Proc. 13 Int. Symposium on High Current Electronics. Tomsk, 2004. P. 218–223.
64. Rostov V.V., et al., In: Proc. 13 Int. Symposium on High Current Electronics. Tomsk, 2004. P. 250–253.

5

Generators of microsecond pulses

Introduction

In this chapter, we consider the principle of operation, electrical circuits and designs of generators of microsecond voltage pulses (GMP) using the technology of linear induction accelerators on magnetic elements. Output parameters of the generators: voltage 450–1000 kV, current 1–2 kA, duration of the flat part of the pulse 1 µs, pulse repetition frequency up to 1 kHz. Due to the appropriate selection of elements, it is possible to form voltage pulses of a linearly increasing or linearly decreasing shape. Generators of microsecond pulses allow for an on-line change in the polarity of the output pulses with the same amplitude and time characteristics. Such power supply parameters are unique. They are not achievable with the help of traditional pulse generation circuits, which can not provide a rectangular shape of the microsecond voltage pulses and such a high repetition frequency. In the present chapter, engineering and design problems associated with the creation of GMP have been studied and the calculations and computer simulations based on equivalent circuits have been carried out. In addition, estimates of the thermal conditions of the elements of a particular generator have been made, formulas have been given for selecting the cross-section of the windings of pulse transformers and saturation chokes. One of the paragraphs is devoted to calculations of the leakage inductance, magnetization inductance, dynamic capacitance and ohmic loss equivalents. Such material can be useful in the design of high-voltage devices using magnetic elements, including linear induction accelerators. The original generator power circuit permitting its operation with the pulse repetition frequency up to 1 kHz in the long packet mode (to 5000) is presented in detail.

Generators of microsecond voltage pulses with similar unique output characteristics can find application in terms of forming electron beams of relativistic microwave devices of both O- and M-types. In this case, at relatively low levels of generated power, it is possible to obtain microwave pulses of high energy due to the increased duration. Thus, the high average power of the microwave radiation of the installation is attainable. As a result, there are prospects for its practical use. In this case, the limitations associated with heating the surface of collectors of O-type relativistic devices, anode blocks of relativistic magnetrons and with their possible destruction are partially removed, since the amplitude of the pulse heating temperature of the surface is proportional to the square root of the pulse duration of the electron beam (see (4.3)).

5.1. Design and elements of microsecond pulse generators

The design of the MPG was developed at the TPU [1−4], it is protected by patents for the invention [5−8] and is described in [9−11]. The principal difference between the MPG of high-current electron accelerators of microsecond duration is the use of magnetic elements performed in the form of saturation chokes with a ferromagnetic core with a winding wrapping it. Such elements are capable of switching an unlimited resource with a current of hundreds of kiloamperes with a frequency of one kilohertz. Another important advantage is the high stability of the amplitude–time characteristics of the pulses being formed. An important advantage is the simplicity of the design, the lack of preventive operations, adjustments and other problems typical for gas spark discharger. The magnetic elements were used in the construction of the LIAs, described in Ch. 2 and section 4.8.1.

The main drawback of the known designs of high-voltage devices, in particular LIA and high-current electronic accelerators (HCEA), is the short duration of the output voltage pulse (no more than 200 ns). This is due to the following reasons: 1) the use of forming lines having a limited electric length; 2) using magnetic commutators capable of providing charging the forming lines with an electrical length of 0.5−1 μs only for a considerable mass of the ferromagnetic material, and therefore for large dimensions and high inductance of the winding, which makes it impossible to form a rectangular pulse of the output voltage; 3) using high-voltage transformers (induction systems) capable of transforming the voltage pulse for a limited time to the saturation of the ferromagnetic cores. To reduce the size of the induction system, it would be possible to use multi-turn primary and secondary windings. However, an increase in the number of to two leads to an increase of about 4 times of the inductance of the discharge circuit, formed by the inductance of the forming line, the winding of the magnetic commutator, and the magnetizing turns of the induction system. The duration of the output voltage pulse with a proportional decrease in its amplitude parameters was doubled. The output voltage pulse of a high-voltage transformer acquires a bell-shaped form.

The design of the generator of microsecond voltage pulses is based on the original idea, consisting in the sequential discharge of several magnetic pulse generators synchronized in a certain way to the primary winding of a high-voltage pulse transformer.

The generator of microsecond voltage pulses consists of the following main blocks:

–*a high voltage unit*, containing several magnetic pulse generators, each of which consists of the storage capacitors and the saturable chokes, a pulse transformer (PT), sources of demagnetization of the PT, magnetic chokes and commutators (see section 5.1.1.);

–*a power source* comprising a charging device for the primary storage, a primary storage in the form of separate blocks of capacitor batteries, a protective device, an oscillation charging device, a thyristor pulse generator and a demagnetization source (see section 5.2);

–*a high-voltage insulator* for separating the housing filled with transformer oil and a vacuum chamber (see section 5.3);

–*a high-voltage transformer* (HVT), which is a set of inductors with multi-turn primary and secondary windings (see section 5.4);

–*a control system* (see section 5.2.4).

5.1.1. High-voltage unit

The main idea used in the design of the generator is that the generation of a voltage output pulse of microsecond duration is performed by consecutively discharging several (two or more) synchronized magnetic pulse generators to the magnetizing turns of the ferromagnetic cores of a high-voltage transformer manufactured using the technology of the induction system of the LIA. Each magnetic pulse generator is a sequence of $N \geq 2$ compression links with increasing natural frequency. Each of these links consists of a capacitor with lumped parameters and a saturation choke. The operating principle of an individual MPG was discussed in detail in Ch. 2. The number of compression links is determined by the duration of the supply pulse at the input of the MPG and the required amount of energy compression. The first compression links of all MPGs are in parallel connected to the thyristor pulse generator of the power supply. The last compression links of all MPGs are connected to the terminals of the magnetizing coils of the inductors (primary winding) of the high-voltage transformer.

At the same time, the flux linkage values of the saturation chokes of the last compression links (magnetic commutators) of each magnetic pulse generator differ from each other by some amount in order to ensure a consecutive MPG discharge. Possible schemes for implementing this idea are shown in Figs. 5.1 and 5.2. They show high-voltage blocks of the generator of the microsecond pulses consisting of three MPGs, each of which has two compression links.

Figure 5.1 shows a scheme in which the magnetic pulse generators used have a traditional circuit for the LIA on magnetic elements developed at the TPU.

Magnetic pulse generators 5–7 have the same electrical circuit and operating principle. They consist of successive *LC*-links of compression. Unlike the known designs of LIAs on magnetic

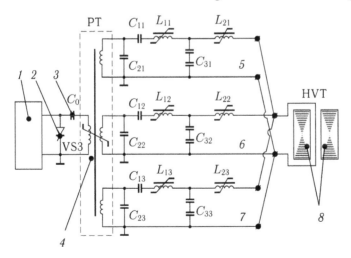

Fig. 5.1. Electrical diagram of the GMP: *1* – power supply; *2*– thyristor primary storage (elements VS3); *3*– primary storage C_0 ; *4* – pulse transformer (PT); *5–7* – magnetic pulse generators consisting of capacitors C_{11}–C_{31}, C_{12}–C_{32}, C_{13}–C_{33} and saturation chokes L_{11}–L_{21}, L_{12}–L_{22}, L_{13}–L_{23}; *8* – high-voltage transformer (HVT).

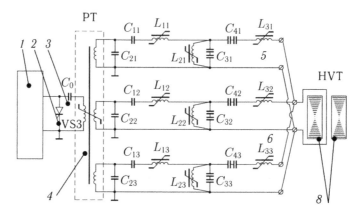

Fig. 5.2. Scheme of GMP with increased discharge voltage of MPG.

elements, there is no forming line in this installation. Its role is played by capacitors C_{31}, C_{32}, C_{33} of the last links of MPG compression. The difference is also that the capacitive storage devices used in the first MPG compression link are divided into three pairs of storage units ($C_{11}-C_{21}$, $C_{12}-C_{22}$, $C_{13}-C_{23}$). The charge of capacitors $C_{11}-C_{13}$, $C_{21}-C_{23}$ of the first compression links of all MPGs from the common pulse transformer PT *4* occurs simultaneously, so that the output pulses are synchronized. The capacitance of the capacitors C_{11}, C_{12}, C_{13} and C_{21}, C_{22}, C_{23}, the number of turns of the secondary winding of the pulse transformer *4* and the parameters of the saturation chokes L_{11}, L_{12}, L_{13} are chosen the same to ensure an equal time interval of the charging processes and equal to the amplitude of the charging voltage of the capacitors. In this case, energy transfer processes are eliminated between the elements of the magnetic pulse generators *5–7*. Proceeding from this, the condition on the number of links for the compression of MPG (at least two) is imposed.

The second variant of the possible electrical circuit of the high-voltage unit is shown in Fig. 5.2. The difference between the first and second options is that the last compression links have different electrical circuits. In the second variant, the discharge voltage of MPG *5–7* is doubled, which makes it possible to obtain a higher voltage at the input of a high-voltage transformer. In addition, the demagnetization of the cores of the magnetic commutators L_{31}, L_{32}, L_{33} and the inductors of the high-voltage transformer occurs when the capacitors C_{41}, C_{42}, C_{43} immediately before the operating impulse. (In the first MPG scheme, the demagnetization of the cores of the magnetic commutator and the inductors of the high-voltage

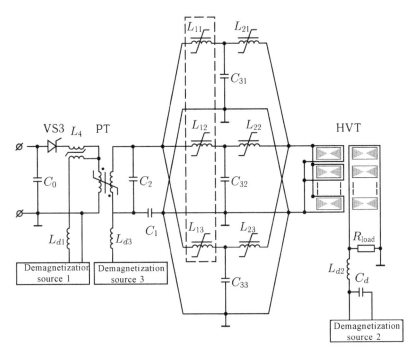

Fig. 5.3. Scheme of high-voltage unit of the MPG: *1* – the source of demagnetization of the cores of the saturation choke L_4, connected through the separating inductance L_{d1}; *2* – the source of demagnetization of the cores of the high-voltage transformer, connected through the separating inductance L_{d2} (terminal L_{d2} is shunted by the capacitor C_d); the source of demagnetization of the cores of the pulse transformer PT, connected through the separating inductance L_{d3}.

transformer occurs simultaneously with the cores of the saturation chokes L_{11}, L_{12}, L_{13}, L_{21}, L_{22}, L_{23} when the capacitors C_{11}, C_{12}, C_{13} are charged.) In addition, the first scheme uses a smaller number of compression stages and, correspondingly, fewer elements. This reduces the energy loss. Nevertheless, the second circuit of the GMP also has its advantages due to the increased discharge voltage.

Let's consider in detail the operating principle of the high-voltage unit of the GMP using a more complete electrical circuit (Fig. 5.3), which corresponds to the first variant (see Fig. 5.1) and is supplemented with devices that ensure its functioning. In addition, significant improvements have been made to it, allowing accurate synchronization of the discharge of all MPGs to the winding of a high-voltage transformer.

The diagram shows thyristors VS3 of the primary storage devices, primary energy storage C_0, pulse transformer PT, three parallel switched magnetic pulse generators, C_1, C_2, C_{31}, C_{32}, C_{33} capacitors and saturation chokes L_{11}–L_{21}, L_{12}–L_{22}, L_{13}–L_{23}, and the high-voltage transformer HVT. The saturation chokes L_{11}, L_{12}, L_{13} have a common ferromagnetic core with three identical windings. The first compression links of all MPGs are connected to the secondary windings of the common pulse transformer PT. The capacitors of the first MPG links are electrically connected. They form the storages C_1 and C_2. The parameters of the windings of the saturation chokes L_{11}, L_{12}, L_{13} are chosen to be equal to ensure an equal time interval of the charge–discharge processes and equal to the amplitude of the charging voltage on the capacitors C_{31}, C_{32}, C_{33}.

5.1.2. Calculation of the high-voltage unit

Here is an example of calculating a generator of microsecond voltage pulses with an amplitude of 450 kV, a current of 1 kA and a duration of the flat part of the pulse of 1 μs. Note that the following formulas can be used to calculate and design a generator with any output parameters.

Before the generator is turned on, it is required to transfer the cores of the magnetic elements to the saturation state $(-B_S)$. Therefore, it is necessary to pass constant demagnetization currents. Initially, the required pulse current I_{d3}, determining the magnetic state of the core of the saturation chokes, is set in the circuit of the PT pulse transformer, and a rectifier for charging the capacitor C_0 is connected. The demagnetization source *3* connected through the separating inductance L_{d3} to the secondary winding of the pulse transformer generates a demagnetization current I_{d3} of the last one (1.5–2 A). The demagnetization source *1*, connected through the separating inductance L_{d1}, forms the current of the saturation choke L_4 (70 A). The cores of the high-voltage transformer are demagnetized by the demagnetization source *2* connected to the secondary winding of the high-voltage transformer through the inductance L_{d2}, shunted by the high-voltage capacitor C_d. All sources of demagnetization currents have the same electrical circuits and are connected to an AC power source.

Discharge of primary storage C_0. With the arrival of the control pulse on the thyristor unit VS3 (time t_0, the voltage changes on the circuit elements are shown in Fig. 5.4), the capacitor C_0 is connected to the primary winding of the pulse transformer. The capacitors C_2 are charged directly from the secondary winding of the PT, and the windings of the saturation chokes L_{11}, L_{12}, L_{13}, L_{21}, L_{22}, L_{23} and magnetizing coils of the induction system of the high-voltage transformer HVT (the cores of which are additionally demagnetized) are connected in the C_1 charge circuit.

If we do not take into account the energy losses on the elements of the generator, then the voltage at the capacitors $U_{C_1}(t)$, $U_{C_2}(t)$ and the current I_1 can be written as follows:

$$U_{C_1}(t) = \frac{U_{C_0}}{1+\lambda_1}(\lambda_1 + \cos \omega_0 t); \tag{5.1}$$

$$U_{C_1}(t) = \frac{k_{PT}\lambda_1 U_{C_0}}{1+\lambda_1}(1 - \cos \omega_0' t) - \frac{I_{d3}}{C_1}t; \tag{5.2}$$

$$U_{C_2}(t) = \frac{k_{PT}\lambda_1 U_{C_0}}{1+\lambda_1}(1 - \cos \omega_0 t) - \frac{I_{d3}}{C_1}t; \tag{5.3}$$

$$I_1 = \frac{U_{C_0}}{2\omega_0 L_0}\sin \omega_0 t + \frac{U_{C_0}}{2\omega_0'(L_0 + L_1 / k_{PT}^2 + L_2 / k_{PT}^2)}\sin \omega_0 t - I_{d3}; \tag{5.4}$$

$$\omega_0 = \sqrt{\frac{1+\lambda_1}{L_0 C_0}}\omega_0' = \sqrt{\frac{1+\lambda_1}{(L_0 + L_1 / k_{PT}^2 + L_2 / k_{PT}^2)C_0}}, \; \lambda_1 = \frac{C_0}{k_{PT}^2(C_1 + C_2)}, \tag{5.5}$$

where k_{PT} is the transformer ratio of the pulse transformer; L_1, L_2 is the inductance of the windings of the saturation chokes L_{11}, L_{12}, L_{13} and L_{21}, L_{22}, L_{23}, connected in series-parallel, through which the charge C_1 occurs. The demagnetization current I_{d3} is much lower than the discharge current of the capacitors, so it can be neglected in calculations.

At the time $t = t_1$, the voltage at C_1 and C_2 reaches a maximum and the process of energy transfer to the capacitors C_1 and C_2 can be considered complete. Voltage from the capacitor C_2 is applied to the secondary winding of the transformer in the time interval t_0–t_1. The induction in the transformer core during this period increases in accordance with the following expression:

$$B(t) = -B_S + \frac{1}{\omega_{PT2}S_{PT}}\int_{t_0}^{t_1} U_{C_2}(t)dt, \tag{5.6}$$

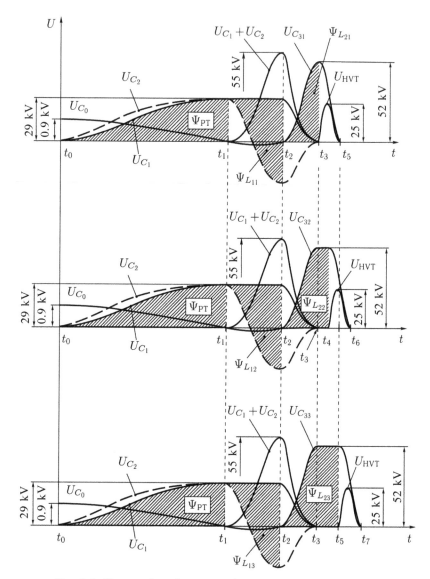

Fig. 5.4. Changes in voltages on elements of the GMP scheme.

where ω_{PT2} and S_{PT} are the number of turns of the secondary winding and the cross-sectional area of the steel of the pulse transformer.

At the time t_1, the saturation induction in the pulse transformer reaches the value B_S, i.e., the transformer core is saturated. Thus, the magnitude of the flux linkage of the core of the pulse transformer should be chosen such that the discharge time of the capacitor C_0 corresponds to the time of magnetization of the core of the pulse transformer:

$$t_1 - t_0 = \pi \sqrt{\frac{L_0 C_0}{2}} \approx \frac{\Psi_{PT}}{\langle U_{C_0} \rangle}, \tag{5.7}$$

where L_0 is the discharge circuit inductance; $\Psi_{PT} = \omega_{PT1} S_{PT} \Delta B$ is the magnitude of the PT flux linkage; $\langle U_{C_0} \rangle \approx U_{C_0} / 2$ is the average operating voltage on the transformer windings; U_{C_0} is the amplitude of the charging voltage C_0. The discharge time of the primary storage device according to the calculations given below is $t_1 - t_0 = 50$ μs. Therefore, the flux linkage of the transformer must be at least

$$\Psi_{PT} = \langle U_{C_0} \rangle (t_1 - t_0) = \omega_{PT1} S_{PT} \Delta B. \tag{5.8}$$

From the expression (5.8) we can determine the necessary size of the cross section of the steel of the core of the pulse transformer $S_{PT} = 90$ cm² at the chosen number of primary turns ($\omega_{PT1} = 1$). The transformer can be manufactured, for example, using two cores with the following parameters: outer diameter $D_{PT} = 1.1$ m; internal diameter $d_{PT} = 0.5$ m; width $l_{PT} = 0.018$ m; the coefficient of filling the steel volume of the core $K = 0.8-0.85$; material – permalloy 50 NP; thickness of the tape 0.02 mm; cross section of steel of one core $S_{core} = (43-46)$ cm².

The primary winding of the pulse transformer is produced either as a solid circular turn having 24 terminals for connection to the capacitors of the primary storage through a saturation choke L_4 and switching thyristors VS3 or from 12 parallel single-turn windings. The secondary winding is made of two parallel sections of 33 turns, each of which is connected to capacitors C_1 and C_2.

The inductance of the secondary winding of a pulse transformer for a saturated state of the core ($\mu \to 1$) is

$$L_{PT2} = \frac{\mu_0}{2\pi} l_{PT2} \omega_{PT2}^2 \ln \frac{D_{PT2}}{d_{PT2}} = 15.4 \ \mu H, \tag{5.9}$$

where $D_{PT2} = 1.14$ m, $d_{PT2} = 0.46$ m and $l_{PT2} = 0.078$ m are the external, internal diameters and length of the secondary winding; $\omega_{PT2} = 33$ is the number of turns of the secondary winding.

The equivalent capacitance in the secondary circuit of the transformer must be

$$C_1 + C_2 = \frac{C_0}{k_{PT}^2} = 2.2 \ \mu F, \tag{5.10}$$

where $C_1 + C_2$ is the total capacity of the capacitors of the first MPG compression link, which are switched on in parallel. Storages C_1 and C_2 are produced from 48 industrial capacitors with a capacity of 0.047 μF. Thus, the total capacity of the storage units is $C_1+C_2 = 2.256$ μF, which is higher than the value calculated from formula (5.10). In this case, the polarity of the voltage at C_0 after the discharge will be negative, which makes it easier to turn off the thyristors, prevents re-saturation of the cores of the pulse transformer and the occurrence of additional pulses.

Recharge of the capacitor C_2 through the secondary winding of the pulse transformer. The voltage on the capacitor and the current in the secondary of the transformer change as follows:

$$U_{C_2}(t) = \frac{U_{C_2}}{2}(1+\cos\omega_1 t); \qquad (5.11)$$

$$I_2 = \frac{U_{C_2}}{2\omega_1 L_{PT2}}\sin\omega_1 t; \qquad (5.12)$$

$$\omega_1 = \sqrt{\frac{2}{L_{PT2}C_2}}, \qquad (5.13)$$

where U_{C_2} is the amplitude of the charging voltage of the capacitor C_2.

During the recharging process, the sum of the voltages at $C_1 = 1.128$ μF and $C_2 = 1.128$ μF is applied to the turns of the saturation chokes L_{11}, L_{12}, L_{13}. The values of the flux linkage of the chokes L_{11}, L_{12}, L_{13} are chosen such that at the time t_2, when C_2 is completely recharged, the common core of the saturation chokes L_{11}, L_{12}, L_{13} was saturated, that is, the following condition is satisfied

$$\pi\sqrt{L_{PT2}C_2} \approx \frac{\Psi_{L_{11}}}{(U_{C_1}+U_{C_2})/2}, \qquad (5.14)$$

where $\Psi_{L_{11}} = \Psi_{L_{12}} = \Psi_{L_{13}} = \omega_{11}S_1\Delta B$; ω_{11} and S_1 are the number of turns and the cross-sectional area of the ferromagnetic core of the saturation chokes L_{11}, L_{12}, L_{13}; $(U_{C_1}+U_{C_2})/2 \approx 27.5$ kV is the average operating voltage on the windings of the chokes L_{11}, L_{12}, L_{13} is the sum of the energy losses during the recharging of the capacitors.

The beginning of the windings of the saturation chokes L_{11}, L_{12}, L_{13}, connected to the capacitors C_1, C_2, can have galvanic connection between themselves. It is important that the opposite ends of the windings connected to the capacitors C_{31}, C_{32}, C_{33}, are not connected. The saturation chokes L_{11}, L_{12}, L_{13} each have three three-turn windings consisting of two parallel sections (6 sections in total).

In order that no transfer of energy occurs between the capacitors of the first links of the individual magnetic pulse generators, it is necessary that the processes in them proceed simultaneously. Therefore, the first links use elements with the same parameters. In this case, the flux linkages and inductances of the saturation chokes are ($L_{11} = L_{12} = L_{13}$) and are composed in the saturated state

$$L_{11} = L_{12} = L_{13} = \frac{\mu_0}{2\pi} l_{11} \omega_{11}^2 \ln\left(\frac{D_{11}}{d_{11}}\right) = 0.4 \ \mu H, \qquad (5.15)$$

where $D_{11} = 1.14$ m, $d_{11} = 0.46$ m are the outer and inner diameters of the winding; $l_{11} = 0.25$ m is the length of the winding.

The recharging time of the capacitor C_2 is

$$t_2 - t_1 = \pi \sqrt{L_{PT2} C_2} = 13.1 \ \mu s. \qquad (5.16)$$

Discharge of capacitors C_1 and C_2 on capacitors C_{31}, C_{32}, C_{33}. At the time t_2, the core of the saturation choke of the first MPG compression links is saturated (the inductance of its winding drops sharply to the value determined by (5.15)) and the capacitors C_1 and C_2, connected in series, start discharging to the capacitors C_{31}, C_{32}, C_{33}, connected in parallel. For effective energy transfer, the following relationship between capacitor capacitors should be fulfilled:

$$\frac{C_1 C_2}{C_1 + C_2} = C_{31} + C_{32} + C_{33}. \qquad (5.17)$$

This process is described by the following expressions for the total voltage:

$$U_{C_1 + C_2}(t) = \frac{U_{C_1} + U_{C_2}}{2}(1 + \cos\omega_2 t); \qquad (5.18)$$

$$U_{C_{31}} = U_{C_{32}} = U_{C_{33}} = \frac{U_{C_1} + U_{C_2}}{2}(1 - \cos\omega_2 t); \qquad (5.19)$$

$$I_2 = \frac{U_{C_1} + U_{C_2}}{2\omega_2 L_{11}'} \sin\omega_2 t, \qquad (5.20)$$

$$\omega_2 = \sqrt{\frac{2(C_1 + C_2)}{L_{11}'(C_1 C_2)}}. \qquad (5.21)$$

In the time interval between t_2 and t_3, the cores of the saturation chokes (magnetic commutators) L_{21}, L_{22}, L_{23}. are magnetized. The magnitude of the flux linkage of the saturation choke L_{21} must be sufficient to complete the discharge of the capacitors C_1, C_2 to capacitors C_{31}, C_{32}, C_{33}, that is, to correspond to the condition

$$\Psi_{L_{21}} = \frac{1}{2} U_{C_{31}} (t_3 - t_2), \qquad (5.22)$$

where $U_{C_{31}} = U_{C_{32}} = U_{C_{33}} \approx 52$ kV is the amplitude of the charging voltage of the capacitors C_{31}, C_{32}, C_{33} (the decrease in the amplitude of the voltage is determined by the losses during energy compression, which can be accurately estimated only in computer simulation); $t_3 - t_2$ is the discharge time of capacitors C_1 and C_2, connected in series, to capacitors C_{31}, C_{32}, C_{33}, connected in parallel:

$$t_3 - t_2 = \pi \sqrt{\frac{L'_{11}(C_1 + C_2)(C_{31} + C_{32} + C_{33})}{C_1 + C_2 + C_{31} + C_{32} + C_{33}}} = 1.18 \ \mu s. \qquad (5.23)$$

The formulas (5.20), (5.21) and (5.23) include the discharge circuit inductance L'_{11} = 0.5 µH, which is composed of the inductance of the winding of the saturation chokes, determined by (5.15), from the intrinsic inductance of the assemblies of the capacitors C_1, C_2, C_{31}, C_{32}, C_{33} and the inductance of the connecting wires.

Capacitor capacitances must be the same (C_{31} = C_{32} = C_{33} = 0.188 µF) in order to ensure a high efficiency of energy transfer from the first MPG compression unit to the second and the equality of charge voltage amplitudes of each capacitor and the duration of charge-discharge processes.

Saturation choke L_{21} can be made on the basis of four cores with the following parameters: $D_{L_{21}}$ = 0.5 m; $d_{L_{21}}$ = 0.22 m; $l_{L_{21}}$ = 0.025 m; K = 0.8; material – permalloy 50 NP, with a solid copper single-turn winding (ω_{21} = 1). The inductance of a winding of a choke at the saturated state of the cores

$$L_{21} = \frac{\mu_0}{2\pi} l_{21} \omega_{21}^2 \ln \frac{D_{21}}{d_{21}} = 0.027 \ \mu H, \qquad (5.24)$$

where D_{21} = 0.525 m, d_{21} = 0.194 m are the outer and inner diameters of the winding; l_{21} = 0.134 m is the axial length of the winding.

Discharge of the capacitors C_{31}, C_{32}, C_{33} *through the primary winding of the high-voltage transformer.* In order to obtain a flat part of the output pulse of the GMP with duration of 1 μs, it is necessary to generate three pulses of ~0.7–0.8 μs duration each with a time delay of ~0.35 μs between the first and second pulses and ~0.7–0.8 μs between the first and third pulse. The output pulse of the second MPG of 0.35 μs can be delayed by increasing the flux linkage of the saturation choke L_{22}, and the output pulse of the third MPG by 0.75 μs by an additional increase in the flux linkage of the saturation choke L_{23}.

$$\Psi_{L_{22}} = \frac{1}{2}U_{C_{32}}\,(t_3 - t_2 + 0.35\ \mu s). \tag{5.25}$$

Saturation choke L_{22} can be made of five cores with the following parameters: $D_{L_{22}} = 0.5$ m; $d_{L_{22}} = 0.22$ m; $l_{L_{22}} = 0.025$; $K = 0.8$; material – permalloy 50 NP, with solid copper single-turn winding ($\omega_{22} = 1$). Thus, the use of an additional core allows to achieve the necessary time delay. With the indicated core sizes, the inductance of the magnetizing coil of the saturation choke L_{22} in the saturated state of the core is

$$L_{22} = \frac{\mu_0}{2\pi}l_{22}\omega_{22}^2\ln\left(\frac{D_{22}}{d_{22}}\right) = 0.032\ \mu H, \tag{5.26}$$

where $D_{22} = 0.525$ m, $d_{22} = 0.194$ m are outer and inner diameters of the windings; $l_{22} = 0.162$ m is the axial length of the winding.

The flux linkage of the saturation choke L_{23} is selected from condition

$$\Psi_{L_{23}} = \frac{1}{2}U_{C_{33}}\,(t_3 - t_2 + 0.75\,\mu s). \tag{5.27}$$

The saturation choke can be produced from six cores with the following parameters: $D_{L_{23}} = 0.5$ m; $d_{L_{23}} = 0.22$ m; $l_{L_{23}} = 0.025$; $K = 0.8$; material – permalloy 50 NP, with a solid copper single-turn winding ($\omega_{23} = 1$).

The inductance of the magnetizing turns of the saturation choke L_{23} in the saturated state of the core is

$$L_{23} = \frac{\mu_0}{2\pi}l_{23}\omega_{23}^2\ln\frac{D_{23}}{d_{23}} = 0.038\ \mu H, \tag{5.28}$$

where $D_{23} = 0.525$ m, $d_{23} = 0.194$ m are the outer and inner diameters of the winding; $l_{23} = 0.19$ m is the axial length of the winding.

In the time interval between t_2 and t_3 (Fig. 5.4), the core of the magnetic commutator L_{21} is magnetized. Capacitor C_{31} is discharged (time interval t_5–t_3) through the primary winding of the HVT and generates a high-voltage pulse. The discharge of the second MPG (capacitor C_{32}) should begin at time t_4 (the core of the magnetic commutator L_{22} is saturated), which corresponds to approximately half the duration of the output pulse of the first MPG. The third MPG (capacitor C_{33}) starts discharging at time t_5 (saturation of the core of the magnetic commutator L_{23}) corresponding to the end of the discharge pulse of the first MPG and half the duration of the discharge pulse of the second MPG. Thus, the delay between the output pulses of the third and first MPG is 0.75 µs.

In the process of tuning the generator of microsecond voltage pulses, it is possible to smoothly change the magnitude of the flux linkage of the magnetic commutators (the moment when individual MPGs are switched on to the primary winding of the high-voltage transformer). For this purpose, additional demagnetization circuits of the saturation chokes L_{21}, L_{22}, L_{23} should be introduced .

Thus, in the considered variant of the high-voltage GMP unit, the delay of the voltage pulse from the second MPG to the primary turns of the high-voltage transformer is accomplished by increasing the linkage of the saturation choke L_{22} by adding one core. The delay of the voltage pulse from the third MPG to the primary turns of the high-voltage transformer is achieved by increasing the linkage of the saturation choke L_{23} by adding one more core. The total duration of the pulses transmitted by the three magnetic pulse generators of the high-voltage unit to the primary winding of the high-voltage transformer is equal to the time interval t_7–t_3. It should be noted that the necessary value of the pulse delay can be achieved by increasing the number of windings of the magnetic commutator coils, and also by a joint variation in the number of turns and the number of cores.

Magnetization reversal of the cores. In order that the magnetic elements have the minimum dimensions, a change in the induction of ΔB by the demagnetized pulse should be maximized, i.e. reversal of the cores should be carried out by limiting the hysteresis loop. In addition to reducing the size, the use of the maximum induction drop allows minimizing the leakage inductance and parasitic capacitance of the elements of the magnetic pulse generator.

Let us consider the process of magnetization reversal of the cores in the initial state. The return of the pulse transformer core to the initial state occurs when C_2 is recharged through its secondary winding. Additional demagnetization of the core of the pulse transformer is carried out by the current I_{d3}, flowing through the secondary winding of the pulse transformer, the winding of the saturation chokes and the primary winding of the high-voltage transformer. Although the magnitude of the current is negligible (1.5−2 A), the ampere turns are sufficient to demagnetize the core of the pulse transformer, since the current flows through the 33-turn secondary winding. The demagnetization of the cores of the saturation chokes, magnetic commutators and high-voltage transformer is performed by the charging current of the capacitor C_1, the value of which exceeds 300 A. However, the current through a separate inductor will be lower by as many times as the number of cores enters the set (i.e. N times). Therefore, for reliable demagnetization of the high voltage the demagnetizing source 2 with a pulsed current of amplitude 50 A is connected to the secondary winding of a high-voltage transformer is included in the circuit.

5.2. Power source

The high-voltage unit of the GMP, containing compression links with magnetic elements, is most advantageous to feed from the circuit with thyristor commutation. Thyristor switches provide switching currents with high repetition rate and with the required impulse and average characteristics.

The functional diagram of the power supply for the GMP is shown in Fig. 5.5. It is based on the scheme of a power source with energy recovery, which is conditionally divided into two parts: the input of energy into the high-voltage GMP unit (A9) and the primary storage device charger (A5−A8). For reliable operation of the circuit elements, the duration of energy input from the thyristor pulse generator (C_0, VS3, L_4) to the high-voltage GMP unit at high average currents should be tens of microseconds.

For a given power level and pulse repetition frequency, the only acceptable way to charge the capacitor C_0 of a thyristor pulse generator is a classical oscillating charge from a constant voltage source comprising a choke L_2 and a thyristor VS1. At the same time, this simple scheme does not ensure the stabilization of the pre-discharge voltage, which varies widely due to the drop in the

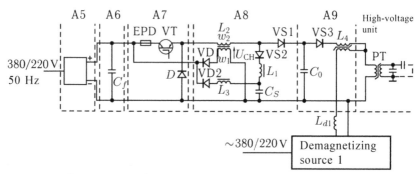

Fig. 5.5. Functional scheme of power supply for GMP.

voltage of the primary storage and the return of part of the energy from the high-voltage unit. Voltage stabilization on C_0 is reached by the interruption of the charging process by switching on the thyristor VS2 on the capacitor C_S. The energy remaining in L_2 after VS1 is turned off, is output to the primary storage A6 (capacitor C_f) via VD1 and the additional coil of the choke L_2. Energy from C_S also returns to C_f through VD2 and L_3. The selected elements allow one to commutate very high currents at high efficiency.

When choosing the charging device A5 of the primary storage A6, it is necessary to take into account the operating mode of the GMP. In the packet mode, it is not expedient to charge the primary storage directly from the AC mains, since it requires expensive and large equipment at an average power of ~1 MW. Pulse power take off of this level creates power surges in the supply network and is highly undesirable for energy suppliers. Therefore, it is more expedient to use the primary capacitive energy storage A6 in the partial discharge mode and replenish its energy with a low-power A5 charger for a long time interval between bursts of pulses. The high energy stored in the primary storage, coupled with the possibility of its rapid withdrawal, causes the catastrophic explosive nature of the emergency process in the event of the breakdown of any element of the scheme or isolation and requires the use of special protective measures. One of these is the inclusion of a chopper circuit breaker on IGBT modules (VT-D, see Fig. 5.5) and an exploding fuse (EPD) (A7).

The charging device A5 of the primary capacitive storage can be made using high-frequency inverting technology, which will allow to regulate charging current and pre-discharge voltage within wide range.

5.2.1. Thyristor pulse generator

The thyristor pulse generator is the first link of the energy compression scheme and realizes the same principle of energy transfer from the capacitor to the capacitor through inductance, which is also used in the MPG links of the high-voltage GMP unit. It is necessary initially to estimate the energy value, $Q = 900$ J, which should be stored in the capacitor C_0 of a thyristor pulse generator. The capacitance C_0 of the primary storage should be such that the energy reserve in it exceeds the energy of the output pulse ($Q = 450$ kV · 1 kA · 1 µs $= 450$ J) approximately in 2 times to compensate losses as a result of compression of energy and extension of pulse fronts. (The coefficient of efficiency of the transfer of energy from the primary storage device to the energy of the output pulse can be estimated with high accuracy in the computer simulation of GMP, taking into account losses in the cores, capacitors and windings, see section 5.8.2.).

The operating voltage C_0 depends on the type of thyristors used and the scheme of their group switching. Since the duration of the energy input is small (given by the input duration $t_1 - t_0 = 50$ µs), it is possible to use only high-speed thyristors, the maximum non-repeating voltage of which does not exceed 2500 V, and the operating voltage is 1200 V. The element base of thyristors use both serial and parallel connections. However, it is known from the experience of designing power supply systems for LIA on magnetic elements (see Chapter 2) that parallel connection provides a higher reliability of the circuit and in this case the modular principle of constructing the feed system of the GMP is simpler to be realized.

In this way,

$$C_0 = 4Q/U_{C_0}^2 = 2.2 \text{ mF}, \tag{5.29}$$

The choice of charge voltage value $C_0 \leq 1000$ V is determined from the condition of reliable operation of thyristors.

The capacitance of C_0 increases to 2.4 mF to collect C_0 from 12 capacitors of 1600 V and 200 µF.

The inductance of the solid toroidal winding (the primary turn of the impulse transformer) will be

$$L_{PT1} = \frac{\mu_0}{2\pi} l_{PT1} \omega_{PT1}^2 \ln \frac{D_{PT1}}{d_{PT1}} = 7.55 \text{ nH}, \tag{5.30}$$

where $D_{PTI} = 1.108$ m, $d_{PTI} = 0.492$ m and $l_{PTI} = 0.046$ m are the external, internal diameters and axial length of the winding. To quantity L_{PTI} it is necessary to add the inductance of mounting $L_{as}^{(0)}$, the inherent inductance of the capacitors L_{C_0}, the inductance L_4 of the thyristor assemblies and the turns of the saturation choke L_4.

The duration of the discharge is

$$t_1 - t_0 = \pi\sqrt{L_0 C_0/2} = 50 \text{ μs},\qquad (5.31)$$

where L_0 is the discharge circuit inductance:

$$L_0 = L_{PTI} + L_4 + L_{C_0} + L_{as}^{(0)} \sim 211 \text{ nH}.\qquad (5.32)$$

The amplitude of the current of the thyristor switch VS3 is

$$I_{dm} = \frac{U_{C_0}}{\sqrt{2 \cdot L_0/C_0}} = 68.5 \text{ kA}.\qquad (5.33)$$

at the maximum rate of current rise

$$\frac{dI_d}{dt} = \frac{\pi I_{dm}}{t_0} = 4.27 \text{ A/μs}.$$

Switching such a current with one thyristor is irrational, since in this mode it is at the limit of its capabilities. It is more reliable to apply parallel connection of the thyristors. We choose for switching C_0 12 thyristors. The current amplitude and the rate of its rise are reduced for one thyristor, respectively, to 5.67 kA and 355 A/μs, which is an easy mode of operation for these thyristors.

For an even distribution of the current between the thyristors, the C_0 storage is divided into 12 blocks of 200 μF. Each block is commutated by a separate thyristor connected to the primary winding of the pulse transformer of the high-voltage unit of the GMP through the winding of the saturation choke L_4. The choke is a set of 12 ferrite rings with windings of 3–5 turns. Ferrite rings are covered by a common demagnetizing coil connected to a demagnetization source. The purpose of the saturation choke is synchronization of the discharge of all power supply units to the primary winding of the pulse transformer, since the time variation of the activation of individual thyristors can reach hundreds of nanoseconds. A choke should compensate for this spread by equalizing the thyristor currents during the discharge pulse, thereby preventing their failure. If one or more thyristors are switched on in advance, the current in them is set at the level of the magnetization reversal of the ferrite core. At the same time, due to the transformer coupling through the demagnetizing winding of the saturating ferrite core to the unconnected thyristors, the emf is transformed, which leads to the switching on of devices, naturally in the presence of a control pulse on them. Until all the devices turn on, the current in them is limited by the magnetization reversal of the cores of the saturation choke.

Thus, the pulse generator consists of 12 identical blocks. In this case thyristor pulse generator entails the separation into 12 parallel blocks: 1) charging circuits A8; 2) protection schemes for the primary storage A7; 3) the primary storage A6. As a result, the power supply system of the GMP should be made of 12 modules with a common charging device for A5 primary storage devices (see Fig. 5.5). The voltage diagrams for the elements of the power circuit are shown in Fig. 5.6.

5.2.2. Oscillating charging device

The circuit of this device implements the principle of oscillation charging of a capacitor from a constant-voltage source through a choke and a switching element. Such a scheme allows, in the limiting case (without losses), to charge the capacitor to a double voltage of a source with a high efficiency. When choosing the charging time, it is necessary to take into account that after the energy input through a pause $t_{pause} \approx t_1 - t_0$, a part of the energy can be returned back to C_0 in a time $t \approx t_1 - t_0$ (voltage pulse U_{rec} in Fig. 5.6). Such regimes are possible when the GMP is not fully matched with the load. Thus, the complete operating process of discharge C_0 can occupy the time interval ≈ 3 $(t_1 - t_0)$ and the charging circuit's thyristor should be switched off after an interval $t \approx 300$ μs .

In this case, the charging process duration C_0 is

$$t_{charge} = 1 / F - 3t_{pause} - t = 1000 \times 3.50 - 300 = 0.55 \text{ ms,}$$

where F is the repetition rate of the GMP pulses.

The inductance of the charge choke must be

$$L_2 = \frac{t_{charge}^2}{\pi^2 C_{0m}} = 153 \cdot 10^{-6} \text{H,}$$

where $C_{Om} = 200$ μF is the capacitance of the capacitor block.

The oscillatory process of charging the capacitor C_0 is described by equation

$$U_{C_0}(t) = E\left[1 - \left(1 + \frac{U_0}{E}\right)\exp\left(-\frac{\pi t}{2Qt_{charge}}\right)\cos\left(\frac{\pi t}{t_{charge}}\right)\right], \qquad (5.34)$$

where E is the voltage of the primary energy storage C_f (A6); U_0 is the voltage remaining from the previous packet; Q is the quality factor of the charging circuit.

Hence, assuming $U_0 = 0$, $t = t_{charge}$, we find the minimum voltage to which the primary storage can be discharged at the end of the burst packet:

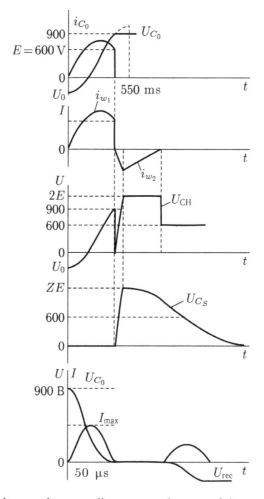

Fig. 5.6. Voltage and current diagrams on elements of the power circuit.

$$E_{min} = \frac{U_{C_0}}{2 - \pi/2Q} = 487 \text{ V,} \qquad (5.35)$$

where $Q = 10$ is the technically realizable Q-factor. Setting the previously possible level of the reverse voltage, $U_0 = 200$ V, which is 22% of $U_{C_0} = 900$ V, we find the time point at which the voltage reaches the preset level E_{min} and the charging process C_0 must be interrupted:

$$\cos \frac{\pi t_x}{t_{charge}} = \frac{U_{C_0}/E_{min} - 1}{1 + U_0/E_{min}}, \qquad (5.36)$$

from which $t_x = 0.72 t_{charge}$ (in this case, for simplicity of calculation, the energy losses were ignored).

If you select the maximum (initial) voltage of the primary storage $E_{max} = 600$ V, and the voltage drop on it after generating a burst of pulses $\approx 20\%$, then the time at which the charging should be interrupted is found from the following relationship:

$$\cos \frac{\pi t_x}{t_{charge}} = \frac{U_{C_0}/E_{max} - 1}{1 + U_0/E_{max}}, \qquad (5.37)$$

It is fulfilled at $t_x = 0.62 t_{charge}$ and at $t_x = 0.67 t_{charge}$, if we assume $U_0 = 0$. Thus, in order to compensate for the voltage drop of the primary storage device, taking into account the residual voltage of the capacitor C_0, due to the return of energy from the high-voltage unit of the GMP, the time required to adjust the charge time $t_x = (0.6-1.0) t_{charge}$, which does not exceed the technically achievable level $t_p = (0.5-1.0) t_{charge}$.

The maximum amplitude of the charging current of the capacitor of the block is

$$I_{chargem} = \frac{U_0 + E_{max}}{\sqrt{L_2/C_{0m}}} = 915 \text{ A} \qquad (5.38)$$

at a current rise rate

$$\frac{dI_{charge}}{dt} = \frac{\pi I_{chargem}}{t_{charge}} = 5.2 \text{ A / ms.} \qquad (5.39)$$

Elements of the charging interruption circuit are chosen for the heaviest mode, $t = 0.5 t_{charge}$, i.e. for the case when the current in the choke L_2 is maximal. The capacitance of the capacitor C_S is

$$C_S = \frac{t_q I_{chargem}}{U_{C_0}} = 51 \ \mu F, \tag{5.40}$$

where t_q is the turn-off time of the thyristor. C_S is increased to 60 μF to compose it from three capacitors of 1600 V and 20 μF.

The voltage amplitude on the capacitor is limited to the level $U_{C_S} = 2E_{max} = 1200$ V, if the number of turns of the main and additional windings of the choke L_2 are equal.

The rate of rise of current through the thyristor of the charging interruption circuit VS2 is limited to a level of 500 A/μs by the inductance L_1 of the value

$$L_1 = \frac{U_{C_0}}{di/dt} = 1.8 \ \mu H.$$

The duration of the current through the thyristor VS2 is

$$t_{VS2} = \frac{2E_{max}C_S}{I_{chargem}} = 80 \ \mu s.$$

After charging the capacitor C_S to $2E_{max}$, the diode VD1 opens and the energy output from the choke L_2 begins. The amplitude of the diode current $I_{VD1} = I_{chargem} = 915$ A for the duration of the output energy $t_{del \ L2} = I_{chargem}L_2/E_{max} = 233$ μs, that does not limit the maximum frequency of repetition of charging cycles ($F = 1000$ Hz).

Since always $U_{C_S} = 2E$, the circuit L_3–VD2 completely removes energy from C_S to the storage unit A6. The inductance of the choke L_3 is found from the condition that the duration of the energy output should be $t_{del \ C_S} = 800$ μs. In this case, the inductance of the choke L_3 must be equal to

$$L_3 = \frac{t_{del C_S}^2}{\pi^2 C_S} = 10^{-3} H. \tag{5.41}$$

The current amplitude in the coil of the choke is

$$I_{L_3m} = \frac{E_{max}}{\sqrt{L_3/C_S}} = 147 \ A. \tag{5.42}$$

We estimate the instrumental error of voltage stabilization on the capacitor of the block C_{0m} by the charging interruption circuit. The error is maximal in the mode $t_x = 0.5t_{charge}$, when the maximum rate of voltage rise at C_{0m} is:

$$\frac{dU_{C_{0m}}}{dt} = \frac{(E_{max} + U_0)\pi}{t_{charge}} = 4.66 \text{ V/}\mu\text{s.} \tag{5.43}$$

The error of the circuit is determined mainly by the delay of the switching on of the thyristor VS2 ($t_{delay} = (1-2)$ µs). In this case

$$\frac{\Delta U}{U} = \frac{(dU_{C_{0m}}/dt)t_{delay}}{U_{C_{0m}}} \approx 0.01. \tag{5.44}$$

The total error in stabilizing the charging voltage level of the capacitors of the primary storage device C_0, taking into account the non-ideality of the voltage divider (reference voltage source), the comparator and the delay of the control pulse generation circuit may increase to 5%.

5.2.3. Primary capacitive energy storage

The initial and final voltages of the primary storage C_f are defined above. It is necessary to determine the capacity of an individual unit:

$$C_f = \frac{2Q_f n_{pul}}{(E_{max}^2 - E_{min}^2)\eta_{charge}} = 6.8 \text{ F,}$$

where Q_f is the energy of the capacitor of the module C_{0m}; $n_{pul} = 5000$ is the number of pulses in the packet; η_{charge} is the efficiency of the charging circuit. Thus, the primary storage unit A6 can be assembled from 12 individual modules of the 16 series-connected capacitor banks (ultracapacitors) BMOD0115PV, produced by the company Maxwell Technologies with the capacitance $C_{BMOD} = 145$ F at a voltage of <42 V. Then the voltage on each module $U_{BMOD} = 600 \text{ V}/16 = 37.5 \text{ V} < 42 \text{ V}$.

The capacitance of the separate module of the primary storage will equal $C_f = C_{BMOD}/16 = 9.1$ F. Each module must be provided with the diode-resistor circuit and the voltage equalization overcharge protection. The average current drawn from one primary storage during a burst of pulses is

$$I_0 = \frac{2E_{min}t_{charge}F}{\pi\sqrt{L_2/C_{0m}}} = 195 \text{ A.} \qquad (5.45)$$

This is substantially less than the current of a five-second battery discharge to E_{min} equal to 600 A, according to the specifications of BMOD0115PV. Thus, the operating mode of the molecular capacitor battery is chosen relatively easy and it can be expected that they will work with high reliability.

The value of the source current for the charge of the primary storage is determined by the duration of the pause t_{pause} between the GMP pulse packets:

$$I = \frac{(E_{max} - E_{min})C_f n_{module}}{t_{pause}}, \qquad (5.46)$$

where C_f is the capacity of the primary storage of the module; $n_{module} = 12$ is the number of modules of the feed circuit of the GMP. A 20-minute pause between bursts requires a 5–10 kW charger (600 V, 8.5 A). Reducing the pause between bursts requires an increase in the capacity of the primary storage. It should work in the mode of the current stabilizer before reaching the set voltage level, and then maintain this level. The device must be connected via the isolation diodes to the primary drives C_f and through the serial communication channel to the controller.

5.2.4. Control systems of the power source

Each power supply unit is controlled by its controller. Its input receives analogue signals of current, voltage and temperature. The controller controls via drivers the switching on of thyristors and IGBT chopper. All 12 block controllers are connected by a serial data channel and a clock channel with a central controller that controls the charging device of the primary storage, turning on the power and cycling start pulses to the power supply units. The central controller is connected to the computer, where it transmits information about the status of the blocks.

The control system of the GMP should monitor the value of the charging voltage of the primary storage devices, set the necessary pulse repetition frequency, set the number of pulses in the packet, automatically or manually start the packet and the time interval between the pulse packets. The control system must be equipped with interlocks which exclude incorrect operation.

So, the technical data of the feed system of the GMP are as follows.

1. Method of power input to the high-voltage unit: discharge of the storage capacitor C_0.

2. The energy of the storage capacitor $C_0 = 900$ J.

3. The maximum voltage of the storage capacitor $U_{C_0} = 900$ V.

4. Voltage stabilization error <5%.

5. The duration of the energy input is 50 µs.

6. The amplitude of the current is 68.5 kA.

7. The current rise rate is 4.27 kA/µs.

8. The maximum pulse repetition frequency is 1000 Hz.

9. Operating mode (number of pulses): packet (5000).

10. Power consumed from a three-phase network to charge the primary storage device in a pause between bursts of pulses, 5–10 kW.

5.3. High-voltage insulator

Insulators of radial and cylindrical geometry can be used for GMP. In the first case, the maximum operating voltage of the generator depends on the outer radius of the radial insulator, limited by the size of the tank in which the high-voltage transformer is placed. In addition, it is difficult to solve the problem of uniform distribution of voltage at the insulator surface. It is necessary to use insulators of complex shape (see, for example, Fig. 3.30). Therefore, for a generator of microsecond voltage pulses one should choose a cylindrical structure. This design allows to increase the distance between the ground and high-voltage electrodes to the value providing the necessary electrical strength. Since the output pulse of the GMP has a microsecond duration, the effect of the inductance of the cathode holder of increased length positioned along the axis of the high-voltage insulator on the duration of the front and the amplitude characteristics of the output pulse is small.

Here is an example of calculating a high-voltage insulator for the output voltage of the GMP of 450 kV. The calculation is carried out for the purpose of optimizing the design for the most uniform distribution of the electric field strength and its tangential component along the external surface facing the vacuum. Calculations determine the geometric dimensions and the effect of the metallic elements located inside the vacuum chamber on the resulting picture of the electric field is estimated. Numerical experiments are carried out with the help of the ERA modernized software intended for calculating electric fields in complex electrophysical devices, the dynamics of charged particles in them, taking into account electrostatic fields induced on isolated conductive elements, as well as external and intrinsic magnetic fields. The software is made up of ready-made modules linked together by control and service tools and implementing certain algorithms to describe a mathematical model that corresponds to the physical content of specific tasks. At the same time, the spectrum of physical productions described by the same mathematical model can be quite broad. The modular structure of the software makes it possible to compile a final accounting program from the ready-made modules using the editor of the monitor system for specific tasks.

The mathematical formulation of the problem of computing the distribution of fields in the vicinity of a high-voltage insulator looks as follows. It is necessary to find a solution of the system of field equations in the closed region: $\Delta\varphi = 0$, where the symbol Δ denotes the Laplace operator, which for the case of a cylindrical axisymmetric system takes the form

$$\Delta \equiv \frac{1}{x}\frac{\partial}{\partial x}(x\frac{\partial}{\partial x}) + \frac{\partial^2}{\partial y^2}. \tag{5.47}$$

The following boundary conditions can be set:

1) $\varphi|_r = g\,(r)$, that is, the potential distribution on the boundary is given in the form of a function. For a single electrode this is a constant value. The condition $\left.\dfrac{\partial\varphi}{\partial n}\right|_r = 0$ of the vanishing of the normal derivative with respect to the potential is given at the symmetry boundary and in the 'closure' regions, sufficiently far from the investigated part of the region;

2) $\varphi|_{r+} = \varphi|_{r-}$, $\varepsilon + \dfrac{\partial\varphi}{\partial n} = \varepsilon - \dfrac{\partial\varphi}{\partial n}$ – the conjugation conditions given by on the interface of the media with different dielectric constants ε_+ and ε_-.

All three types of boundary conditions are present in the design of a high-voltage insulator. The conjugation conditions are set twice, since there are two media interfaces (oil/dielectric and dielectric/vacuum). The initial design was chosen a cylindrical insulator, previously successfully used for the LIA on magnetic elements. The design of the insulator allows repair, prompt replacement, as well as an increase, if necessary, of its length. In addition, it makes it possible to arrange a spiral on the surface of the demagnetization of the cores of a high-voltage transformer and magnetic commutators. The high-voltage insulator of a cylindrical type is made of polymethyl methacrylate (plexiglass resistant to mineral oils, gasoline and alkalis, which generates gases during the formation of an electric arc, which help to extinguish this arc). The insulator is fastened between two metal flanges by means of eight dielectric studs 25 mm in diameter located at a diameter of 560 mm and intended to seal vacuum connections. For mechanical strength, the thickness of the insulator wall is 22–25 mm, the outer diameter is 650 mm, the inner diameter is 600 mm in the region of the high-voltage electrode and 606 mm in the region of the zero electrode. The inner surface of the insulator is made conical to remove air bubbles when the tank is filled with transformer oil (dielectric permeability $\varepsilon = 2.2$). The insulator is installed in a vacuum chamber with a diameter of 1200 mm. One flange, which is a ground (zero) electrode, is under the same potential as the chamber, and a high voltage is induced on the second (high-voltage electrode with an external diameter of 870 mm), since the terminals of the secondary winding of the high-voltage transformer are attached to this flange.

The design of an insulator with a central electrode in the form of a metal tube with an external diameter of 200 mm, placed in oil and passing along the axis of the system, was investigated. Computational results (Fig. 5.7 *a*) indicates that for a smooth surface of the insulator the distribution of the electric field is highly non-uniform on the vacuum side. In the direction from the zero electrode to the high-voltage electrode the field strength increases monotonically. At the same time, the change in the field distribution practically does not reflect the change in the permittivity of polymethyl methacrylate from $\varepsilon = 2.2$ to $\varepsilon = 2.6$, and also the location of the copper spiral for the demagnetization circuit with a cross section of 1.6×6 mm^2 installed on the rib in a rectangular groove 3 mm deep. Along the length of the insulator, equal to 450 mm, it was possible to arrange ~90 turns with a pitch of 5 mm.

Preliminary numerical experiments made it possible to draw the following conclusion: to uniformly distribute the electric field strength and its tangential component over the insulator surface in order to avoid breakdowns and increase its reliability, it is necessary to introduce metal gradient rings into the structure. The next stage of

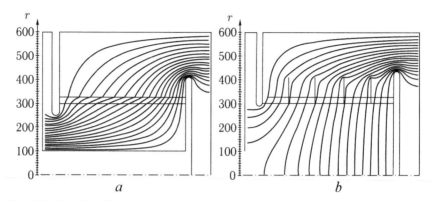

Fig. 5.7. The distribution of the electric field strength in a smooth cylindrical insulator (*a*) and in an insulator with gradient rings (*b*).

numerical calculations was connected with the determination of the number of gradient rings, their location and geometric dimensions. At the same time, in order to reduce the number of vacuum seals and increase the mechanical strength of the structure, the task was to achieve an even division of the potential based on the minimum possible number of rings. Numerical experiments with a sequential increase in the number of rings from one to three did not achieve the desired result, since the partial 'improvement' of the field distribution remained unsatisfactory. At the same time, the size of the high-voltage electrode was used as a criterion for limiting the maximum ring size (diameter) to avoid electrical breakdowns. The required result was obtained using four gradient rings with an external diameter of 818 mm. The resulting picture with a practically uniform distribution of the tangential component of the electric field strength along the surface of the insulator for four rings is given in Fig. 5.7 *b*. The gradient rings 2 mm thick are located relative to the zero electrode at distances of 90, 180, 280 and 370 mm, respectively.

Thus, numerical calculations allow optimizing the construction of a high-voltage insulator and ensuring a uniform distribution of the electric field strength and its tangential component along the external surface facing the vacuum. The design of the high-voltage insulator makes it possible to arrange on its outer surface a demagnetization coil of the cores of a high-voltage transformer and magnetic commutators. If necessary, it is possible to quickly repair a high-voltage insulator, replacing its individual sections (insulator elements located between gradient rings and electrodes).

5.4. High-voltage transformer

The high-voltage transformer GMP is manufactured according to the principle of the induction system of LIA in the form of a set of individual inductors – ferromagnetic cores covered by magnetizing coils.

Such an arrangement makes it possible to minimize the leakage inductance of the magnetizing turns of a high-voltage transformer. The number of turns can be increased several times to reduce the amount of ferromagnetic material of the cores (output pulse duration ~1 μs allows to increase the discharge inductance in contrast to the traditional design of LIA). The outputs of the primary turns of the inductors are assembled into groups and connected to all MPGs. To increase the output voltage, the secondary winding of the high-voltage transformer must also have several turns. The chapter deals with the case where the primary winding of each inductor of a high-voltage transformer is structurally made in the form of three three-turn windings, each consisting of two sections connected in parallel (6 sections in all), and the secondary winding consists of 6 sections of three turns made of high-voltage cable, and covers all the inductors from the outside.

For toroidal ferromagnetic cores of rectangular cross section, the leakage inductance of the magnetizing coil is minimal at the maximum of the flux linkage, if the ratio of the outer diameter of the core to the internal diameter is $e \approx 2.2$–2.4. For GMPs cores with dimensions $D_{HVT} = 1.1$ m, $d_{HVT} = 0.5$ m, $l_{HVT} = 0.018$ m, material – permalloy 50 NP have been selected. The cross section of the steel of one core is $S_{HVT} = 43$–46 cm². For a rectangular pulse of duration $\tau \approx 1$ μs the permissible value of the voltage on the core winding is determined from the following formulas:

$$\Delta B\omega_{HVT1}S_{HVT} = \int_0^\tau U_{C_{31}}(t)dt, \ U_{C_{31}} = \frac{\Delta B\omega_{HVT1}S_{HVT}}{\tau} = 32 \text{ kV.} \quad (5.48)$$

Here $\omega_{HVT1} = 3$ is the number of turns of the primary winding. By making an adjustment of the pulse duration associated with the presence of fronts, the maximum permissible value of the voltage acting on each core is limited to a value of 25 kV.

The number of inductors of the high-voltage transformer N is determined by the requirement for good matching of the wave impedance $\rho \approx \sqrt{L/C_{31}}$ of the discharge circuit with load R_{load}/N^2

($L \sim 0.2$ μH is the inductance of the discharge circuit). Dependences of the amplitude and duration of the front of the output pulse of the GMP on the number of inductors are calculated using the Mathcad mathematical software. From Fig. 5.8 it follows that the number of inductors of a high-voltage transformer should be 18–21. It is advisable to choose a smaller number of inductors, i.e. 18. As a result, the dimensions of the transformer are reduced, the duration of the front of the output pulse is shortened, the energy losses for magnetization of the inductor cores decrease, and the cost of the transformer is reduced.

The following expressions were used for calculating the value of the current on the load:

$$I = \frac{U_{C_{31}}}{2aL}(-e^{p_1 t} + e^{p_2 t});$$

$$p_{1,2} = -\frac{R_{\text{load}}}{2L} \pm \sqrt{\frac{R_{\text{load}}^2}{4L^2} - \frac{1}{LC_{31}}}; \qquad (5.49)$$

$$a = \sqrt{\frac{R_{\text{load}}^2}{4L^2} - \frac{1}{LC_{31}}},$$

where $L = L_{21} + L_{\text{HVT1}} + L_{\text{load}} + L_{C_{31}} \sim 0.2$ μH is the inductance of the discharge circuit; $C_{31} = 0.188$ μF is the capacitance of the last MPG compression link capacitor; $U_{C_{31}} = 52$ kV is the discharge voltage of the capacitor C_{31}; $R_{\text{load}} = 450$ Ohm/$N^2 = 1.389$ Ohm; $N = 18$ is the coefficient of transformation (number of inductors) of a high-voltage transformer. The characteristics of the output pulse of a high-voltage

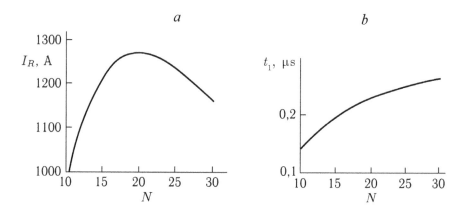

Fig. 5.8. Dependences of the amplitude of the output current of the GMP (*a*) and duration of the front edge of the output pulse (*b*) on the number of inductors *N*.

transformer are: current 1263 A; voltage of 568 kV (Fig. 5.9 shows a discharge current pulse of one of the three capacitors of the last compression links of MPG). In engineering calculations, the losses in the high-voltage transformer, which lead to a decrease in the output characteristics of the generator, were not taken into account.

The axial length of the transformer made of 18 inductors is approximately 0.84 m. The induction system is fixed between the flanges and is tightened by studs. Insulating disks are installed between the inductors and the magnetizing coils.

5.5. Calculation of currents in GMP elements and selection of the cross-section of windings

Pulse transformer. The use of a primary winding of a pulse transformer with a minimum number of turns leads to a reduction in the duration of the discharging pulse of the primary storage and its front, and to a decrease in the energy losses in the winding. Therefore, the number of turns in the primary winding of the PT is set to one. Taking into account the charging voltage of the primary storage device $U_{C_0} = 0.9$ kV and the need to obtain the required value of the voltage of the capacitors of the first link of the high-voltage unit C_1 and C_2, $U_{C_1} = U_{C_2} \approx 30$ kV the PT transformation ratio should be

$$k_{PT} = \frac{U_{C_1}}{U_{C_0}} = 33.33. \tag{5.50}$$

Assume $k_{PT} = 33$, i.e. the secondary winding PT must have 33 turns.

We define the currents in the windings of the pulse transformer. The pulse transformer performs two functions: it raises the voltage during the time interval $t_1 - t_0 = 50$ μs, and at saturation its secondary

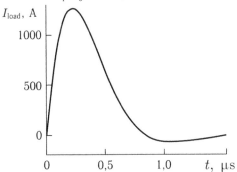

Fig. 5.9. The impulse of the discharge current of the capacitors C_{31}, C_{32}, C_{33} through the primary winding of the high-voltage transformer.

winding plays the role of inductance for the capacitor C_2 recharge (the duration of the recharge process is $t_2-t_1 = 13.1$ μs according to (5.16)). The primary storage device C_0 is discharged to the primary winding by a current whose amplitude value is

$$I_{\mu PT1} = \frac{U_{C_0}C_0}{(t_1-t_0)}\frac{\pi}{2} = 67.86 \text{ kA}. \tag{5.51}$$

The effective value of the current in the primary winding at a repetition rate of $F = 1000$ Hz

$$I_{ef\ PT1} = I_{\mu PT1}\sqrt{\frac{(t_1-t_0)F}{2}} = 10.73 \text{ kA}. \tag{5.52}$$

At the allowable current density $j = 5$ A/mm² the required primary winding conductor section is

$$S_{PT1} = \frac{I_{ef\ PT1}}{j} = 2146 \text{ mm}^2. \tag{5.53}$$

Structurally, the primary winding of the pulse transformer is made in the form of a hollow toroid of a rectangular cross section with an open lateral part (Fig. 5.10). On top there are disks are made of an insulating material, on which two 33-turn sections of the secondary winding are wound, connected in parallel.

The amplitude value of the current in the secondary winding of the PT at a discharge of C_0 is

$$Im'_{PT2} = \frac{I_{\mu PT1}}{k_{PT}} = 2.06 \text{ kA},$$

and the effective current through the secondary winding

$$I'_{ef\ PT2} = \frac{I_{ef\ PT1}}{k_{PT}} = 325 \text{ A}. \tag{5.54}$$

The amplitude value of the current in the secondary winding PT when recharging C_2 is

$$Im''_{PT2} = U_{C_2}\sqrt{\frac{C_2}{L_{PT2}}} = 8 \text{ kA},$$

where $U_{C_2} = U_{C_0}k_{PT} = 29.7$ kV is the maximum voltage on the secondary winding.

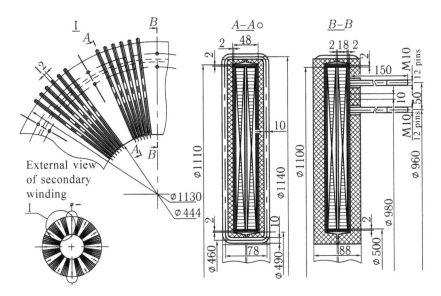

Fig. 5.10. Design of pulse transformer.

The effective value of the current in the secondary winding of the PT when recharging C_2

$$I''_{\text{ef PT2}} = \text{Im}''_{\text{PT2}} \sqrt{\frac{(t_2 - t_1)F}{2}} = 649 \text{ A.}$$

Thus, the total effective current through the secondary winding PT, consisting of two parallel sections, with a charge of C_1 and C_2 and recharge of C_2 will be equal to

$$I_{\text{ef PT2}} = I'_{\text{ef PT2}} + I''_{\text{ef PT2}} = 974 \text{ A.} \tag{5.55}$$

The required cross-section of the conductor of one section of the secondary winding must be at least

$$S_{\text{PT2}} = \frac{I_{\text{ef PT2}}/2}{j} = 97.4 \text{ mm}^2. \tag{5.56}$$

We will choose a wire with a copper rectangular core and fiberglass insulation of 3.5 mm × 1.4 mm² with the number of conductors in one section equal to 18.

Saturation chokes L_{11}, L_{12}, L_{13}. According to (5.23), the discharge time of the series-connected capacitors C_1 and C_2 through the windings of the saturation chokes L_{11}, L_{12}, L_{13} is $t_3 - t_2 = 1.18$ μs. The amplitude value of the discharge current of the capacitances C_1 and C_2 into three parallel windings of the saturation choke is given by

$$I_{\mu l_{11}-l_{13}} = U_{C_1+C_2}\sqrt{\frac{(C_1+C_2)}{2L_1'}} = 40.55 \text{ kA}, \qquad (5.57)$$

where $U_{C_1+C_2} = 54$ kV is the amplitude of the voltage on the capacitors C_1 and C_2 connected in series; $L_1' = L_{11} + L_{as}^{(1)}$ is the inductance of the discharge circuit, equal to the sum of the inductances of the windings of the chokes L_{11}, L_{12}, L_{13} (denoted as L_{11}), the inductances of their connection to the capacitors and the inductances of the capacitors C_1, C_2, C_{31}, C_{32}, C_{33}.

The amplitude value of the current flowing in the coil of one choke L_{11} (L_{12}, L_{13}) when the capacitors C_{31} (C_{32}, C_{33}) are charged, is equal to

$$I_{\mu l_{11}} = \frac{I_{\mu l_{11}+l_{13}}}{3} = 13.52 \text{ kA}. \qquad (5.58)$$

Effective total current

$$I_{ef\,L_{11}+L_{13}} = \frac{I_{\mu l_{11}+l_{13}}}{\sqrt{2}}\sqrt{(t_3-t_2)F} = 985 \text{ A}, \qquad (5.59)$$

and the effective current through one winding L_{11} (L_{12}, L_{13})

$$I_{ef\,L_{11}} = \frac{I_{ef\,L_{11}+L_{13}}}{3} = 328 \text{ A}. \qquad (5.60)$$

The windings of the saturation chokes L_{11}, L_{12} and L_{13} are wound on a common ferromagnetic core consisting of 11 cores of the type K1100×500×18. Structurally, the windings consist of 6 sections, in which the conductor is uniformly located along the toroidal magnetic system (Fig. 5.11).

The required cross section of the conductor in one section can be determined by the allowable current density $j = 5$ A/mm²:

$$S_{l_{11}} = \frac{I_{ef\,L_{11}+L_{13}}/6}{j} \approx 33 \text{ mm}^2. \qquad (5.61)$$

The conductor of each section consists of 12 copper wires of rectangular cross-section with dimensions of 3.55 × 1.4 mm² (dimensions of the wire in the insulation 3.95 × 1.8 mm²). With this design, the inductance of the windings of the chokes L_{11}, L_{12}, L_{13} at the saturated core will be equal to each other and equal to

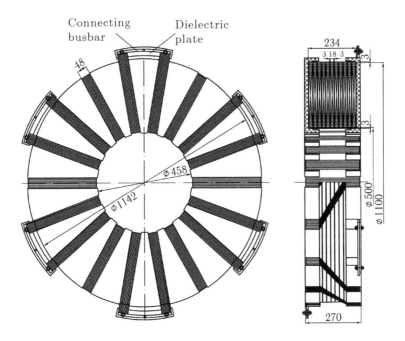

Fig. 5.11. The design of saturation chokes L_{11}, L_{12} and L_{13} with a common ferromagnetic core.

the equivalent inductance L_1, the value of which is calculated by formula (5.15).

Magnetic commutators L_{21}, L_{22}, L_{23}. Let's calculate the currents for the circuit $C_{31}-L_{21}$ – the primary winding of the HVT. Values of currents in circuits $C_{32}-L_{22}$ – primary winding of the HVT and $C_{33}-L_{23}$ – the primary winding of the HVT differs little from the current in the primary circuit because of the insignificant difference in the inductances of the switch windings when the cores are saturated. When the capacitor is discharged through a circuit consisting of a series-connected inductance and resistance, the current in the circuit is a damped sinusoid if the energy stored in reactive elements, exceeds the energy dissipated by active elements.

The nature of the current in the circuit depends on the value of d_d, called the damping constant:

$$d_d = \frac{R^2 C}{4L}. \qquad (5.62)$$

At $0 < d_d < 1$ the discharge is oscillatory, and when $d_d > 1$ it is aperiodic. In the present scheme, $d_d = \dfrac{R_{load}^2 C_{31}}{4L} = 0.453$ where R_{load} is the load resistance applied to the primary side of the high-voltage transformer; L is the discharge circuit inductance consisting of the leakage inductance of the magnetic commutator L_{21}, the inductance of the assembling, the leakage inductance of the primary winding HVT, and the inductance of capacitors C_{31}.

The amplitude value of the current with a damped oscillatory discharge is

$$I_{\mu L_{21}} = U_{C_{31}} k(d_3) \sqrt{C_{31}/L} = 22.6 \text{ kA}, \qquad (5.63)$$

where $U_{C_{31}}$ = 50 kV is the charging voltage of the capacitor C_{31}; k (d_d) = 0.466 is the attenuation coefficient, determined from the graph for the given function.

The effective currents in the conductors of the three discharge capacitor circuits C_{31}, C_{32} and C_{33} are approximately equal:

$$I_{ef\, L_{21}} = I_{ef\, 22} = I_{ef\, 23} = 423 \text{ A},$$

where $t_5 - t_3$ = 0.7 μs – is the duration of the discharge pulse.

The required cross section of the coils of switches at a desired current density j = 5 A/mm² should be

$$S_{L_{21}} = S_{L_{22}} = S_{L_{23}} = 85 \text{ mm}^2. \qquad (5.64)$$

Given the influence of the skin effect and mechanical stress, the single-turn winding of the switches L_{21}, L_{22} and L_{23} should be made as a solid copper disk covering the cores (Fig. 5.12). The thickness of the disk is 5 mm, and the average cross-sectional area of the conductor is $S_{21} = S_{22} = S_{23}$ = 5655 mm².

High-voltage transformer. The cross section of the primary winding of one inductor of the HVT depends on the magnitude of the total effective current flowing in the magnetizing windings of the core from three MPGs. For a quasi-rectangular pulse shape, the effective current is

$$I_{ef\, HVT1} = I_{\mu HVT1} \sqrt{(t_7 - t_3)F} \approx 32 \text{ A}, \qquad (5.65)$$

where $I_{\mu\, HVT1}$ is the amplitude value of the current in the primary winding of one inductor of the HVT; $t_7 - t_3$ = 1 μs is the duration of the current pulse. At the allowable current density j = 5 A/mm² the conductor cross-sectional area will be

$$S_{\text{HVT1}} = I_{\text{ef HVT1}}/j = 6.4 \text{ mm}^2. \qquad (5.66)$$

The primary winding of the HVT was in the form of a copper tape having a width of 40 mm and a thickness of 0.5 mm with a cross-sectional area of 20 mm². Structurally, the primary winding has 3 turns and consists of 6 parallel sections. In this case, the total

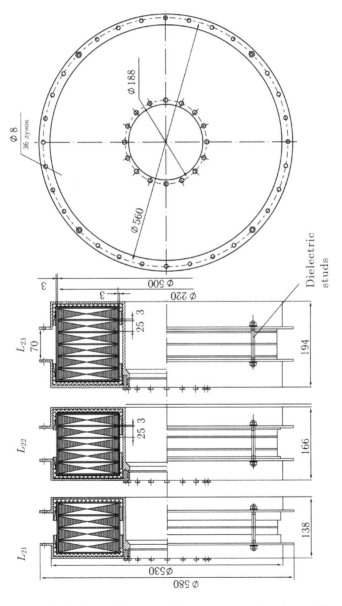

Fig. 5.12. The design of the magnetic commutators L_{21}, L_{22} and L_{23}.

cross-section of the winding conductor coiled on one core will be $S_{HVT1} = 120$ mm^2.

The current in the secondary winding of a high-voltage transformer is equal to the current in the primary winding. Consequently, in accordance with (5.65), the effective value of the current in the secondary winding will be $I_{ef\,HVT2} = I_{ef\,HVT1} = 32$ A. Based on these conditions, the secondary winding of the HVT was a copper wire in enamel insulation with a diameter of 0.6 mm. The conductor of the secondary winding contained 44 wires twisted together, with a total cross-sectional area of 12.6 mm^2. With regard to insulation, the outer diameter of the conductor is 5 mm. Structurally, the secondary winding contains 6 parallel sections, each of which has 3 turns each. The construction of a high-voltage transformer is shown in Fig. 5.13.

5.6. Thermal calculation of GMP elements

The energy losses in the GMP elements occur during the passage of pulsed current that causes their heating. Providing the normal thermal regime is the most important requirement for trouble-free operation. Therefore, it is necessary that the temperature of all heat-loaded elements does not exceed the permissible limits. The maximum permissible temperature [T] of magnetic elements and their windings is limited by the operating temperature of the insulation (95°C for transformer oil, 105°C for textolite, 155°C for fiberglass). For capacitor assemblies [T] is 70°C or 55°C. The ambient temperature T_0 for the calculations was chosen to be 20°C.

The tasks of thermal calculation are: 1) evaluation of losses in the elements of the installation, determination of the temperature of their heating per pulse and per packet of pulses; 2) calculation of the operating time at which the temperature of the GMP elements exceeds the permissible limits.

At a high repetition rate of pulses during a relatively short time interval, the GMP elements rapidly become hot. The heat released during this period is so insignificant that it can be ignored when calculating the heating. Such regimes in which all the heat released is concentrated within the element are called adiabatic and described by the equation

$$Q = c_t \gamma_t V_m \Delta T, \tag{5.67}$$

where Q is the amount of the released heat; c_t, γ_t and V_m are the

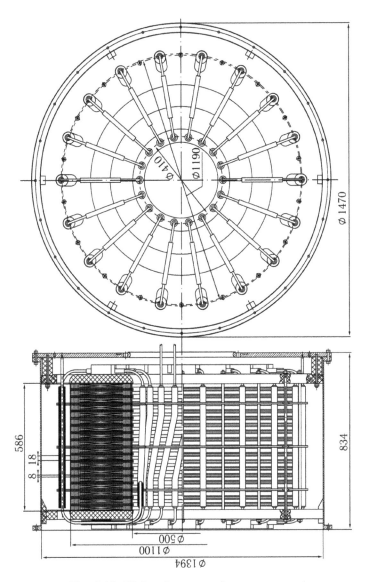

Fig. 5.13. High-voltage transformer construction.

specific heat, density and volume, in which heat release takes place; ΔT is the temperature difference. From the equation of adiabatic heating regime we determine temperature ΔT. The conditions for allowing the propagation of the packet of pulses is given by $\Delta T <$ $[\Delta T]$, where $[\Delta T]$ is the maximum permissible heating which is the difference between the maximum allowable operating temperature of the element and the ambient temperature T_0: $[\Delta T] = [T] - T_0$.

The operating time interval is the time during which the temperature of the heating elements of the generator does not exceed the permissible limits. Since the installation works in the packet mode, the operating interval is more convenient to express through the number of consecutive packets, which can tolerate installation. The number of packets is determined from the ratio of temperature differences and is calculated from the most heat-stressed element:

$$n_{calc} = [\Delta T]/\Delta T. \qquad (5.68)$$

The magnetic elements of the generator consist of a magnetic system (MS) and a winding. The mechanism of energy loss in them is different. Heating of the MS in the range of the effect of the pulse τ occurs when the cores are reversed, the heating of the winding is due to the flow of currents. Taking this remark into account, it is expedient to calculate the loss of energy into two parts: the calculation of the energy losses in the MS and the calculation of the ohmic energy losses in the winding.

5.6.1. Calculation of energy losses in magnetic systems of GMP elements

The magnetic systems of GMP elements consist of a set of permalloy cores with a tape thickness of 10, 20 μm. When magnetization is reversed by short pulses of tape cores, the energy loss consists from the losses due to the effect of magnetic viscosity and the action of eddy currents. Specific losses (J/m³) per magnetization per pulse are calculated by the following formula (corresponds to (1.25)):

$$q = 2B_S H_0 + \frac{2B_S}{\tau}\left(S_{\omega 0}\lambda^2 + S_{\omega e}\lambda^3\right), \qquad (5.69)$$

where H_0 is the start field; B_S is saturation induction; τ is the pulse duration; $S_{\omega 0}$ is the switching factor due to the influence of eddy currents (it does not depend on the duration τ and the form of the magnetization reversal current pulse); $S_{\omega e}$ is the switching factor due to the effect of magnetic viscosity (a constant value for the given material and the thickness of the tape, independent of the duration and shape of the magnetizing current pulse); $\lambda = \Delta B/(2B_S)$ is the load factor of the core by the flux linkage ($\lambda = 1$ at $\Delta B = 2B_S$).

The data for calculating the magnetization reversal losses in permalloy 50 NP are given in Table 1.2.

After simplifications, expression (5.69) is transformed to the form $q = 2B_S (H_0 + S_\omega/\tau)$. Specific energy losses in the core of the MS are added from the losses $q_{\tau 1}$ to the magnetization reversal by the pulse τ_1 (magnetization reversal of the cores to prepare them for the operating pulse) and $q_{\tau 2}$ to the magnetization by the operating impulse τ_2:

$$q_\tau = q_{\tau 1} + q_{\tau 2} = 2B_S \left[2H_0 + S_\omega \left(\frac{1}{\tau_1} + \frac{1}{\tau_2} \right) \right]. \qquad (5.70)$$

For the packet, the specific losses are $q = n_{pul} q_\tau$ [J/m³].

The complete loss of per pulse and packet make up in one core $Q1_\tau = q_\tau V_c$ and $Q_1 = qV_c$, and in the magnetic system as a whole $Q_\tau = Q1_\tau n_{cor}$ and $Q = Q1 n_{cor}$ (where $V_c = \pi (D^2-d^2)hk/4$ is the volume of the core; D, d is the outer and inner diameters of the core; h is the width of the core; k is the filling factor; n_{cor} is the number of cores).

Calculation of energy losses in the cores of the magnetic elements of the GMP shows that the highest of them occur in the magnetic system of a high-voltage transformer and constitute approximately 55% of the energy loss in all elements. In turn, the total losses are approximately 40% of the energy stored in the primary storage.

5.6.2. Calculation of ohmic energy losses in the windings of GMP elements

The windings of the elements of the GMP are heated due to ohmic losses in the flow of the two current pulses: current I_1 of duration $\tau_1 = 50$ microsecond at the discharge capacitor C_0 and the charge of the capacitor C_2 and current I_2 of duration τ_2 when discharging the respective capacitors (other than primary PT winding in which only current I_1 of the discharge of the capacitance C_0 flows). The total amount of heat released in the winding is equal to the sum of Q_1 and Q_2, separated from the action of each of these currents. For one pulse, it will be $Q_\tau = Q_{\tau 1} + Q_{\tau 2}$, where $Q_{\tau 1} = P_1 \tau_1$; $Q_{\tau 2} = P_2 \tau_2$; P_1 and P_2 are the power losses during the flow of currents I_1 and I_2, respectively. The amount of heat released in the winding during a burst of pulses is $Q = n_{pul} Q_\tau$.

The energy losses in the winding are determined by the ohmic resistance of the winding wires and the effective value of the current flowing through the winding. For short pulses, it is necessary to take into account the surface effect in the wires and the proximity effect. The essence of the skin effect is that no current flows through the entire cross-section of the conductor, and in the thin surface layer of the cross-section equal to $\Delta_{ef} = \sqrt{\dfrac{2\tau \rho_{Cu}}{\pi \mu_0}}$, where $\rho_{Cu} = 1.78 \cdot 10^{-8}$ ohm $\cdot$ m is the specific electrical resistance of copper. The quantity Δ_{ef} is the effective depth of penetration of the pulse current, and the cross section S_{ef}, through which the current flows is the effective cross section. The proximity effect is manifested in the displacement of the current under the action of eddy currents and a magnetic field to the periphery of the winding adjacent to the frame. As a result, the cross section through which the current flows becomes sickle-shaped, which leads to an additional increase in resistance. The proximity effect is considered by the coefficient k_b, selected in the range of 1.5–2.5 (smaller values are accepted for the windings of the well insulated wires, and higher values – for windings with the the small thickness of the 'turn to turn' insulation). For the primary windings of pulse and high-voltage transformers we take $k_b = 2.5$ as for the windings made 'turn to turn'. For the other windings of the magnetic elements, we choose $k_b = 1.5$ as for the windings with reinforced insulation.

The power loss per one pulse in a copper wire with a specific resistance ρ_{Cu} and the length l during the packet of a pulsed current I of duration τ with a pulse repetition rate F is calculated from $P = I_{ef}^2 \rho_{Cu} \dfrac{l}{S_{ef}} k_b$, where I_{ef} is the effective value of the impulse current equal to $I_{ef} = I\sqrt{\tau F / 2}$. For multi-core windings, the power loss is the sum of the losses in each core. After all transformations, the expressions for calculating the amount of heat released per pulse will have the form

$$Q_{\tau 1} = I_1^2 \frac{l\tau_1^2 F}{2mn_t S_{ef 1}} k_b;$$ (5.71)

$$Q_{\tau 2} = I_2^2 \frac{l\tau_2^2 F}{2mn_t S_{ef 2}} k_b,$$ (5.72)

where m is the number of sections; n_t is the number of cores in the winding; S_{ef1} and S_{ef2} are the effective cross sections for the pulsed currents I_1 and I_2. The effective cross sections are calculated by the following formulas:

– for the uniform winding $S_{ef} = 2\pi \bar{D} \Delta_{ef}$, where $\bar{D}$ is the average diameter of the turn;

– for a rectangular bar $S_{ef} = 2(a + b)\Delta_{ef}$;
– for round wire $S_{ef} = \pi (D\Delta_{ef} - \Delta_{ef}^2)$, where D is the diameter wires.

The length of the winding is $l = \bar{p}\omega$, where ω is the number of turns; $\bar{p}$ is the average perimeter of the turn, calculated as $\bar{p} = 2[D - d/2 + h]$ for a pulse transformer, chokes L_{11}, L_{12}, L_{13}, L_{21}, L_{22}, L_{23} and the primary winding of the HVT and $\bar{p} = 2[D - d/2 + h] + r(\pi - 4)$ for the secondary winding of the HVT (r is the bending radius of the secondary winding of the HVT).

Analysis of the results of calculating the energy losses in the windings of the magnetic elements of the GMP shows that the highest losses for a packet of 5000 pulses, 848 J (less than 1% of total energy losses) occur in the windings of a pulsed transformer, which is associated with a long duration of the current pulse. In the windings of the magnetic elements, installed closer to the output of the GMP, the losses are greatly reduced.

Calculation of temperature heating when a burst of pulses passes. In view of equation (5.67) heating of the MS per pulse and a pulse packet equals, respectively, $\Delta T_\tau = \dfrac{q_\tau}{c_{50}\gamma_{50}}$ and $\Delta T = \dfrac{q}{c_{50}\gamma_{50}}$, where $c_{50} = 500$ J/(kg · deg) is the specific heat of permalloy 50 NP; $\gamma_{50} = 8.2 \cdot 10^3$ kg/m^3 is the density of permalloy 50 NP.

Heating the coils $\Delta T_\tau = \dfrac{Q_\tau}{c_{Cu}\gamma_{Cu} V_{vol}}$ and $\Delta T = \dfrac{Q}{c_{Cu}\gamma_{Cu} V_{vol}}$, where $\gamma_{Cu} = 8.94 \cdot 10^3$ kg/m^3 – density of copper; $c_{Cu} = 385$ J/(kg · deg) is the specific heat of copper.

The volume V_{vol} occupied by the solid copper windings, is calculated by formula $V_{vol} = \pi \bar{D}al$.

For a multicore winding made of a rectangular bar, the volume of copper is calculated as $V_{vol} = mn_t abl$. If the winding is made of a circular wire, then $V_{vol} = \dfrac{\pi D^2}{4} mn_t l$.

Calculation of the operating interval should be carried out on the most heat-loaded element, which will also determine the thermal regime of the GMP in general. Calculations show that the losses in the windings are extremely small and, with the correct choice of the cross section, have practically no effect on the thermal state of the installation.

5.6.3. Calculation of energy losses in capacitors

GMP designs use modern capacitors based on a combined weakly polar dielectric (condenser paper and a non-polar impregnated lavsan film): K75-74 (with foil plates) and K75-81 (with metallized plates). The specificity of the operation of capacitors is the presence of two modes of operation (charging and discharging), characterized by the time of effect of the voltage on the insulation of the capacitors, the time and amplitude values of the currents flowing through the capacitor. The energy loss in the capacitor when a half-sinusoidal current pulse passes through it is approximately equal to

$$Q_c \approx \frac{\pi}{8} C U^2 \tan\delta(\omega), \tag{5.73}$$

where C is the capacitance of the capacitor; U is the voltage at the capacitor; $\tan\delta(\omega)$ is the tangent of the loss angle, which depends on the angular frequency ω of the flowing current.

For capacitors based on a weakly polar dielectric, the temperature and frequency dependence of $\tan\delta$ is weakly expressed. Therefore, the value of heat release over the whole range of operating temperatures and frequencies can be considered constant, and $\tan\delta$ – independent of temperature and frequency. Hence, the total losses per charge–discharge cycle will be $Q1_\tau = 2Q_c$ in one capacitor.

In operation of the capacitor C_2, in addition to a charge–discharge cycle, there is a stage of overcharging from $+U_{C_2}$ to $-U_{C_2}$. At this stage the losses in the capacitor will be equal to

$$Q_{rec} = \frac{\pi}{8} C_2 \left(2U_{C_2}\right)^2 \tan\delta, \tag{5.74}$$

general losses

$$Q1_\tau^{\langle C_2 \rangle} = Q1_\tau + Q_{rec}. \tag{5.75}$$

The total losses in the assembly of capacitors are determined by the formula

$$Q_\tau = Q1_\tau n_{cap}, \tag{5.76}$$

where n_{cap} is the number of capacitors in the assembly. The energy released for a packet in one capacitor is $Q1 = Q1_\tau n_{pul}$, and in the capacitor assembly $Q = Q_\tau n_{pul}$.

Calculation of the heating temperature of capacitors. Heating of a capacitor per cycle amounts to $\Delta T_r = \dfrac{Q1_r}{c_c \gamma_c V_{cap}}$, wherein $Q1_r$ is the amount of heat released in the capacitor in one cycle; V_{cap} is the volume of the capacitor; c_c is the specific heat of the capacitor; γ_c is the density of the capacitor. Since the thermal losses in the pulse capacitor are determined by the losses in the dielectric of the capacitor, which amount to ~97% of the total losses, the specific heat of the capacitor can be replaced by the specific heat of the dielectric c_d. Specific heat capacities of lavsan and paper are the same and equal to 1.5 kJ/(kg · deg). The capacitor density is $\gamma_c = m_{cap}/V_{cap}$ (where m_{cap} is the mass of the capacitor). After the transformation $\Delta T_r = \dfrac{Q1_r}{c_d m_{cap}}$. Heating the capacitor with a packet of pulses will be

$$\Delta T = \Delta T_r n_{pul}.$$

The number of pulse packets that a capacitor allows in the operating temperature range is determined by a formula similar to (5.68) for the most heat-loaded capacitor. The losses in the capacitor directly depend on its capacitance. If storage units are assembled from capacitors of a smaller capacity, the energy losses in each individual capacitor and, correspondingly, their heating decrease. In this case, the total losses in the entire assembly remain unchanged. Calculations show that the losses in the MPG capacitors account for about a quarter of the total energy losses of the generator. The largest losses are observed in the capacitors of the primary energy storage, since these capacitors are low-voltage and have the highest density of stored energy. Estimates of the thermal state of capacitor assemblies show that they allow continuous generation of 28 packets of pulses, i.e., 140 000 pulses.

In general, the results of the thermal calculation of all the elements of the generator allow us to state that:

1) the thermal resource of all the elements of the installation allows 28 packets of 5000 pulses with a duration of 1 μs with a frequency of 1 kHz to be continuously transmitted;

2) the main energy losses occur in the cores of the magnetic elements of the GMP;

3) the most heat-loaded elements are capacitors C_0 of the primary energy storage and capacitors C_2 of the first compression link, which limit the duration of continuous operation of the GMP.

It should be noted that the thermal calculations were performed in the adiabatic approximation, that is, under conditions where the heat sink is neglected. The allowance for the heat sink leads to a decrease in the design heating temperature of individual elements, which should lead to an increase in the permissible operating time intervals of the GMP.

The use in the GMP of the elements that have a large mass and high heat capacity, as well as the choice of windings of the corresponding structures and wire cross-sections eliminates the need for cooling the elements of the high-voltage unit of the GMP.

5.7. Computer modelling of the GMP

Since the generator contains magnetic elements in its composition, it can be argued that it is an expensive unit and also labor-intensive in manufacture. Thus, the stage of computer modelling is very important. Elements of the installation are represented by complex equivalent circuits. Therefore, the calculation of GMP is possible only with the use of computer tools, which allows one to take into account almost all physical effects when compressing energy.

The aim of this work is to create a computer model for calculating the processes taking place in the high-voltage unit of the GMP and for promptly selecting the parameters of the elements for tuning to the maximum output power, efficiency, and for generating a pulse with the required amplitude and time characteristics. To solve this problem, the actual electrical circuit (see Fig. 5.3) is represented by an equivalent circuit for which the parameters of the elements are determined. Then, a circuit is prepared for calculations using the Electronic Workbench software. The processes in the computer model are considered from the time t_0 (see Fig. 5.4), when capacitors C_0 are charged, the switch K_0 is closed (thyristors VS3 are turned on) and C_0 begins to be discharged through the primary winding of the pulse transformer PT.

5.7.1 . Equivalent schemes of elements in GMP

The discharge circuit of the primary storage unit C_0 includes a pulse transformer, primary storage capacitors C_0, inductance elements and capacitors of the high-voltage unit C_1 and C_2. It has the equivalent circuit shown in Fig. 5.14. The circuit includes a pulse transformer, which is a complex electromagnetic system with magnetic and electric fields distributed between the structural elements. As a result of simplifications, the equivalent circuit of a pulse transformer can be represented with the help of lumped elements:

– C_{PT1} and C_{PT12} – energetically equivalent distributed so-called dynamic capacitance of the primary winding of the transformer relative to the magnetic system and the capacitance between the windings;

– L_{SPT} – leakage inductance of the winding;

– R_{PT1} and R_{PT2} – ohmic equivalents of losses in primary and secondary winding circuits;

– $L_{\mu PT}$ – magnetization inductance;

– $R_{\mu PT}$ – equivalent resistance of the losses in the magnetic system;

– $L_{as}^{(0)}$ and $L_{as}^{(1)}$ – inductance of mounting of elements in the primary and secondary circuits of the pulse transformer.

From the presented scheme of substitution it follows that the discharge circuit of the primary storage contains eight independent elements capable of accumulating energy. The transient process in such a scheme will be described by solving a differential equation of the eighth order. Finding its solution for known parameters of elements and initial conditions is possible only numerically using computer facilities.

1. The leakage inductance of the windings of the pulse transformer L_{SPT}, represented in an equivalent circuit in the form of a lumped element, is determined from the value of the magnetic energy stored in the winding leakage fluxes. In the developed PT design, with a good winding density, the magnetic leakage flux is concentrated in the space between the primary and secondary windings. The intensity of the magnetic field in this space can be considered equal to the field strength between two equidistant surfaces, streamlined by a current of constant density. Under such assumptions, the leakage inductance is

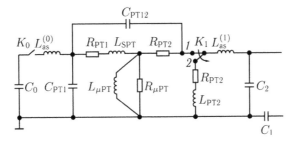

Fig. 5.14. Equivalent circuit of a pulse transformer.

$$L_{SPT} = \frac{\mu_0 \omega_{PT1}^2 (p_1 + p_2)}{2l_0}\left(\Delta_{12} + \frac{d_1 + d_2}{3}\right), \tag{5.77}$$

where p_1 and p_2 are the lengths of the turns of the primary and secondary windings; $l_0 = \pi(D_{PT} + d_{PT})/2$ is the length of the core windings; Δ_{12} is the thickness of the insulation layer between the windings; d_1, d_2 is the thickness of conductors of primary and secondary windings; D_{PT}, d_{PT} are the outer and inner diameters of the core. For the GMP under consideration the leakage inductance of the primary winding will be $L_{SPT} \approx 6 \cdot 10^{-9}$ H.

2. The ohmic equivalent of the losses R_{PT1} includes the resistance of the primary winding of a pulse transformer–current leads and contact connections, as well as losses in the dielectric of capacitors C_0 when they are discharged. The value of $Q_{\tau_1} = 12.4$ J is determined by the thermal calculations described above. We estimate the loss resistance R_{PT1} from the following considerations. The energy losses are equal to

$$Q_{\tau_1} = \int_{t_0}^{t_1} R_{PT1} I_{PT1}^2(t)\, dt, \tag{5.78}$$

where $t_1 - t_0 = 50$ μs is the discharge time of the capacitor C_0 on the capacitors C_1 and C_2; $I_{PT1}(t) = \dfrac{U_{C_0}}{\omega_1 L_0}\sin \omega_1 t$ is the discharge current; $U_{C_0} = 0.9$ kV is the amplitude of voltage on the capacitor C_0; $\omega_1 = 1/\sqrt{L_0 C_0}$.

Integrating (5.78) and substituting the value of Q_{τ_1}, we have

$$R_{PT1} = \frac{2Q_{\tau_1}\pi^2 L_0^2}{U_{C_0}^2 (t_1 - t_0)^3} = 0.11 \cdot 10^{-3} \text{ ohm.} \tag{5.79}$$

3. Magnetization inductance $L_{\mu\,PT}$ is related to the magnitude of the magnetizing current of the pulse transformer during the time $t_1 - t_0$ of the discharge C_0. Let us determine the magnitude of the magnetization inductance through the flux linkage of a pulsed transformer:

$$L_{\mu PT} = \frac{\psi_{PT}}{\Delta I_{\mu PT}}, \tag{5.80}$$

where $\Delta I_{\mu PT}$ is the magnetizing current; $\psi_{PT} = l_{PT} B_s K \omega_{PT1}(D_{PT} - d_{PT})n_{PT}$; l_{PT} is the width of the steel of one core; n_{PT} is the number of cores.

Using equation (5.69) for pulsed magnetization reversal of the permalloy and accepting $\lambda = 1$ (full reversal of the core from $-B_S$ to $+B_S$) gives

$$\Delta I_{\mu PT} = \frac{\pi(D_{PT} + d_{PT})\left[H_0(t_1 - t_0) + 2S_{\omega e} + S_{\omega 0}\right]n_{PT}}{2(t_1 - t_0)\omega_{PT1}}. \tag{5.81}$$

After substituting in (5.81) the data from Table 1.2 we get

$$L_{\mu PT} = \frac{2\omega_{PT1}^2 B_S l_{PT} K(t_1 - t_0)(D_{PT} - d_{PT})n_{PT}}{\pi(D_{PT} + d_{PT})[H_0(t_1 - t_0) + 2S_{\omega e} + S_{\omega 0}]} = 127 \; \mu H. \tag{5.82}$$

4. Dynamic capacitances C_{PT1}, C_{PT12}. The replacement of the distributed capacitances of the pulsed transformer windings with lumped ones is based on the energy principle. If the geometry of the windings and voltage distribution on them are known, then it is possible to calculate the energy of the electric field concentrated between the corresponding elements of the design of the pulse transformer. Equating the energy thus calculated to the energy $W = CU^2/2$ expressed by capacitance C and voltage U, we can define a lumped capacitance, energetically equivalent to the distributed capacitance.

A. The dynamic capacitance of the primary winding of a pulse transformer depends on its type. For the construction shown in Fig. 5.10, we can assume that the primary winding covers almost the entire surface of the core, forming an uniform magnetizing coil. In this case, the capacitance consists of one component between the magnetizing coil and the inductor core. The voltage $U(D)$ in the gap l_{10} (coil–inductor core) varies from 0 at the inner diameter d_{PT} to the value $U_{C_0}/2$ at the outer diameter D_{PT} according to the law

$$U(D) = \frac{U_{C_0}}{2}\frac{(D - d_{PT})}{(D_{PT} - d_{PT})}. \tag{5.83}$$

By dividing into elementary capacitances, finding the energy concentrated in them, and then integrating in the range from D_{PT} to d_{PT} with the two sides of the inductor taken into account, we obtain

$$C_{10}^{(1)} = \frac{\pi \varepsilon_{10} \varepsilon_0}{8 l_{10} (D_{PT} - d_{PT})^2} \times$$

$$\left[\frac{D_{PT}^4 - d_{PT}^4}{2} - \frac{4 d_{PT} (D_{PT}^3 - d_{PT}^3)}{3} + d_{PT}^2 (D_{PT}^2 - d_{PT}^2) \right] = 2.6 \text{ nF}, \quad (5.84)$$

where ε_{10} = 5.5 is the dielectric constant of the insulation material (fiberglass) between the primary winding and the core; l_{10} is the thickness of the insulation.

B. The dynamic capacitance C_{PT12} between the primary and secondary windings of the pulse transformer is [11]

$$C_{PT12} = \frac{\varepsilon_{12} \varepsilon_0 p h_{PT}}{3 \Delta_{12}} \left(\frac{k_{PT}}{m} + 1 \right)^2 = 3.6 \text{ μF}, \quad (5.85)$$

where ε_{12} = 5.5 is the dielectric constant of the insulation material between the windings (fiberglass); Δ_{12} is the distance between windings; k_{PT} = 33 is the transformation ratio; m = 1 is the number of secondary winding layers; h_{PT} is the height of the windings of the pulse transformer; p is the average perimeter between the windings.

5. The ohmic equivalent of losses consists of two parts: 1) resistance R_{PT2} of the secondary winding of a pulse transformer connecting current leads, contact connections, losses in the dielectric of capacitors C_1 and C_2 when they are charged; 2) the resistance R'_{PT2} of the winding of a pulse transformer connecting current leads, contact connections, losses in the dielectric of capacitors during the recharging of C_2 through the secondary winding of a pulse transformer. The data for calculating these losses, Q_{τ_2} = 4.8 J and Q'_{τ_2} = 9.6 J were determined as a result of thermal calculations. We estimate the loss resistance in the same way as for the primary winding of the pulse transformer:

$$R_{PT2} = \frac{2 Q_{\tau_2} \pi^2 \left(L_{PT2} + L_{as}^{(1)} \right)^2}{U_{C_2}^2 (t_1 - t_0)^3} = 0.22 \text{ ohm}, \quad (5.86)$$

$$R'_{PT2} = \frac{2 Q'_{\tau_2} \pi^2 \left(L_{PT2} + L_{as}^{(1)} \right)^2}{U_{C_2}^2 (t_2 - t_1)^3} = 24 \text{ ohm}, \quad (5.87)$$

where U_{C_2} = 29 kV is the amplitude of the voltage on the capacitor C_2; $t_2 - t_1$ = 13.1 μs is the duration of the capacitor recharge, determined by the relation (5.16); L_{PT2} = 15.4 μH is the inductance of the secondary winding of a pulse transformer with the saturated core (5.9); $L_{as}^{(1)} = 0.068$ μH (see piece 9 of this section).

Triggering of the switch K_1 (see Fig. 5.14) from position *1* to position *2* simulates the saturation of the core of the pulse transformer and starts recharging the capacitor C_2 through its secondary winding with the inductance L_{PT2}.

6. The ohmic equivalents $R_{\mu PT}$ of the losses in the magnetic system of a pulsed transformer associated with eddy currents and magnetic viscosity are determined using the results of thermal calculations: $Q_{rPT} = 4.36$ J. The energy loss is

$$Q_{rPT} = \frac{U_{C_0}^2}{R_{\rho PT}}(t_1 - t_0). \tag{5.88}$$

From here

$$R_{\mu PT} = \frac{U_{C_0}^2}{Q_{rPT}}(t_1 - t_0) = 9.3 \text{ ohm.} \tag{5.89}$$

7. The inductance $L_{as}^{(0)}$ of mounting the elements in the primary circuit of the pulse transformer includes: 1) the inductance of the connecting wires between the elements of the discharge circuit 'capacitors C_0 – thyristors – saturation choke L_4 – the primary winding of the pulse transformer'; 2) intrinsic inductance of the capacitors C_0; 3) inductance of thyristors VS3; 4) the inductance of the turns of the saturation choke L_4. The total inductance of the discharge circuit of the primary storage device $L_0 = L_{as}^{(0)} + L_{SPT} = 211$ nH, which ensures its discharge in 50 μs.

8. The inductance $L_{as}^{(1)}$ of mounting of the elements in the secondary circuit of the pulse transformer includes the inductance of the connecting wires between the capacitors C_2 and the secondary winding of the pulse transformer and the inherent inductance of the assembly of capacitors C_2. The installation inductance of the discharge circuit, assembled from 24 parallel lines, is $L_{as}^{(1)} = L_{wir}^{(2)} + L_{K75-74}/24 = 74$ nH.

The magnetic pulse generator consists of two compression links. The first link contains the saturation chokes L_{11}, L_{12}, L_{13}, the capacitors C_1, C_2, C_{31}, C_{32}, C_{33} and mounted inductance $L_{as}^{(2)}$, and the other link (three parallel lines) – saturation chokes (magnetic commutators L_{21}, L_{22}, L_{23}), capacitors C_{31}, C_{32}, C_{33} and inductance of mounting L. It has the equivalent circuit shown in Fig. 5.15.

The first link of the compression of the magnetic pulse generator contains:

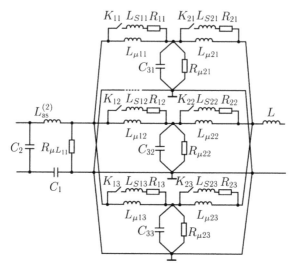

Fig. 5.15. Equivalent MPG scheme.

a) the leakage inductance $L_{S11} = L_{S12} = L_{S13} = 0.4$ μH of the windings of the chokes L_{11}, L_{12}, L_{13} when the core is saturated, the value of which is determined by formula (5.15);

b) ohmic equivalents R_{11}, R_{12}, R_{13}, which take into account the total losses in the coils, in the supply circuits and in the contact connections, losses in the discharged capacitors C_1 and C_2, and the loss of the recharged capacitors C_{31}, C_{32}, C_{33}. The energy loss, taking into account the thermal calculations, is $Q_{r3} = 8.66$ J. By analogy with (5.79), we can write

$$R_{11} = \frac{2Q_{r3}\,\pi^2(L_{S1}+L_{as}^{(2)})^2}{(U_{\tilde{N}_1}+U_{C_2})^2(t_3-t_2)^3} = 8.6 \text{ mohm}, \qquad (5.90)$$

where $U_{C_1} = U_{C2}$ is the amplitude of the charging voltage on the capacitors C_1 and C_2 ; $t_3-t_2 = 1.18$ μs is the discharge time of the series-connected capacitors C_1 and C_2 to parallel-connected capacitors C_{31}, C_{32}, C_{33}, determined by the relation (5.23).

Since the saturation chokes L_{11}, L_{12} and L_{13} have the same parameters, and the capacitances of capacitors C_{31}, C_{32}, C_{33} are equal to each other, they are expressed as losses in three MPGs $R_{11} = R_{12} = R_{13}$;

c) equivalent magnetization inductance $L_{\mu 11}$ of the saturation choke L_{11} is related to the magnetizing current during the recharging of C_2. The formula for calculating the quantity $L_{\mu 11}$ is similar to (5.82):

$$L_{\mu 11} = \frac{2\omega_{11}^2 B_S l_{L1} K (t_3 - t_2)(D_{L1} - d_{L1}) n_{11}}{\pi (D_{L_1} + d_{L_1})[H_0 (t_3 - t_2) + 2S_{\omega e} + S_{\omega 0}]} = 4 \text{ mH},$$

where ω_{11} = 3 is the number of turns of the winding L_{11}, L_{12}, L_{13}; l_{11} is the width of the steel of one core; D_{L_1}, d_{L_1} are the outer and inner diameters of the core; n_{11} = 11 is the number of cores of the saturation choke; $t_2 - t_1$ = 13.1 μs is determined according to (5.16). Since the parameters of the windings are chosen to be the same, and the windings themselves are located on the common core, $L_{\mu 11} = L_{\mu 12} = L_{\mu 13}$;

d) inductance of mounting $L_{as}^{(2)}$ includes the inductance of connecting wires between the elements C_1-C_2-L_{11}-L_{13}-C_{31}-C_{33} and the inherent inductance of assemblies of capacitors K75-74. It is equal to $L_{as}^{(2)}$ = 100 nH;

e) The ohmic equivalents of the losses $R_{\mu L_{11}}$ in the magnetic system of the saturation chokes L_{11} are determined using the results of thermal calculating by the formula similar to (5.89):

$$R_{\mu L_{11}} = \frac{4U_{C_2}^2}{Q_{\tau L_{11}}}(t_2 - t_1) = 811 \text{ ohm}, \tag{5.91}$$

where $Q_{\tau L_{11}}$ = 45 J is the energy loss in the magnetic system of the saturation choke L_{11}.

The switching of all the switches, K_{11}, K_{12} and K_{13}, takes place simultaneously at the moment of saturation of the core of the chokes L_{11}, L_{12}, L_{13} and simulates the beginning of the process of energy transfer from the series-connected capacitors C_1 and C_2 to the parallel capacitors C_{31}, C_{32}, C_{33}.

The second link of compression of the magnetic pulse generator contains :

a) the leakage inductance L_{S21} = 0.027 μH, L_{S22} = 0.032 μH, L_{S23} = 0.038 μH, the winding of magnetic commutators at the saturated state of the cores, which are calculated from formulas (5.24), (5.26) and (5.28);

b) ohmic equivalents R_{21}, R_{22}, R_{23}, which take into account the total losses in the windings of the chokes and primary winding of the high-voltage transformer, in the supply circuits and in the contacts, the losses in discharged capacitors C_{31}, C_{32}, C_{33}. According to thermal calculations, the losses for magnetic commutators L_{21}, L_{22},

L_{23} are $Q_{\tau 4}^{(L21)} \approx Q_{\tau 4}^{(L22)} \approx Q_{\tau 4}^{(L23)} \approx 1.3$ J. On the other hand, the losses are expressed by the equation

$$Q_{\tau 4}^{(L21)} = \int_{t_3}^{t_5} R_{21} I_{21}^2(t)dt, \tag{5.92}$$

where t_5-t_3 is the discharge time of the capacitors C_{31} to the primary winding of the high-voltage transformer; I_{21} is the discharge current. The discharge of capacitor C_{31} has an oscillatory character, since the following relation is satisfied

$$r = \frac{R_{\text{load}}}{C^2} < 2\sqrt{\frac{L}{C_{31}}}, \tag{5.93}$$

where $L = L_{21} + L_{\text{HVT1}} + L_{\text{load}} \sim 0.2$ µH is the inductance of the discharge circuit; $C_{31} = 0.188$ µF is the capacitance of the last MPG compression link capacitors. Therefore, for the discharge current of the capacitor C_{31},

$$I_{21}(t) = -\frac{U_{C_{31}}}{\omega_{31}L}e^{-R_{\text{load}}t/2L}\sin\omega_{31}t, \tag{5.94}$$

where $\omega_{31} = \sqrt{\dfrac{1}{LC_{31}} - \dfrac{R_{\text{load}}^2}{4L^2}}$ is the natural frequency of the circuit.

Numerically integrating $\int_{t_3}^{t_5} I^2(t)dt$ (Fig. 5.16) and substituting the value found into (5.93), we obtain

$$R_{21} = R_{22} = R_{23} = 7.7 \text{ mOhm}. \tag{5.95}$$

c) the equivalent magnetization inductances $L_{\mu 21}$, $L_{\mu 22}$, $L_{\mu 23}$ of the saturation chokes L_{21}, L_{22}, L_{23} reflect the process of the current flow during magnetization reversal of the magnetic commutator cores. Similarly to (5.82), we can write

$$L_{\mu 21} = \frac{2\omega_{21}^2 B_S l_{21} K(t_5 - t_3)(D_{L_{21}} - d_{L_{21}})n_{21}}{\pi(D_{L_{21}} + d_{L_{21}})[H_0(t_5 - t_3) + 2S_{\omega e} + S_{\omega 0}]} = 40.3 \text{ µH}, \tag{5.96}$$

where $\omega_{21} = 1$ is the number of magnetizing turns of the windings L_{21}, L_{22}, L_{23}; l_{21} is the width of steel of one core; $D_{L_{21}}$, $d_{L_{21}}$ are the outer and inner diameters of the core; $n_{21} = 4$, $n_{22} = 5$, $n_{23} = 6$ are the number of cores of saturation chokes L_{21}, L_{22}, L_{23} respectively.

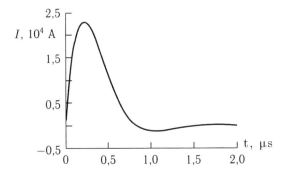

Fig. 5.16. Results of the numerical calculation of the current through the magnetic commutator L_{21}.

Due to the use of a different number of cores the magnetization inductance of the chokes differs in magnitude. They are calculated using formulas similar to (5.97): $L_{\mu22}$ = 50.4 nH, $L_{\mu23}$ = 60.5 nH;

d) the turning on the switches K_{21}, K_{22}, K_{23} (Fig. 5.15) with the necessary delay occurs at the moments of saturation of the cores of the chokes L_{21}, L_{22}, L_{23}. It simulates the beginning of the process of consecutive discharge of capacitors C_{31}, C_{32} and C_{33} through the primary winding of the high-voltage transformer;

e) The ohmic loss equivalents $R_{\mu l_{21}}$, $R_{\mu l_{22}}$, $R_{\mu l_{23}}$ in the magnetic systems of the saturation chokes L_{21}, $L_{22}0$, L_{23} are determined using the results of thermal calculations by a formula similar to (5.89):

$$R_{\mu21} = \frac{(U_{C_1} + U_{C_2})^2}{Q_{\tau l_{21}}} (t_3 - t_2) = 181 \text{ ohm,}$$

$$R_{\mu22} = \frac{(U_{C_1} + U_{C_2})^2}{Q_{\tau l_{22}}} (t_3 - t_2) = 147 \text{ ohm,} \qquad (5.97)$$

$$R_{\mu23} = \frac{(U_{C_1} + U_{C_2})^2}{Q_{\tau l_{23}}} (t_3 - t_2) = 124 \text{ ohm,}$$

where $Q_{\tau l_{21}} = 19.7$ J, $Q_{\tau l_{22}} = 24.3$ J, $Q_{\tau l_{23}} = 28.9$ J are the energy losses in the magnetic systems of the saturation chokes L_{21}, L_{22}, L_{23}.

The high-voltage transformer and the load have a common equivalent circuit (Fig. 5.17), containing:

a) leakage inductance of the primary winding. Figure 5.18 shows the location of the magnetizing coils on the inductor surface and in the space between the inductors.

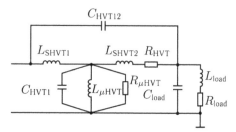

Fig. 5.17. Equivalent circuit of high-voltage transformer.

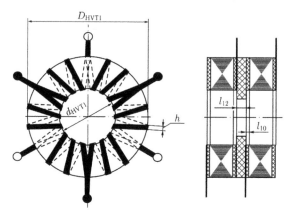

Fig. 5.18. The location of the magnetizing coils on the surface of the inductor and in the space between the inductors: D_{HVT1} – the outer diameter of the turn; d_{HVT1} – inner diameter of the turn; h – the width of the turn; l_{12} – the distance between turns in the transverse direction; l_{10} – thickness of insulation between the turn and the magnetic core; $d = l_{12} + 2l_{10}$ – distance between the cores (number of turns in the gap between the cores $n = 36$),

The distribution of the leakage magnetic field in the gap between the cores of the high-voltage transformer and the load is shown in Fig. 5.19 *a* . The superposition principle of magnetic fields was used to calculate it.

We consider Fig. 5.19 *b* (*1*). We assume that the magnetic field in the gap d within the coil h is uniform and equal

$$H_d = \frac{2I\omega_{HVT1}}{hn}, \qquad (5.98)$$

where $\omega_{HVT1} = 3$ is the number of turns of the primary winding; I is the current in the primary winding.

Then the energy of the leakage fluxes will be

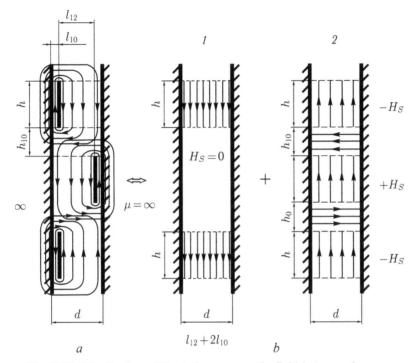

Fig. 5.19. Distribution of the leakage magnetic field between the cores.

$$W'_\mu = \frac{\mu_0 H_d^2}{2} V_{\text{leak}} = \frac{\mu_0}{2} \left(\frac{4I^2 \omega_{HVT1}^2}{h^2 n^2} \right) \cdot \left[\left(\frac{D_{HVT1} - d_{HVT1}}{2} \right) dh \frac{n}{2} \right]$$
$$= \frac{\mu_0 I^2 \omega_{HVT1}^2 (D_{HVT1} - d_{HVT1}) d}{2hn},$$

(5.99)

where V_{leak} is the volume of the leakage field. The leakage inductance is defined as

$$L'_S = \frac{2W'_\mu}{I^2} = \frac{\mu_0 \omega_{HVT1}^2 (D_{HVT1} - d_{HVT1}) d}{hn}.$$

(5.100)

Figure 5.19 *b* (*2*) represents the second component of the magnetic leakage field. Here, the opposing magnetic fields ($-H_S$ and $+H_S$) are mutually compensated. Field H_S in *n* gaps between turns the current flows I/n. In this case, the induction of the magnetic field in the gaps is expressed as

$$B = \frac{\mu_0 I \omega_{HVT1}}{n(l_{12} + 2l_{10})} = \frac{\mu_0 I \omega_{HVT1}}{nd}.$$

(5.101)

The energy of the leakage field in the gaps is

$$W''_{\mu} = \frac{B^2}{2\mu_0} V_g =$$

$$\frac{1}{2\mu_0} \frac{\mu_0^2 I^2 \omega_{HVT1}^2}{n^2 d^2} \left[\frac{D_{HVT1} - d_{HVT1}}{2} d \left(\pi \frac{D_{HVT1} + d_{HVT1}}{2} - hn \right) \right] = \qquad (5.102)$$

$$\frac{\mu_0 I^2 \omega_{HVT1}^2 \left(D_{HVT1} - d_{HVT1} \right)}{4n^2 d} \left[\frac{\pi \left(D_{HVT1} + d_{HVT1} \right) - 2hn}{2} \right].$$

Leakage inductance

$$L''_S = \frac{2W''_{\mu}}{I^2} = \frac{\mu_0 \omega_{HVT1}^2 \left(D_{HVT1} - d_{HVT1} \right)}{4n^2 d}$$

$$\left[\pi \left(D_{HVT1} + d_{HVT1} \right) - 2hn \right]. \qquad (5.103)$$

The total leakage inductance of the primary winding of one core is:

$$L_{S1}^{(1)} = \frac{2(W'_{\mu} + W''_{\mu})}{I^2} L'_S + L''_S =$$

$$\frac{\mu_0 \omega_{HVT1}^2 (D_{HVT1} - d_{HVT1})}{n} \left[\frac{d}{n} + \frac{\pi (D_{HVT1} + d_{HVT1}) - 2hn}{4nd} \right] \qquad (5.104)$$

$$= 275.6 \text{ nH},$$

Taking into account the parallel connection of inductors, we obtain the general leakage inductance of primary winding of high-voltage transformer:

$$L_{SHVT1} = \frac{L_{S1}^{(1)}}{N} = 15.3 \text{ nH}; \qquad (5.105)$$

b) leakage inductance L_{SHVT2} of the secondary winding of the HVT. In a high-voltage transformer, the turns of the secondary winding are relatively sparse along the length of the magnetic system, so using a formula similar to (5.77) leads to an underestimated value of L_{SHVT2}. Let us single out in the transformer three zones in which the magnetic leakage field, caused by the counter-current currents, is concentrated between the turns of the primary and secondary windings, and we determine the leakage inductance by parts. In this case we neglect the intervals l_{12} between magnetizing coils due to their smallness in comparison with the dimensions of the secondary winding (Fig. 5.19).

We calculate using the known formulas for a coaxial conductor. *Zone I* (Fig. 5.20). The inductance of the cable is

$$L_s^1 = \frac{\mu_0 \omega_{HVT2}^2 l_1}{2\pi} \times$$

$$\times \left[\ln\left(\frac{R_1}{p_1}\right) + \frac{n_1^4}{(p_1^2 - n_1^2)^2} \ln\left(\frac{p_1}{n_1}\right) - \frac{1}{4}\frac{3n_1^2 - p_1^2}{p_1^2 - n_1^2} + \frac{1}{m}\ln\left(\frac{R_1}{mp}\right) + \frac{1}{4m} \right] = \tag{5.106}$$

$$= 269.5 \text{ nH},$$

where n_1 and p_1 are the radii of the inner wire (the conductor of the primary winding); ρ is the radius of the strand (conductor of the secondary winding); R_1 is the radius of the circle on which the centres of the strands are located; $m = 18$ is the number of cores of the outer wire; $\omega_{HVT2} = 3$ is the number of turns of the secondary winding; l_1 is the length of the windings conductors; d is the distance between centres of the windings.

Zone II (Fig. 5.21). The cable inductance is defined as

$$L_s^2 = \frac{\mu_0 \omega_{HVT2}^2 l_1}{2\pi} \times$$

$$\times \left[\ln\left(\frac{q}{R_2}\right) + \frac{r^4}{(r^2 - q^2)^2} \ln\left(\frac{r}{q}\right) - \frac{1}{2}\frac{r^2}{r^2 - q^2} - \frac{1}{4}\left(1 - \frac{1}{m}\right) + \frac{1}{m}\ln\left(\frac{R_2}{mp}\right) \right] = \tag{5.107}$$

$$= 328.65 \text{ nH},$$

where q and r are the inner and outer radii of the outer wire (primary winding); ρ is the radius of the strand (secondary winding); R_2 is the

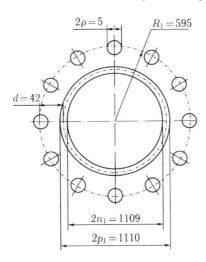

$2\rho = 5$

$R_1 = 595$

$d = 42$

$2n_1 = 1109$

$2p_1 = 1110$

Fig. 5.20. Scheme of coaxial cable with multistrand outer and hollow internal wires.

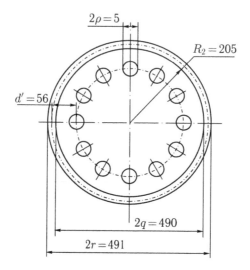

Fig. 5.21. Scheme of coaxial cable with multistrand inner and hollow outer wires.0

radius of a circle on which the centres of the strands are located; $m = 18$ is the number of strands of the internal wire; l_1 is the length of the windings; d' is the distance between the centres of the windings.

Zone III (area at the ends of the transformer). To calculate the inductance of the winding turns in the ends of the transformer, we transform this region into an equivalent coaxial one.

In this case, the leakage inductance is defined similarly to the calculation in zone I:

$$L_S^3 = \frac{\mu_0 \omega_{HVT2}^2 l_2}{2\pi} =$$

$$= \left[\ln\left(\frac{R_3}{p_3}\right) + \frac{n_3^4}{(p_3^2 - n_3^2)^2} \ln\left(\frac{p_3}{n_3}\right) - \frac{1}{4}\frac{3n_3^2 - p_3^2}{p_3^2 - n_3^2} + \frac{1}{m}\ln\left(\frac{R_3}{mp}\right) + \frac{1}{4m} \right] = \quad (5.108)$$

$$= 174.37 \text{ nH},$$

The total value of the leakage inductance of the secondary winding of the HVT will be $L_{SHVT2} = L_S^1 + L_S^2 + 2L_S^3 = 946.9$ nH. After reducing to the primary winding of the HVT, we obtain the value $L'_{SHVT2} = 2.92$ nH;

c) the dynamic capacitance of the primary winding of the inductor C_{HVT1} of a high-voltage transformer, which depends on the type of inductors used. To simplify the calculations, we assume that the primary winding of a high-voltage transformer covers almost the entire surface of the core, forming a uniform magnetizing coil. In this case, the capacity of the inductor consists of two components: $C_{10}^{(1)}$ – the magnetizing coil–the inductor core and the $C_{12}^{(1)}$ – magnetizing turn – the magnetizing turn of a neighbouring inductor.

The voltage $U(D)$ in the gap l_{10} (turn–core of the inductor) varies from 0 at the inner diameter d_{HVT} up to the value $U_{C31}/2\omega_{\text{HVT1}}$ on the outer diameter D_{HVT} according to the law

$$U(D) = \frac{U_{C31}}{2\omega_{\text{HVT1}}}\frac{(D - d_{\text{HVT}})}{(D_{\text{HVT}} - d_{\text{HVT}})}. \qquad (5.109)$$

By subdividing into elementary capacitances, finding the energy concentrated in them, and then integrating in the range from d_{HVT} to D_{HVT} taking into account the two sides of the inductor, we obtain

$$C_{10}^{(1)} = \frac{\pi\varepsilon_{10}\varepsilon_0}{8l_{10}\omega_{\text{HVT1}}^2(D_{\text{HVT}} - d_{\text{HVT}})^2} \times$$

$$\times\left[\frac{D_{\text{HVT}}^4 - d_{\text{HVT}}^4}{2} - \frac{4d_{\text{HVT}}(D_{\text{HVT}}^3 - d_{\text{HVT}}^3)}{3} + d_{\text{HVT}}^2(D_{\text{HVT}}^2 - d_{\text{HVT}}^2)\right] =$$

$$= 0.267 \text{ nF},$$

where $\varepsilon_{10} = 5.5$ is the dielectric constant of the insulation material between the primary winding and the core (fiberglass); l_{10} is the thickness of the insulation.

The capacitance $C_{12}^{(1)}$ is determined on the basis that the electric field strength between the coils of adjacent inductors is uniform. Then, using the formula of a flat capacitor

$$C_{12}^{(1)} = \frac{\varepsilon_{12}\varepsilon_0\pi(D_{\text{HVT}}^2 - d_{\text{HVT}}^2)}{4l_{12}\omega_{\text{HVT}}^2} = 0.51 \text{ nF}, \qquad (5.110)$$

where $\varepsilon_{12} = 5.5$ is the dielectric constant of the insulation material between the primary windings the of neighbouring inductors (fiberglass); l_{12} is the thickness of the insulation. The total dynamic capacity of the primary winding is

$$C_{\text{HVT1}} = N(C_{10}^{(1)} + C_{12}^{(1)}) = 14 \text{ nF}. \qquad (5.111)$$

d) the dynamic capacitance C_{HVT12} between the primary and secondary windings of the high-voltage transformer. The energy in the space between the windings is stored in the following areas:

– between the coils of the primary and secondary windings located on the outer diameter of the inductors;

– between the coils of the primary and secondary windings located on the internal diameter of the inductors;

– between the turns of the primary and secondary windings, located in the two-rod array of inductors.

To simplify the calculations, we assume that:

– coils of the primary winding, located on the outer and inner diameters of the inductors, form the outer and inner cylinders with diameters D_{HVT1}, d_{HVT1};

– the turns of the secondary winding form the outer and inner cylinders with diameters D_{HVT2}, d_{HVT2};

– in the end regions of the inductors, the primary and secondary windings form two flat cylindrical capacitors, uniformly charged over the entire surface. The voltage on the outer and inner cylinders varies according to the law

$$U(l) = \frac{NU_{C_{31}} l}{\omega_{HVT1} l_{HVT}};$$

– the electric field strength in the space between the cylinders formed by the primary and secondary windings is changed as follows:

$$E(l,r) = \frac{NU_{C_{31}} l}{l_{HVT} \omega_{HVT1} \ln(D_{HVT2}/d_{HVT2})}, \tag{5.112}$$

where l_{HVT} is the length of the high-voltage transformer.

In this case, the total stored energy in the first two regions is

$$W = 2 \int_V \frac{\varepsilon_m \varepsilon_0 E^2(l,r)}{2} dV = \frac{2\varepsilon_m \varepsilon_0 N^2 U_{C_{31}}^2}{2l_{HVT}^2 \omega_{HVT1}^2 \ln^2(D_{HVT2}/D_{HVT1})} \times$$

$$\int_0^{l_{HVT}} l^2 dl \int_{D_{HVT1}/2}^{D_{HVT}/2} \frac{2\pi r}{r^2} dr = \frac{2\pi \varepsilon_m \varepsilon_0 N^2 U_{C_{31}}^2 l_{HVT}}{3\omega_{HVT1}^2 \ln(D_{HVT2}/D_{HVT1})}, \tag{5.113}$$

where ε_m is the permittivity of the transformer oil. In deriving the formula we used the simplification

$$\ln \frac{D_{HVT2}}{D_{HVT1}} \approx \ln \frac{d_{HVT1}}{d_{HVT2}},$$

since the diameters of the primary and secondary windings are large in comparison with the distance between them.

On the other hand,

$$W = \frac{C_{SN(1+2)}^{(2)}N^2U_{C_{31}}^2}{2\omega_{HVT1}^2},$$

Consequently,

$$C_{SN(1+2)}^{(2)} = \frac{4\pi\varepsilon_m\varepsilon_0 l_{HVT}}{3\ln(D_{HVT2}/D_{HVT1})} = 0.293 \text{ nF.} \qquad (5.114)$$

The capacity of the third region is equal to the capacity of two cylindrical capacitors. It is

$$C_{SN(3)}^{(2)} = \frac{\pi\varepsilon_m\varepsilon_0(D_{HVT1}^2 - d_{HVT1}^2)}{2\Delta_{12HVT}} = 0.755 \text{ nF.} \qquad (5.115)$$

Thus, the dynamic capacitance between the primary and secondary windings is

$$C_{HVT12} = C_{SN(1+2)}^{(2)} + C_{SN(3)}^{(2)} = 1.05 \text{ nF.} \qquad (5.116)$$

It should be noted that the simplification used (representation of the primary and secondary windings of a high-voltage transformer in the form of a solid metal cylinder) leads to an overestimation of the dynamic capacitance;

e) magnetization inductance $L_{\mu HVT}$ of one inductor of a high-voltage transformer, which is determined in the same way as for the saturation chokes:

$$L_{\mu HVT}^{(1)} = \frac{2\omega_{HVT1}^2 B_S l_c K (t_7 - t_3)(D_{HVT} - d_{HVT})}{\pi(D_{HVT} + d_{HVT})[H_0(t_7 - t_3) + 2S_{\omega e} + S_{\omega 0}]} = 56 \text{ }\mu H, \qquad (5.117)$$

where $t_7 - t_3 = 1$ μs is the duration of the output pulse of the GMP.

The equivalent inductance of magnetization of the magnetic system of a high-voltage transformer will be

$$L_{\mu HVT} = \frac{L_{\mu HVT}^{(1)}}{N} = 3.1 \text{ }\mu H. \qquad (5.118)$$

f) Ohmic losses in the cores of the induction system $R_{\mu HVT}$, which are determined using the results of thermal calculations by a formula similar to (5.89):

$$R_{\mu HVT} = \frac{U_{C_{31}}^2}{Q_{THVT}}(t_7 - t_3) = 3.64 \text{ ohm,} \qquad (5.119)$$

where Q_{rHVT} = 206.1 J are the energy losses in the magnetic system of the HVT;

g) ohmic loss equivalent R_{HVT}, which determines the loss in the secondary winding of a high-voltage transformer made of a set of high-voltage cables. The formula for its calculation is similar to (5.79):

$$R_{\text{HVT}} = \frac{2Q_{\tau 5}\pi^2 L^2}{n^2 U_{C_{31}}^2 (t_7 - t_3)^3} = 1.7 \cdot 10^{-2} \text{ohm}, \qquad (5.120)$$

wherein $Q_{\tau 5}$ = 1.76 · 10^{-6} J;

h) the inductance of the discharge circuit of the magnetic pulse generator to the primary winding of the high-voltage transformer, which consists of three parts:

– the inductance of conductors connecting the capacitors C_{31}–C_{33}, the magnetic commutators L_{21}–L_{23} and the primary winding of the high voltage transformer;

– the inductance of each assembly of capacitors C_{31}, C_{32}, C_{33}, equal to $L_{C_{31}} = L_{C_{32}} = L_{C_{33}} = 36$ nH ;

– the inductance of the wires connecting the primary winding of a high-voltage transformer.

The assembling inductance of the discharge circuit is $L_{\text{wir}}^{(3)} + L_{C_{31}} + L_{\text{wir}}^{(4)}$ = 152 nH.

The load comprising a high-voltage insulator is represented by the following elements. Load capacitance C_{load} consists of two components: the capacitance $C_k^{(1)}$, formed by high-voltage cables located inside the insulator, and the spiral of the insulator, and the capacitance $C_k^{(2)}$ between the outer surface of the spiral and the body of the vacuum chamber. According to the calculations, the dynamic load capacity, reduced to the primary side of the high-voltage transformer, is $C_{\text{load}} = N^2 \left(C_k^{(1)} + C_k^{(2)} \right) = 2.2$ nF; ;

i) the inductance of the load formed by the inductances of the high-voltage cables passing inside the high-voltage insulator, which is calculated by a formula similar to (5.108) for a coaxial cable with multistrand inner and hollow outer wires:

$$L_{\text{load}} = \frac{\mu_0 l_{\text{INS}}}{2\pi} \times$$

$$\times \left[\ln\left(\frac{D_{\text{VC}}}{d_{\text{HVT2}}} \right) + \frac{r^4}{(r^2 - q^2)^2} \ln\left(\frac{r}{q} \right) - \frac{1}{2}\frac{r^2}{r^2 - q^2} - \frac{1}{4}\left(1 - \frac{1}{m} \right) + \frac{1}{m}\ln\left(\frac{d_{\text{HVT2}}}{m\rho} \right) \right] + \qquad (5.121)$$

$$= 128 \text{ nH,}$$

where q, r, D_{VC} are the internal, external and average diameters of the vacuum chamber; ρ is the radius of a strand of a high-voltage cable; $m = 6$ is the number of cores of the internal wire; l_{INS} is the length of the insulator.

j) load resistance, selected as $R_{load} = 450$ ohm.

The general equivalent scheme of the GMP is shown in Fig. 5.22.

5.7.2. Computer model of GMP

Computer simulation of the generator of microsecond voltage pulses was performed using the Electronic Workbench software. A special feature of the program is the approximation of the process of studying the work of various electrical devices to the conditions of laboratory studies of practical schemes. In this case, the system for recording parameters consists of instrumentation, appearance, controls and characteristics as close as possible to industrial analogues. The simulation scheme for the physical processes occurring in the GMP for generating high-voltage pulses, represented in the elements of the Electronic Workbench program, corresponds to the equivalent circuit diagram of the device shown in Fig. 5.22. The difference is that to simulate the process of flow of current through the GMP circuit it is necessary to use commutating devices – switches K which at a certain moment of time switch the current from one circuit to the other or disconnect the circuit. This is due to the fact that GMP uses magnetic elements – chokes and transformers with a saturable magnetic system. The saturation choke is a non-linear inductive switching element. The process of transition of the core material of the choke from the unsaturated state to the saturated state takes place at a certain time. In the model, the process of switching the circuit with the inductance of magnetization to a circuit with linear inductance is produced by the switch K 'instantly'. The energy consumed during the real choke switching period is taken into account in the model by connecting equivalent loss resistances calculated analytically. It should be noted that in the simulated GMP scheme, the calculation parameters of the PT pulse transformer must be reduced to its secondary circuit, and the high-voltage transformer HVT and the load to the primary side of the HVT. In this case:

1) for the circuit C_0–PT: $U'_{C_0} = U_{C_0} k_{PT} = 29.7$ kV; $C'_0 = C_0/k^2_{PT} = 2.204$ µF; $C'_{PT1} = C_{PT1}/k^2_{PT} = 2.2$ pF; $L^{(0)}_{as} = L^{(0)}_{as} k^2_{PT} = 223.25$ mH (corresponds to $L^{(0)}_{as}$ in Fig. 5.22); $L'_{SPT} = L_{SPT} k^2_{PT} = 6.53$ µN; $L'_{\mu PT} = L_{\mu PT} k^2_{PT} = 138$ mH (corresponds to $L_{\mu PT}$ in Fig. 5.22); $R'_{PT1} =$

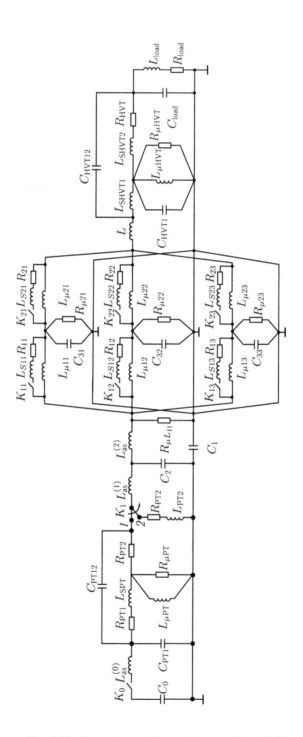

Fig. 5.22. General equivalent scheme of the GMP.

$R_{PT1}k^2_{PT} = 0.12$ ohm; $R'_{\mu PT} = k^2_{PT} = 10.13$ kohm (corresponds to $R_{\mu PT}$ in Fig. 5.22);

2) for the HVT–load circuit: $R'_{load} = R_{load}/N^2 = 1.389$ ohm; $L'_{load} = L_{load}/N^2 = 0.4$ nH; $C'_{load} = C_{load}N^2 = 2.2$ nF; $L'_{SHVT2} = L_{SHVT2}/N^2 = 2.92$ nH; $L_{SHVT} = L_{SHVT1} + L'_{SHVT2} = 18.2$ nH.

The electrical diagram of the GMP for computer simulation is shown in Fig. 5.23. The work of the GMP model proceeds as follows. The capacitance C_0 is charged to the voltage $U_{C_0} = 0.9$ kV (the reduced value $U'_{C_0} = 29.7$ kV). At time t_0, the switch K_0 is switched on. The capacitance C_0 is discharged via PT into parallel connected capacitors C_1 and C_2. The switching time of K_0 is 1 μs (it is chosen arbitrarily). In the future, the time interval for triggering keys in the circuit is a fixed value that corresponds to the calculated parameters of the magnitude of the flux linkage of magnetic elements. The switch K_0 and diode D_1 simulate the operation of thyristors VS3 of the thyristor pulse generator. Diode D_1 does not allow C_0 to be recharged (in the real circuit this process is prevented by magnetization reversal of the core of the pulse transformer). Charging current C_1 flows through the circuit $(+)\ C_0-K_0-D_1-L^{(0)}_{as} - (R_{PT1}-L_{SPT}-R_{PT2}$ parallel to $C_{PT12})-K_1-C_1-K_5-L_{HVT}-K_1$ $(K_{52},\ K_{53})-L_{as21}(L_{as22},\ L_{as23}-L_{21}(L_{22},\ L_{23})-D_{21}(D_{22},\ D_{23})-K_{31}(K_{32},K_{33})-L_{11}(L_{12},\ L_{13})-D_{11}\ (D_{12},\ D_{13})-K_2-(-)C_0.$

The charge current C_2 flows through the circuit $(+)C_0-K_0-D_1-\ L^{(0)}_{as}-(R_{PT1}-L_{SPT}-R_{PT2}$ parallel to $C_{PT12})-K_1$ – current shunt $R_{SC1}-C_2-K_2\ -(-)C_0$. After 50 μs (current time 51 μs), the charge process of C_1 and C_2 ends. At this time, the cores of the pulse transformer PT are saturated (K_1 switches and closes the secondary circuit of the pulse transformer L_{PT2}), the C_2 capacitor starts to be recharged through the secondary winding of the pulse transformer L_{PT2}. This circuit includes the resistance R'_{PT2}, which takes into account the energy losses in the capacitors C_2, the connecting wires and the conductor of the secondary winding PT.

To record current and voltage signals, an 'oscilloscope' is used with adjustable parameters for the sensitivity and duration of the signals, as well as the accuracy of the calculation. The oscilloscope is connected to the circuit elements via voltage dividers: $R_{VD11}-R_{VD12}$ for current, $R_{VD21}-R_{VD22}$ for voltage. The current multiplication factor for $R_{VD11}-R_{VD12}$ is $k_J = (1$ MOhm + 1 MOhm) /1 MOhm = 2. The coefficient of multiplication by voltage for $R_{VD21}-R_{VD22}$ is equal to $k_U = (950$ kOhm + 50 kOhm)/50 kOhm = 20. To measure the currents in the circuit, the current shunts R_{SC1}, R_{SC2}, R_{SC3} with a resistance of 0.0001 ohm are used. The sequence of signal recording in different

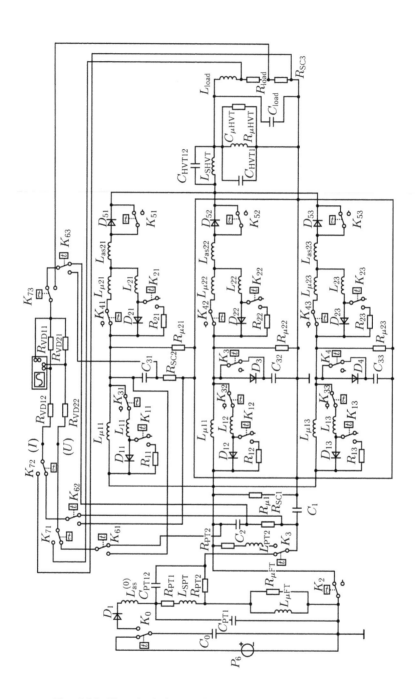

Fig. 5.23. Electrical circuit of the GMP for computer modelling.

electrical circuits of the GMP is provided by switching the measuring conductors with the keys K_{61}, K_{62}, K_{63} and K_{71}, K_{72}, K_{73}.

After recharging the capacitor C_2 the capacitors C_1 and C_2 are connected in series and have an equivalent capacitance of 0.564 µF. Until this time, the discharge of C_1 and C_2 into three parallel MPGs is impossible, since the chokes L_{11}, L_{12}, L_{13} are in an unsaturated state, which in the model is represented by the counter-switched diodes D_{11}, D_{12}, D_{13}. The magnetization inductances $L_{\mu11}$, $L_{\mu12}$, $L_{\mu13}$ are high and only the magnetization current of the cores is passed.

We especially note the following. It was previously indicated that in order to synchronize the operation of all MPG windings of the chokes L_{11}, L_{12}, L_{13}, situated on a common magnetic core and have a magnetic coupling. However, it is impossible to show this in the model scheme. The chokes L_{11}, L_{12}, L_{13} are represented therein as three separate inductances, L_{11}, L_{12} and L_{13}, which are connected in parallel. At the same time, the calculated inductances of these chokes must be increased three-fold in the model. Thus, $L_{\mu11} = L_{\mu12} = L_{\mu13} = L_{\mu11 \, (12.13)} \cdot 3 = 12$ mH; $L_{11} = L_{12} = L_{13} = L_{11 \, (12.13)} \cdot 3 = 1.5$ mH.

At the moment of saturation of the chokes L_{11}, L_{12}, L_{13} (the magnetization reversal time of the core is equal to 13.1 ms) instead of the high magnetizing inductance $L_{\mu11}$, $L_{\mu12}$, $L_{\mu13}$ small leakage inductance of the winding L_{11}, L_{12}, L_{13} are connected. In the model this is done by including K_{11}, K_{12} and K_{13} (the current time is 64.1 µs). There is a transfer of energy from the capacitors C_1 and C_2 to the capacitors C_{31}, C_{32} and C_{33}. The switches simultaneously connect the loss resistances R_{11}, R_{12}, R_{13}, which simulate energy losses in the discharge circuit. The resistance $R_{\mu1}$ reflects losses in the magnetic system of the chokes L_{11}, L_{12}, L_{13}, caused by the magnetization reversal of the ferromagnetic material. With the appearance of voltage on the capacitors C_{31}, C_{32} and C_{33}, the process of magnetization of the commuting chokes L_{21}, L_{22} and L_{23}, respectively, begins.

Let us consider the operation of MPG after charging capacitors C_{31}, C_{32} and C_{33}. In the first MPG during this time the core of the choke L_{21} is saturated (magnetization time is 1.18 µs). In the model, the switch K_{21} is closed and the K_{41} switch is opened (the current time is 65.28 µs). Note that at the same time switches K_{31}, K_{32} and K_{33} open, as well as K_3, K_4 and K_{51}, K_{52}, K_{53}. In the measuring circuits, the switches K_{71}, K_{72} and K_{73} are switched. There is a discharge of the capacitor C_{31} on the load in the circuit $C_{31} - R_{21} - K_{21} - L_{21} - L_{as21} -$

D_{51} −(L_{SHVT} parallel to C_{HVT12})−(L_{load}−R_{load}−R_{SC3} parallel to C_{HVT}, $L_{\mu HVT}$, $R_{\mu HVT}$ parallel to C_{load})−R_{SC2}−C_{31}.

Simulation of the discharge delay of the capacitor C_{32} in the second MPG and the capacitor C_{33} in the third MPG is performed by opening the switched K_3 and K_4. In this case, in the circuit C_{32} and C_{33}, the diodes D_3 and D_4 are switched on, which prevent the discharge of these capacitors. The discharge of C_{32} and C_{33} for the load starts after the keys K_3 and K_4 are re-closed with the corresponding shunting of the diodes D_3 and D_4, as well as closing the switches K_{22} and K_{23}. The switch K_3 closes after 0.35 μs after its opening (current time 65.63 μs), and K_4 − after 0.7 μs (current time 65.98 μs). Thus, the operation of real commutating chokes L_{22} and L_{23}, magnetized for a longer time than the choke L_{21}, is modelled, due to the larger number of cores used.

The switches K_{22} and K_{23} at corresponding switching shunt the diodes D_{22} and D_{23} which are included for the discharge current, and simultaneously connect to the loss resistance circuit R_{22} and R_{23}. Opening the switches K_{41}, K_{42} and K_{43}, disconnected from the circuit the magnetization inductances of the commuting chokes $L_{\mu 21}$, $L_{\mu 22}$ and $L_{\mu 23}$. This excludes their influence on the process of discharge of the capacitors C_{31}, C_{32} and C_{33}, respectively. The diodes D_{51}, D_{52} and D_{53}, included in the MPG, prevent the flow of current in the direction opposite to the worker, from one MPG to another during their joint operation. This is ensured by opening the switches K_{51}, K_{52} and K_{53} shunting the diodes, before the process of discharging the capacitor C_{31} of the first MPG to the load begins. The model process in the scheme is similar to the real one. Thus, when the operating current pulse passes, the cores of the commuting chokes L_{21}, L_{22} and L_{23} are saturated in one direction of magnetization. The current from neighbouring MPGs has a direction opposite to the operating current in the circuit and, therefore, the magnetic state of commuting chokes prevents its flow.

In each MPG the resistances $R_{\mu 21}$, $R_{\mu 22}$ and $R_{\mu 23}$ are connected in parallel with the capacitors C_{31}, C_{32} and C_{33}. These resistances take into account the energy loss spent on the magnetization reversal of the core of the chokes L_{21}, L_{22} and L_{23}. The process of discharging the capacitors C_{32} of the second MPG and C_{33} of the third MPG along the corresponding circuits to the load is similar to the discharge of the capacitance C_{31} of the first MPG. The only difference is the time delay of the operation of commutating chokes (by 0.35 μs for L_{22}

and for 0.7 µs for L_{23}). The computer model of GMP is operating in the single mode. Below are the results of the GMP model work.

5.7.3. Results of computer modelling

To correct the shape of the GMP output pulse in modelling, the inductance of the discharge circuits of the second MPG compression links on the primary winding of the high-voltage transformer varied. The best pulse shape was achieved when using mounting inductances $L_{wir}^{(3)}$, equal to 162, 112 and 112 nH, installed in the discharge circuits C_{31}, C_{32} and C_{33} respectively. Practically, these quantities can be realized with different numbers of the used connecting wires.

The preliminary modelling results showed that the losses in the cores of the inductors of the high-voltage transformer are sufficiently high (206.1 J, which corresponds to the ohmic equivalent of the losses $R_{\mu\,HVT}$ = 3.64 Ohm; see (5.121)). In addition, the inductance of the magnetization of the cores, which is $L_{\mu HVT}$ = 3.1 µH, is small (see (5.120)). These factors lead to a decrease in the output parameters of GMP. A solution to the problem may be the use of inductor cores of a high-voltage transformer made of a permalloy strip 10 µm in thickness. In this case, losses associated with the magnetic viscosity are reduced. To calculate the losses, the permalloy characteristics given in Table 1.2 were used.

In this case, the magnetization inductance $L_{\mu HVT}$ of the primary winding of one inductor is determined by the formula (5.117)

$$L_{\mu HVT}^{(1)} = \frac{2\omega_{HVT1}^2 B_S l_c K (t_7 - t_3)(D_{HVT} - d_{HVT})}{\pi(D_{HVT} + d_{HVT})[H_0(t_7 - t_3) + 2S_{\omega e} + S_{\omega 0}]} = 135.88 \ \mu H. \quad (5.122)$$

The equivalent inductance of magnetization of a high-voltage transformer

$$L_{\mu HVT} = \frac{L_{\mu HVT}^{(1)}}{N} = 7.55 \ \mu H. \quad (5.123)$$

The losses in one core and in the magnetic system of a high-voltage transformer, calculated by a formula similar to (5.78), are $Q1_{\tau\,HVT}$ = 6.66 J and $Q_{\tau HVT}$ = 120 J, respectively. Substituting the last value in (5.119), we find

$$R_{\mu HVT} = \frac{U_{C_{31}}^2}{Q_{\tau HVT}}(t_7 - t_3) = 6.25 \text{ ohm}. \qquad (5.124)$$

The calculation using the mentioned values show that the increase in the output power of the generator is 30%.

To reduce losses during energy compression and to increase the output parameters of GMP is possible by using in the cores of magnetic commutators of steel with a rolled stock thickness of 10 μm. The calculation of heat losses in these cores gives values $Q_{\tau L_{31}} = 10.7$ J, $Q_{\tau L_{32}} = 12.75$ J, $Q_{\tau L_{33}} = 15.3$ J. In this case, the ohmic equivalent losses in the magnetic systems of the saturation chokes L_{21}, L_{22}, L_{23} are $R_{\mu L_{21}} = 350$ ohm, $R_{\mu L_{22}} = 280$ ohm, $R_{\mu L_{23}} = 233$ ohm. At the same time, the corresponding magnetization inductance of commutators increase, which in this case is $L_{\mu 21} = 81.6$ μH, $L_{\mu 22} = 120.48$ μH, $L_{\mu 23} = 163.12$ μH. Calculation using the mentioned values shows that the power released on the load increases by 4–6%, and the pulse width increases by 15 ns to 1 μs at the 0.9 amplitude level.

Current and voltage diagrams for various GMP elements with cores made of 10 μm steel are shown in Figs. 5.24–5.27.

Figure 5.24 demonstrates the discharge of the capacitor C_0 through the primary winding of the pulse transformer to the parallel connected capacitors C_1 and C_2. Charging voltage of the capacitors C_1 and C_2 is 29.04 kV, charging current 1011 A, duration 50 μs. In this case, the efficiency of energy transfer from the primary storage device C_0 to the capacitors of the first MPG compression link is

$$\eta_1 = \frac{C_1 U_{C_1}^2 + C_2 U_{C_2}^2}{C_0 U_{C_0}^2} = 0.98. \qquad (5.125)$$

The discharge current of the primary storage C_0 is 66.9 kA.

Figure 5.25 shows the energy compression process (the charge C_2 – recharge of C_2 – charge of C_{31}–C_{33} – formation of the output pulse). The voltage of the capacitor C_2 after its recharge ($U_{C_2 \text{ rec}}$) is 28.34 kV, and the efficiency of energy transfer from primary storage C_0 to C_1 and C_2 after the recharge C_2

$$\eta_1 = \frac{C_1 U_{C_1}^2 + C_2 U_{C_2 \text{rec}}^2}{C_0 U_{C_0}^2} = 0.93. \qquad (5.126)$$

According to calculations, the charging current C_2 is equal to 1011 A, and the recharge current C_2 is 7757 A (duration 13.1 μs).

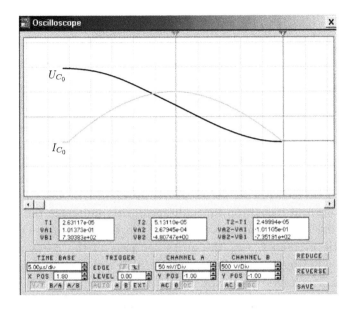

Fig. 5.24. The discharge of the capacitor C_0.

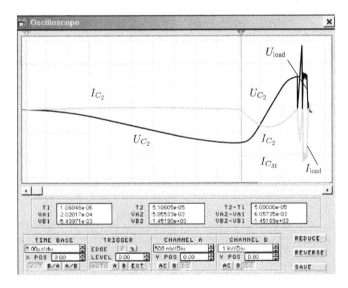

Fig. 5.25. The process of energy compression.

The charging voltage of the capacitors C_{31}, C_{32}, C_{33} is 55.4 kV. The charging current is 13890 A (duration 1.18 µs). At the same time, the efficiency of energy transfer from primary storage C_0 to the capacitors of the last compression link of MPG is

$$\eta_1 = \frac{C_{31}U_{C_{31}}^2 + C_{32}U_{C_{32}}^2 + C_{33}U_{C_{33}}^2}{C_0 U_{C_0}^2} = 0.89. \qquad (5.127)$$

The discharging currents of the capacitors on the winding of the high-voltage transformer are: for C_{31} 28814 A with a duration of 0.53 μs; for C_{32} 29167 A with a duration of 0.53 μs; for C_{33} 29984 A with a duration of 0.73 μs.

As can be seen from Figs. 5.24 and 5.25, the beginning of the discharge of the capacitors of the MPG compression links and recharge C_2 (activation of switches) coincides with the instant of the charge current transition through zero. This indicates the correct choice for the engineering calculation of the flux coupling values of the magnetic elements, i.e., the cross sections of the cores and the number of windings.

Figure 5.26 shows the output pulse of the GMP, which has the following parameters: voltage not less than 450 kV in any part momentum; current is not less than 1 kA in any part of the pulse; pulse saturation is 1 μs at the level of 0.9.

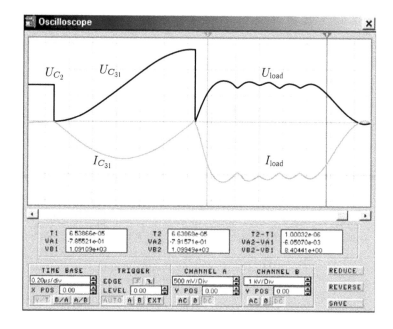

Fig. 5.26. The charge of the capacitor C_{31} and the pulse load.

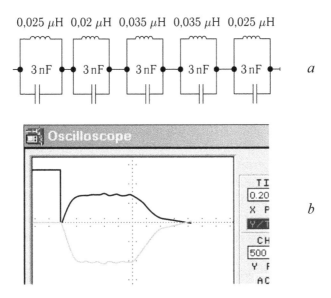

Fig. 5.27. Scheme of the antiresonant circuit (*a*). Oscillograms of the voltage pulse and the current load of the GMP (*b*).

Thus, the presented computer simulation of the parameters of the GMP is carried out derived using an equivalent circuit, which really reflects the physical processes. The correctness of the proposed method has been repeatedly confirmed by the modelling of LIA on magnetic elements. The discrepancy between the results of computer modelling and the experimental data obtained is not more than 10%. The simulation carried out allowed us to determine the elements which limit the output parameters of the GMP, adjust their characteristics and set up the GMP elements to get the required characteristics of the output pulse (voltage 450 kV; current 1 kA; pulse duration 1 µs).

To reduce oscillations on the flat part of the pulse, it is possible to use antiresonant circuits installed before the primary winding of the high-voltage transformer. The circuit and the nominal values of the elements of such a circuit are given in Fig. 5.27. There are also calculated oscillograms of the voltage and current pulse on the ohmic load of the GMP.

5.8. Formation of voltage pulses of special shape

On the basis of this idea it is possible to create generators of high-voltage linearly increasing or linearly decreasing pulses of microsecond duration. The need to form pulses with similar characteristics is caused by the fact that using explosive–emission diodes, relativistic magnetrons, reflex triodes, vircators and other devices, the impedance of the diode is reduced and, consequently, the amplitude of the output voltage decreases during the pulse. This is explained by the expansion of the cathode and anode plasma in the interelectrode gaps of the devices, which leads to a reduction in the accelerating gap during the voltage pulse and a decrease in the energy of the electron beam particles. To correct the voltage drop, a generator with a linearly increasing output pulse can be used, which allows to stabilize the level of the electric field strength on a diode. For some microwave devices, it may be advisable to use generators with linearly decreasing pulses, since this will allow to reduce the rate of expansion of the cathode and anode plasma and thereby eliminate short-circuiting the generator to ground.

As in the GMP considered above, it is necessary to ensure a consistent discharge of capacitors through the windings of the magnetic commutators of the last compression links of several (two or more) magnetic pulse generators to the primary winding of a high-voltage pulse transformer. To form output pulses of a special shape, capacitors and saturation chokes are used in the last stages of MPG compression the capacitances and inductances of which are in a certain relationship with one another. In the generator, to form an output voltage pulse with an increasing amplitude the capacitor capacitances should be in the following ratio:

$$C_{N1} < kC_{N2} < ... < kC_{Nm},$$

where $k = 1.1-2$, and the condition

$$C_{N1}L_{N1} \approx C_{N2}L_{N2} \approx ... \approx C_{Nm}L_{Nm},$$

where L_{N1}, L_{N2}, ..., L_{Nm} are the inductances of the windings of magnetic commutators of the MPG. For a generator with an output pulse of a linearly decreasing shape, the following inverse relationship is satisfied

$$C_{N1} > kC_{N2} > ... > kC_{Nm},$$

where $k = 1.1-2$.

The processes of energy transfer from one link of compression to another are described in detail above. Therefore, here we confine ourselves to describing the processes in the last compression links of the generator, which contains three parallel MPGs. Let us consider the principle of the action of the generator of linearly decreasing pulses using a concrete example. Let the generator contain capacitors of the following capacitances: $C_{31} = 0.282$ µF; $C_{32} = 0.188$ µF; $C_{33} = 0.123$ µF. Thus, the coefficient $k = 1.5$. The capacitors C_{31}, C_{32}, C_{33} are charged to $U_{C_{31}} = U_{C_{32}} = U_{C_{33}} = 50$ kV from the previous compression links in the time interval $\Delta t_{11} = 1$ µs. The values of the flux linkage of the saturation chokes L_{21}, L_{22} and L_{23} must be

$$\langle U \rangle \Delta t_{11} \approx \Psi_{21} = \omega_{21} S_{21} \Delta B;$$

$$\langle U \rangle (\Delta t_{11} + 0.3 \text{ µs}) \approx \Psi_{22} = \omega_{22} S_{22} \Delta B; \qquad (5.128)$$

$$\langle U \rangle (\Delta t_{11} + 0.6 \text{ µs}) \approx \Psi_{23} = \omega_{23} S_{23} \Delta B,$$

where 0.3 µs and 0.6 µs are the delays of switching the second and third magnetic pulse generators on the primary winding of the high-voltage transformer.

The saturable choke L_{21} should be produced using one core ($N_{21} = 1$) of a ring having an external diameter $D_{ext} = 500$ mm, inner diameter $D_{in} = 220$ mm, width $h = 25$ mm with a filling ratio of the steel core $K = 0.8$, wound it from a permalloy 50 NP tape with a thickness of 0.02 mm, the cross section of the steel core will be

$$S_{21} = \frac{D_{ext} - D_{in}}{2} N_{21} h K = 28 \text{ cm}^2.$$

To satisfy the equality in the left-hand side of the first formula (5.128), the number of the turns in the winding of the saturation choke L_{21} must be at least $\omega_{21} = 4$. The inductance of the winding of the choke L_{21} in the saturated state of the cores will be

$$L_{21} = \frac{\mu_0}{2\pi} \omega_{21}^2 a_{21} \ln \frac{D_{ext.wind.}}{D_{in.wind.}} \approx 0.1 \text{ µH}, \qquad (5.129)$$

where $a_{21} = 40$ mm is the linear dimension of the winding; $D_{ext.wind.} = 520$ mm and $D_{in.wind.} = 200$ mm are the external and internal diameters of the saturation choke winding.

The duration of the discharge pulse C_{31} through the inductance of the winding of the saturation choke L_{21} and the inductance of the magnetizing turns of the inductors L_{SHVT1} (approximately equal to 0.15 µH) is equal to

$$\Delta t_{21} = \pi \sqrt{\frac{C_{31}(L_{21} + L_{SHVT1})}{2}} \approx 0.65 \ \mu s. \qquad (5.130)$$

The flux linkage of the saturation choke L_{22} must exceed the value of the flux linkage of the saturation choke L_{21} to delay the discharge of the capacitor of the last compression link of the second MPG. The saturation choke L_{22} is produced from two ferromagnetic rings with outer and inner diameters of 500 mm and 220 mm and 25 mm wide and a three-turn winding is also used. It is then possible to delay the discharge pulse of the capacitor C_{32} to the desired value ($\Delta t = 0.3 \ \mu s$; see the second equation (5.128)).

Thus, the second magnetic pulse generator is connected to the magnetizing coils of the transformer inductors approximately at the maximum of the discharge current pulse of the first magnetic pulse generator.

The inductance of the winding of the choke L_{22} in the saturated state of the cores is

$$L_{22} = \frac{\mu_0}{2\pi} \omega_{22}^2 a_{22} \ln \frac{D_{ext.wind.}}{D_{in.wind.}} = 0.1 \ \mu H, \qquad (5.131)$$

where $a_{22} = 70$ mm; $D_{ext.wind.} = 520$ mm; $D_{in.wind} = 200$ mm.

The duration of the discharge pulse of C_{22} through the inductance of the winding of the saturation choke (L_{22}) and the magnetizing turns ($L_{m.t}$) is

$$\Delta t_{22} = \pi \sqrt{\frac{C_{32}(L_{22} + L_{m.t.})}{2}} = 0.53 \ \mu s.$$

The third magnetic pulse generator increases the delay of the output pulse by another 0.3 μs. To do this, the saturation choke of the last compression link of the third magnetic pulse generator has an increased flux linkage and is made of four ferromagnetic rings with a double-turn winding.

The inductance of the winding of the choke L_{23} in the saturated state of the cores is

$$L_{23} = \frac{\mu_0}{2\pi} \omega_{23}^2 a_{23} \ln \frac{D_{ext.wind.}}{D_{in.wind.}} = 0.08 \ \mu H, \qquad (5.132)$$

where $a_{32} = 120$ mm; $D_{ext.wind.} = 530$ mm; $D_{in.wind.} = 200$ mm.

The duration of discharge pulse C_{33} through the inductance of the saturation choke and the magnetizing turns $L_{m.t.}$ is

$$\Delta t_{23} = \pi \sqrt{\frac{C_{33}(L_{23} + L_{m.t.})}{2}} \approx 0.4 \; \mu s. \qquad (5.133)$$

In this way, choosing the parameters of the magnetic commutators (the number of cores and the number of turns of the windings) of the last compression links of the parallel installed magnetic pulse generators, it is possible to delay the moment of their switching to the primary winding of the high-voltage pulse transformer, and also to realize a different inductance of the discharge circuits for approximate equality of the duration of the discharge processes. Application in this generator capacitors of different capacitances in the last links of the compression of magnetic pulse generators provides different amplitudes of the discharge current, which makes it possible to obtain output pulses of a linearly decreasing shape.

The calculation of the output pulse parameters of the generator with the parameters of the elements given above is performed using the Electronic Workbench software. As a result of consecutive discharge of the capacitors of the last compression links of the three magnetic pulse generators on a load of 100 ohms, a linearly decreasing pulse is formed with an amplitude of 480 kV, a current of 4.8 kA, 1.2 μs long at the base with a leading edge duration of 0.15 μs and the rear edge 0.2 μs. Figure 5.28 shows a linearly decreasing pulse generated by the GMP.

The number of MPGs used in the generator of high-voltage pulses of microsecond duration and hence the possible duration of the output pulses is limited by the following circumstance. At the end of the discharge of the capacitor of the last link of the first magnetic pulse generator, the core of the saturation choke L_{21} is in a saturated state. When switching on the primary winding of the high-voltage pulse transformer of the second and third magnetic pulse generators along the winding of the saturation choke L_{21} the demagnetizing current begins to flow under the action of the potential difference at the terminals of the primary winding. The duration of the process of magnetization reversal of the choke is limited by the magnitude of its flux linkage. In the considered example of a specific embodiment of the generator, the duration of the reverse magnetization process depends on the shape of the voltage pulse at the terminals of the primary winding and the flux linkage of the choke and is ~1.4 μs.

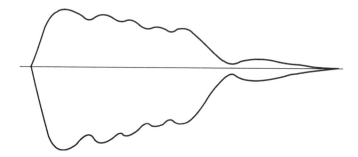

Fig. 5.28. The linearly-decreasing pulse formed by GMP.

Therefore, in this case, no more than three magnetic pulse generators can be connected.

To generate a voltage output pulse with an increasing form of the output pulse of the capacitors of the last compression links of the magnetic pulse generators should be in the reverse ratio:

$$C_{1N} > kC_{2N} > ... > kC_{mN},$$

where $k = 1.1-2$, and the following condition should be satisfied

$$C_{1N}L_{1N} \approx C_{2N}L_{2N} \approx ... \approx C_{mN}L_{mN}.$$

Thus, the generator of high-voltage linearly-increasing (linearly-decreasing) pulses of microsecond duration realizes an original idea that is connected with the supply to the primary winding of a high-voltage pulse transformer of discharge pulses from several magnetic pulse generators switched on with the necessary time delay. To form pulses of a special shape, capacitors of different capacitances are used in the last stages of compression.

Conclusion

This chapter describes the principle of the operation of generators of microsecond voltage pulses, their design, the method of engineering calculation based on analytical expressions, the process of computer modelling using equivalent circuits, and the thermal calculation of elements. A possible power source for the GMP is presented and the design and calculation of the parameters of the high-voltage insulator and transformer are described.

The main element of the GMP – high-voltage unit – is made on the basis of three synchronized parallel magnetic pulse generators, whose discharge on the primary turns of the inductors of the high-voltage transformer is performed with a certain time delay. The synchronization of the three MPGs is achieved by applying to the saturation chokes of the last compression links the common magnetic core and selecting the elements of equal parity in the first links of the MPG compression. The delay in switching on the MPG is achieved by using saturation in the last links of compression with increased flux coupling. The superposition of the output pulses of the three MPGs makes it possible to form a pulse with the required length and amplitude parameters on the load. To reduce the weight and cost of the high-voltage transformer is made on the principle of the induction system of the LIA with multi-turn windings inductors.

The primary storage device can be made on the basis of capacitors BMOD0115 PV (ultracapacitors). For their charge in a pause between bursts of pulses in 10−20 min, it is sufficient to use a source with a power of 5−10 kW (600 V, 8.5 A). To reduce the pause between bursts, it is possible to increase the power of the charging source of the primary storage. Elements of the high-voltage unit can be placed in a cylindrical container with a diameter of 1.4 m with a length of 2.3 m, and elements of the power system − in its stand. The results of engineering calculations and computer simulation allow to conclude that on the basis of the proposed concept it is possible to create generators of microsecond voltage pulses of rectangular shape with the following parameters: 450−1000 kV; 1−2 kA; duration 1 µs with the repetition rate up to 1 kHz.

We note the possibility of inverting the polarity of the output pulses of the GMP by reconnecting the input lines of the high-voltage unit from the power source and changing the polarity of the sources of demagnetization of the magnetic elements. The thermal operating life of the GMP allows to operate in a continuous mode 140 000 pulses. The above advantages substantially expand the capabilities of generators as power sources for various relativistic microwave devices.

References

1. Vintizenko I.I., Furman E.G., Izv. VUZ, Fizika, 1998, No. 4. Appendix. P. 111–119.
2. Butakov L.D., et al., Prib. Tekh. Eksper., 2000, No. 3. P. 159, 160.
3. Butakov L.D., et al., *ibid,* 2001, No. 5. P. 1045110.
4. II Vintizenko, Izv. VUZ, Fizika, 2007, No. 10/2. Pp. 136–141.
5. Vintizenko I.I., Linear induction accelerator, Patent of the Russian Federation for invention No. 2178244. BI. 2002, No. 1.

6.   Vintizenko I.I., Linear induction accelerator, Patent of the Russian Federation for invention No. 2185041. BI. 2002, No. 19.
7.   Vintizenko I.I., Generator of high-voltage linearly-increasing pulses of microsecond duration, Patent of the Russian Federation for invention No. 2305379. BI. 2007, No. 24.
8.   Vintizenko I.I., Generator of high-voltage linearly-decreasing pulses of microsecond duration, Patent of the Russian Federation for invention No. 2303338. BI. 2007, No. 20.
9.   Vintizenko I.I., et al., Izv. VUZ, Fizika, 2006, No. 11. Annex. Pp. 262–265.
10.   Vintizenko I.I., et al., in: Pulsed Power and Plasma Science Conference. Albuquerque, USA, 2007. Digests of Technical Papers. P. 865–868.
11.   Vintizenko I.I., et al., Izv. VUZ, Fizika, 2009, No. 11/2. Pp. 139-144.
12.   Vizir V.A., et al., In: Proc. 13 Int. Symposium on High Current Electronics. Tomsk, 2004. P. 198–200.
13.   Vdovin S.S., Design of pulse transformers, Leningrad, Energoatomizdat, 1991.

Index

A

acceelerator
  'non-iron' LIAs 44, 83, 85, 90, 225, 237
accelerator
  Advanced Test Accelerator (ATA) 34
  Astron 1, 7, 57, 65, 178
  Racetrack induction accelerator 10, 12
  SILUND 2, 33
  two-beam accelerator 183, 184
adgezator 4
amplifier
  relativistic klystron amplifier , 188, 189, 190, 194, 198, 203, 205, 208
antenna amplifier , 240, 241, 242, 248, 253, 258, 263, 264

B

breaker
  plasma current breaker 16, 17
buncher 185, 187, 189, 190, 264, 265, 266, 268

C

Cherenkov , 208, 214, 232, 233, 238, 240, 242, 243, 245, 246, 272
coefficient
  energy transfer coefficient 146
  Townsend ionization coefficient 47
Compact LIA 9, 319, 320
Compact Linear Induction Accelerator (CLIA) 203

D

DFL (double forming line) 8, 35, 40, 41, 42, 43, 44, 45, 61, 87, 98, 99,
       101, 102, 105, 108, 111, 112, 114, 115, 116, 117, 118, 119, 193,
       195, 196, 221, 222, 254, 255, 267, 278
dielectric emitter 124, 127, 129, 132, 194, 195, 198